Geophysical Monograph Series

Including

IUGG Volumes

Maurice Ewing Volumes

Mineral Physics Volumes

GEOPHYSICAL MONOGRAPH SERIES

Geophysical Monograph Volumes

41 **Gondwana Six: Stratigraphy, Sedimentology, and Paleontology** *Garry D. McKenzie (Ed.)*
42 **Flow and Transport Through Unsaturated Fractured Rock** *Daniel D. Evans and Thomas J. Nicholson (Eds.)*
43 **Seamounts, Islands, and Atolls** *Barbara H. Keating, Patricia Fryer, Rodey Batiza, and George W. Boehlert (Eds.)*
44 **Modeling Magnetospheric Plasma** *T. E. Moore and J. H. Waite, Jr. (Eds.)*
45 **Perovskite: A Structure of Great Interest to Geophysics and Materials Science** *Alexandra Navrotsky and Donald J. Weidner (Eds.)*
46 **Structure and Dynamics of Earth's Deep Interior (IUGG Volume 1)** *D. E. Smylie and Raymond Hide (Eds.)*
47 **Hydrological Regimes and Their Subsurface Thermal Effects (IUGG Volume 2)** *Alan E. Beck, Grant Garven, and Lajos Stegena (Eds.)*
48 **Origin and Evolution of Sedimentary Basins and Their Energy and Mineral Resources (IUGG Volume 3)** *Raymond A. Price (Ed.)*
49 **Slow Deformation and Transmission of Stress in the Earth (IUGG Volume 4)** *Steven C. Cohen and Petr Vaníček (Eds.)*
50 **Deep Structure and Past Kinematics of Accreted Terranes (IUGG Volume 5)** *John W. Hillhouse (Ed.)*
51 **Properties and Processes of Earth's Lower Crust (IUGG Volume 6)** *Robert F. Mereu, Stephan Mueller, and David M. Fountain (Eds.)*
52 **Understanding Climate Change (IUGG Volume 7)** *Andre L. Berger, Robert E. Dickinson, and J. Kidson (Eds.)*
53 **Plasma Waves and Istabilities at Comets and in Magnetospheres** *Bruce T. Tsurutani and Hiroshi Oya (Eds.)*
54 **Solar System Plasma Physics** *J. H. Waite, Jr., J. L. Burch, and R. L. Moore (Eds.)*
55 **Aspects of Climate Variability in the Pacific and Western Americas** *David H. Peterson (Ed.)*
56 **The Brittle-Ductile Transition in Rocks** *A. G. Duba, W. B. Durham, J. W. Handin, and H. F. Wang (Eds.)*
57 **Evolution of Mid Ocean Ridges (IUGG Volume 8)** *John M. Sinton (Ed.)*
58 **Physics of Magnetic Flux Ropes** *C. T. Russell, E. R. Priest, and L. C. Lee (Eds.)*
59 **Variations in Earth Rotation (IUGG Volume 9)** *Dennis D. McCarthy and Williams E. Carter (Eds.)*
60 **Quo Vadimus *Geophysics for the Next Generation* (IUGG Volume 10)** *George D. Garland and John R. Apel (Eds.)*
61 **Cometary Plasma Processes** *Alan D. Johnstone (Ed.)*
62 **Modeling Magnetospheric Plasma Processes** *Gordon K. Wilson (Ed.)*
63 **Marine Particles Analysis and Characterization** *David C. Hurd and Derek W. Spencer (Eds.)*
64 **Magnetospheric Substorms** *Joseph R. Kan, Thomas A. Potemra, Susumu Kokubun, and Takesi Iijima (Eds.)*
65 **Explosion Source Phenomenology** *Steven R. Taylor, Howard J. Patton, and Paul G. Richards (Eds.)*
66 **Venus and Mars: Atmospheres, Ionospheres, and Solar Wind Interactions** *Janet G. Luhmann, Mariella Tatrallyay, and Robert O. Pepin (Eds.)*
67 **High-Pressure Research: Application to Earth and Planetary Sciences (Mineral Physics Volume 3)** *Yasuhiko Syono and Murli H. Manghnani (Eds.)*
68 **Microwave Remote Sensing of Sea Ice** *Frank Carsey, Roger Barry, Josefino Comiso, D. Andrew Rothrock, Robert Shuchman, W. Terry Tucker, Wilford Weeks, and Dale Winebrenner*
69 **Sea Level Changes: Determination and Effects (IUGG Volume 11)** *P. L. Woodworth, D. T. Pugh, J. G. DeRonde, R. G. Warrick, and J. Hannah*
70 **Synthesis of Results from Scientific Drilling in the Indian Ocean** *Robert A. Duncan, David K. Rea, Robert B. Kidd, Ulrich von Rad, and Jeffrey K. Weisset (Eds.)*
71 **Mantle Flow and Melt Generation at Mid-Ocean Ridges** *Jason Phipps Morgan, Donna K. Blackman, and John M. Sinton (Eds.)*
72 **Dynamics of Earth's Deep Interior and Earth Rotation (IUGG Volume 12)** *Jean-Louis Le Mouël, D. E. Smylie, and Thomas Herring (Eds.)*
73 **Environmental Effects on Spacecraft Positioning and Trajectories (IUGG Volume 13)** *A. Vallance Jones (Ed.)*
74 **Evolution of the Earth and Planets (IUGG Volume 14)** *E. Takahashi, Raymond Jeanloz, and David Rubie (Eds.)*
75 **Interactions Between Global Climate Subsystems: The Legacy of Hann (IUGG Volume 15)** *G. A. McBean and Michael Hantel (Eds.)*

Maurice Ewing Volumes

1 **Island Arcs, Deep Sea Trenches, and Back-Arc Basins** *Manik Talwani and Walter C. Pitman III (Eds.)*
2 **Deep Drilling Results in the Atlantic Ocean: Ocean Crust** *Manik Talwani, Christopher G. Harrison, and Dennis E. Hayes (Eds.)*
3 **Deep Drilling Results in the Atlantic Ocean: Continental Margins and Paleoenvironment** *Manik Talwani, William Hay, and William B. F. Ryan (Eds.)*
4 **Earthquake Prediction—An International Review** *David W. Simpson and Paul G. Richards (Eds.)*
5 **Climate Processes and Climate Sensitivity** *James E. Hansen and Taro Takahashi (Eds.)*
6 **Earthquake Source Mechanics** *Shamita Das, John Boatwright, and Christopher H. Scholz (Eds.)*

IUGG Volumes

1 **Structure and Dynamics of Earth's Deep Interior** *D. E. Smylie and Raymond Hide (Eds.)*
2 **Hydrological Regimes and Their Subsurface Thermal Effects** *Alan E. Beck, Grant Garven, and Lajos Stegena (Eds.)*
3 **Origin and Evolution of Sedimentary Basins and Their Energy and Mineral Resources** *Raymond A. Price (Ed.)*
4 **Slow Deformation and Transmission of Stress in the Earth** *Steven C. Cohen and Petr Vaníček (Eds.)*
5 **Deep Structure and Past Kinematics of Accreted Terrances** *John W. Hillhouse (Ed.)*
6 **Properties and Processes of Earth's Lower Crust** *Robert F. Mereu, Stephan Mueller, and David M. Fountain (Eds.)*
7 **Understanding Climate Change** *Andre L. Berger, Robert E. Dickinson, and J. Kidson (Eds.)*
8 **Evolution of Mid Ocean Ridges** *John M. Sinton (Ed.)*
9 **Variations in Earth Rotation** *Dennis D. McCarthy and William E. Carter (Eds.)*
10 **Quo Vadimus Geophysics for the Next Generation** *George D. Garland and John R. Apel (Eds.)*
11 **Sea Level Changes: Determinations and Effects** *Philip L. Woodworth, David T. Pugh, John G. DeRonde, Richard G. Warrick, and John Hannah (Eds.)*
12 **Dynamics of Earth's Deep Interior and Earth Rotation** *Jean-Louis Le Mouël, D.E. Smylie, and Thomas Herring (Eds.)*
13 **Environmental Effects on Spacecraft Positioning and Trajectories** *A. Vallance Jones (Ed.)*
14 **Evolution of the Earth and Planets** *E. Takahashi, Raymond Jeanloz, and David Rubie (Eds.)*
15 **Interactions Between Global Climate Subsystems: The Legacy of Hann** *G.A. McBean and Michael Hantel (Eds.)*

Mineral Physics Volumes

1 **Point Defects in Minerals** *Robert N. Schock (Ed.)*
2 **High Pressure Research in Mineral Physics** *Murli H. Manghnani and Yasuhiko Syona (Eds.)*
3 **High Pressure Research: Application to Earth and Planetary Sciences** *Yasuhiko Syono and Murli H. Manghnani (Eds.)*

Geophysical Monograph 76
IUGG Volume 16

Relating Geophysical Structures and Processes
The Jeffreys Volume

Keiiti Aki
Renata Dmowska
Editors

American Geophysical Union
International Union of Geodesy and Geophysics

Published under the aegis of the AGU Books Board.

Library of Congress Cataloging-in-Publication Data

Relating geophysical structures and processes : the Jeffreys volume /
Keiiti Aki, Renata Dmowska, editors.
p. cm. — (Geophysical monograph ; 76) (IUGG ; v. 16)
ISBN 0-87590-467-X
1. Earth sciences—Mathematical models. 2. Geophysics—
Mathematical models.
I. Aki, Keiiti, 1930– II. Dmowska, Renata.
III. Series. IV. Series: IUGG (Series) ; v. 16.
QE43.R44 1993
550'.1'5118—dc20 93-28483
CIP

ISBN: 0-87590-467-X

ISSN: 0065-8448

Printed in the United States of America.

CONTENTS

PREFACE

A successful geophysicist must learn how to create a model simple enough for the application of principles of mathematical physics, and yet capable of capturing the essence of the phenomena in the Earth. In the past, geophysicists tended to restrict themselves to either a study of structure or a study of process. The time has come to combine these two lines of study for better understanding of phenomena in the Earth. This volume represents an initial effort toward this goal by providing a survey of such modeling of geophysical structures and processes covering the whole Earth from its ionosphere to the inner core.

Structures and processes in the Earth are not as clearly defined as, say, structures of organs and physicochemical processes in the human body. Some Earth structures and processes may exist only in the minds of geophysicists, where they are intertwined in such a way that a structure produces a process, and vice versa. The book begins with Bolt's paper "Jeffreys and the Earth," which succinctly reviews Jeffreys' overwhelming work based largely on his mastery of simple representations to produce mathematical models of complicated structures and processes. Following is a survey of how geophysicists, working on various parts of the Earth, followed Jeffreys' tradition in relating structures and processes, thereby achieving simple and effective models of the Earth's phenomena.

The volume presents work on different areas of the Earth's system, starting with the ionosphere, and then descending through the atmosphere into the complex structures and processes of the solid Earth, ending at the core.

Not all of the geophysical areas discussed here are accessible for direct, in situ measurements, perhaps with the exception of the ionosphere and atmosphere. The ionosphere was actually studied long before in situ measurement became possible, using the usual geophysical approach based on indirect measurements--in this case, diurnal geomagnetic variations and sounding by radio waves. A fascinating history of some aspects of the geophysical study of the ionosphere is presented here by D. R. Bates. This study has addressed fundamental physical processes at the atomic level, concerning the recombination of electrons and positive ions. Some of the early modelers' conclusions about the ionosphere were confirmed later by in situ measurements, and some were disproved.

The Jeffreys tradition in relating structures to processes is addressed here perhaps most directly by Y. Kurihara, who defines hurricanes as stiff cyclonic vortices with a compact warm core. He discusses their structures in relation to general atmospheric processes, using abundant observations and well-established basic equations governing the processes. He emphasizes the importance of interactions between the hurricane system and both larger- and smaller-scale systems. As an example, he describes how the consideration of hurricane asymmetrical structure due to an Earth rotation effect improves the prediction of the hurricane's course.

The remaining papers in this volume are concerned with solid Earth structures and processes, starting with M. H. P. Bott's presentation of the interrelation in which the density structure determines the stress distribution in the crust and upper mantle. Using the finite element method, he calculates the detailed distribution of stress in the rift zone, mid-ocean ridge, and subduction zone, showing that the driving forces of plate motion such as ridge push and slab pull can be modeled this way.

A full understanding of complex and diverse structures and processes of convergent margins is not yet within the grasp of today's scientists in spite of many recent achievements associated with better and more accessible instruments, and advances in laboratory experiments. A simple rheological framework for comparative subductology is presented here by T. Shimamoto, T. Seno and S. Uyeda, who attempt to find the interrelation between the thermal structure and seismic behavior of subduction zones, trying to catch the diversity with a single, but perhaps most important parameter: temperature distribution.

Thermal and density structures can be connected to processes more straightforwardly than can the structures determined by seismic methods. The latter methods, on the other hand, can reveal the Earth's internal structure with much higher spatial resolution than any other geophysical method. The seismic structure must, however, be translated into some other structures to link it directly with processes. Thus, B. Kennett asks (rather desperately) what sort of structures these are--referring to the complex behavior of upper mantle transition zones as seismic reflectors. The nature of the transition zone appears to vary from place to place, so how can these complex seismic signatures be related to the effects of pressure, temperature, phase, and composition?

The velocity structures obtained by global seismic tomography are usually translated directly to the thermal structure and consequently to the density structure. Recent developments in derivation of three-dimensional velocity models are presented by Dziewonski et al., along with some examples of the use of these models in explaining several geodynamic observables. Exploring the interrelation between geophysical structure and process was not possible using the one-dimensional Earth model of classical seismology.

The last three papers discuss the structures and processes in the deep Earth. Hide et al. explore the interrelation between the topographic core-mantle coupling and fluctuations in the Earth's rotation. Chemical mantle-core reactions are then summarized by R. Jeanloz, and their geomagnetic implications pointed out. Processes of accretion and core formation in the Earth are discussed by H. Newsom and F. Slane, as well as their bearing on the present major divisions of the Earth into the core, lower mantle, and upper mantle, and also the geophysical structures that may have been present during the earliest history of the Earth, such as magma oceans and a dense high-temperature atmosphere.

This volume provides an in-depth treatment of a representative set of papers from the Union symposium "Jeffreys Symposium: Interrelation Between Geophysical Structures and Processes" during the meeting of the International Union of Geodesy and Geophysics in Vienna in August 1991. This symposium was extraordinary because of the focus on the interrelation between structures and processes in the broad subject areas of geophysics. The symposium was named appropriately after Sir Harold Jeffreys and was attended by Lady Jeffreys, who expressed keen professional interest in all presentations. Significant contributions made by Jeffreys (1891-1989) in a wide variety of fundamental problems of the Earth are described in the first paper by B. A. Bolt.

Keiiti Aki
Renata Dmowska
Editors

FOREWORD

The scientific work of the International Union of Geodesy and Geophysics (IUGG) is primarily carried out through its seven associations: IAG (briefly, Geodesy), IASPEI (Seismology), IAVCEI (Volcanology), IAGA (Geomagnetism), IAMAP (Meteorology), IAPSO (Oceanography), and IAHS (Hydrology). The work of these associations is documented in various ways.

Relating Geophysical Structures and Processes: The Jeffreys Volume is one of a group of volumes published jointly by IUGG and AGU that are based on work presented at the Inter-Association Symposia as part of the IUGG General Assembly held in Vienna, Austria, in August 1991. Each symposium was organized by several of IUGG's member associations and comprised topics of interdisciplinary relevance. The subject areas of the symposia were chosen such that they would be of wide interest. Also, the speakers were selected accordingly, and in many cases, invited papers of review character were solicited. The series of symposia were designed to give a picture of contemporary geophysical activity, results, and problems to scientists having a general interest in geodesy and geophysics.

In view of the importance of these interdisciplinary symposia, IUGG is grateful to AGU for having put its unique resources in geophysical publishing expertise and experience at the disposal of IUGG. This ensures accurate editorial work, including the use of peer reviewing. So the reader can expect to find expertly published scientific material of general interest and general relevance.

Helmut Moritz
President, IUGG

Jeffreys and the Earth

BRUCE A. BOLT

Seismographic Station and Department of Geology and Geophysics, University of California, Berkeley

The work of Harold Jeffreys related to geophysical processes is described and its interactions with other research of this century on the Earth and planets explored. The extent of his influence on the style of the subject by his insistence on mathematical treatment, order of magnitude arguments, and probability assessments is evaluated. For forty years, he had no peer in his wide application of wave theory to seismological problems. Jeffreys's greatest lasting success was his construction of global travel-time tables, with uncertainty estimates, for seismic waves and their inversion to a seismic velocity structure in a radially symmetric Earth model. He also contributed significantly to a wide variety of fundamental geodynamical problems, especially the Earth's rotation, its figure and gravity field, and thermal history. His development of a theory of inference applicable to geophysics was profound; but his assessment of posterior inference in relation to global tectonic evolution gave little weight to the importance of the geomagnetic observations on the ocean floor. His conclusions on the long-term mechanics of the Earth were contrary to the now widely held theory of plate tectonics. Yet his work on convection and energy dissipation remains central to the study of Earth processes.

JEFFREYS'S CAREER

It is appropriate at the outset to give a brief account of Jeffreys's life. Detailed memoirs have now been published by the Royal Society of London [*Cook*, 1990, including a microfiche with a list of publications and other material], the Seismological Society of America [*Bolt*, 1989], and elsewhere.

Harold Jeffreys was born 100 years ago on 22 April 1891 at Fatfield, England. His parents were school teachers. He obtained a B.Sc. degree in 1910 with distinction in mathematics at Armstrong College, at that time a College of the University of Durham; the site is now the University of Newcastle-upon-Tyne. He went on to study mathematics at Cambridge University and was elected a fellow at St. John's College in 1914, the first year of World War I. He remained a Fellow until his death in 1989 — some 75 years. A 1929 photograph including Jeffreys is reproduced in Figure 1.

Jeffreys published his first investigations in geophysics as early as 1915, inspired by the major geophysical endeavors — not so well known today — of G. H. Darwin [see *Darwin*, 1907-16]. From 1917 to 1921 he worked at the Meteorological Office in London, tackling some hydrodynamical problems that came from the war front in France. He returned to Cambridge in 1922, where he stayed for his life's work and where in 1946 he was elected Plumian Professor of Astronomy and Experimental Philosophy. He retired from his chair in 1958 but vigorously continued his research for many years afterwards.

With the pioneer seismologist H. H. Turner (Director of the *International Seismological Summary*) and Robert Stoneley, he promoted geophysics at the Royal Astronomical Society, recommending the publication of a *Geophysical Supplement* — later the *Geophysical Journal* and now the *Geophysical Journal International*. His papers dominated its numbers for more than thirty years. He was President of IASPEI (1957-1960) and attended the first assembly of the IUGG in 1922 and every meeting of the IUGG from 1933 to 1967.

In 1940 Jeffreys married Bertha Swirles, an atomic physicist. In 1929, she had received a Cambridge Ph.D. for a thesis on "Some Applications of the Theory of Perturbations in Quantum Mechanics." She had been a lecturer in Applied Mathematics at Manchester University when she returned in 1938 to Girton College as Fellow and Lecturer. She retired from lecturing in 1969 and has now been a Fellow for over fifty years. Papers concerning her work will be deposited with those of Sir Harold in St. John's College Library. Together, they wrote the influential and still widely used treatise *Methods of Mathematical Physics* [*Jeffreys and Jeffreys*, 1972]. For over three decades, visiting mathematicians, astronomers, statisticians, and geophysicists were welcomed by Sir Harold and Lady Jeffreys at their home at 160 Huntingdon Road, Cambridge, recently described by a Russian visitor as "an Institute for the History of Geophysics." Sir Harold died in Cambridge just short of the age of 98.

Relating Geophysical Structures and Processes: The Jeffreys Volume
Geophysical Monograph 76, IUGG Volume 16

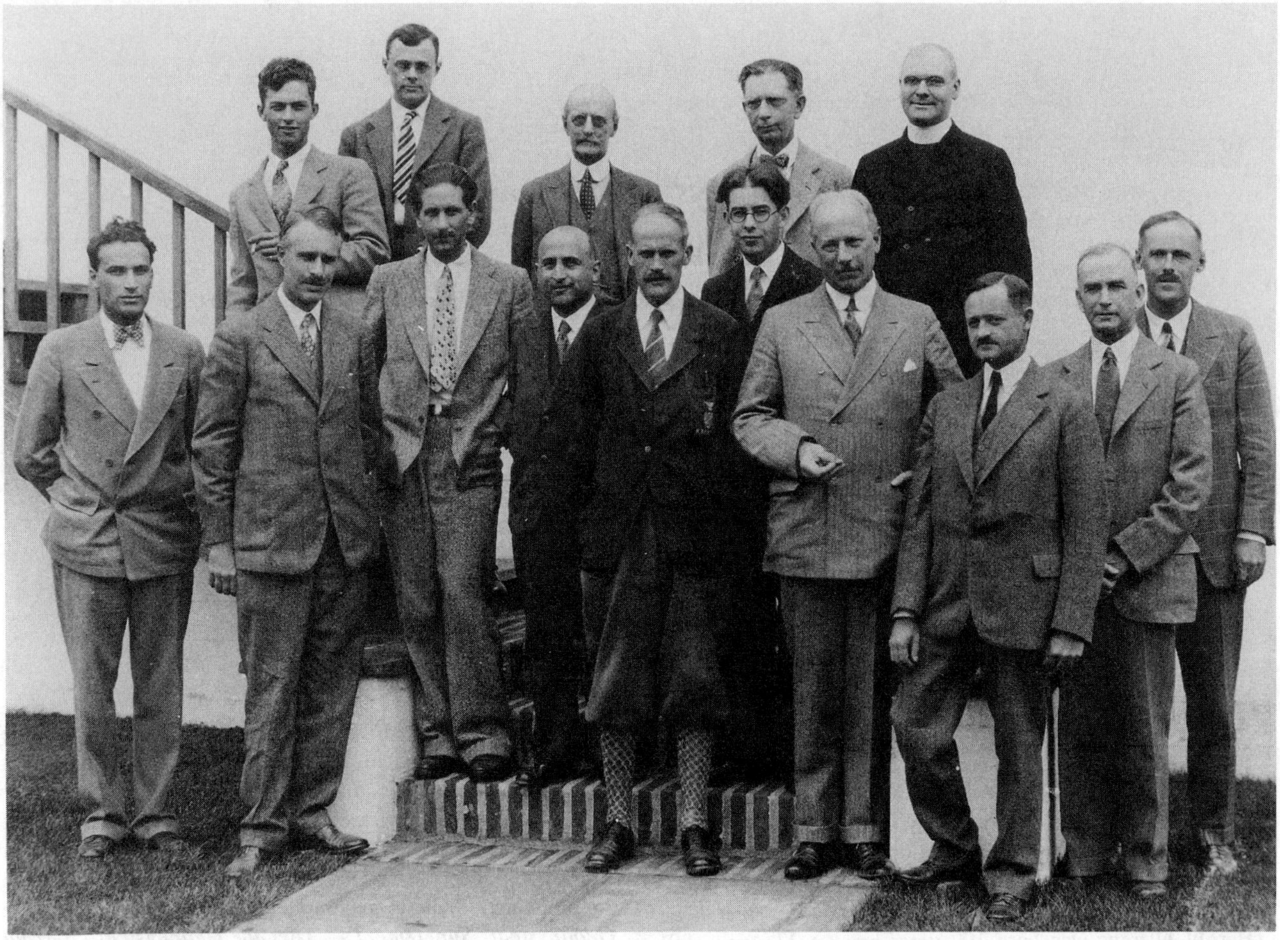

Fig. 1. Conference on the future of seismological studies at CalTech at Pasadena in 1929. Front row, left to right: assistant, Leason Adams, Hugo Benioff, Beno Gutenberg, Harold Jeffreys, Charles Richter, Arthur Day, Harry Wood, Ralph Arnold, John P. Buwalda. Back row: assistant, Perry Byerly, Harry Fielding Reid, John Anderson, Father J. B. Macelwane.

One enduring anecdote of Jeffreys is told in different forms: *Bates et al.* [1982, p. 38, footnote 27] have written "Within the fields of seismology and gravitation, it is difficult to find a research topic that Sir Harold has not already written about. However, when geophysicists from British Petroleum Ltd. approached him with the hope of learning some new survey techniques, they first explained at length the difficulties in finding new oil-bearing structures in Iran. He replied, 'I'm glad it's your problem, not mine!'"

THE TREATISE *THE EARTH*

In recent times there has been wide discussion that contemporary progress will soon bring theoretical physics to an end with a final discovery of "A Theory of Everything" (acronym A TOE). This idea of finality has been around for a while. At the end of the last century, Lord Kelvin believed that all that remained to be done in physics was to measure the physical constants more accurately. In the 1920's, after Einstein's relativity theory, Born suggested that physics would be over in six months. Afterwards, physics got busy again.

With the publication of the first edition of *The Earth* in 1924, Jeffreys brought to geophysics a large part of Kelvin's optimism concerning the power of classical dynamics. Yet Jeffreys never did claim, except perhaps in his formulation of a probability methodology, that he was describing "the geophysical TOE." Each succeeding edition of *The Earth* through the sixth (and last) in 1976, incorporated changes and iterations towards more complete and realistic models and processes. This seminal work demonstrated how comprehensive and quantitative mechanical arguments can define the physical properties of the whole Earth. Fundamental theories were laid down concerning the Earth's genesis, its internal history, its structure and composition, and its external and internal dynamical interactions. Rather than building upon the preconceptions and assumptions from antiquity, *The Earth* altered geophysics permanently by setting up a consistent critical methodology in which physical theory and observations were explicitly interdependent. An exception in Jeffreys's multifaceted canvas which may have affected his views on plate tectonics (see the later section on geodynamics) was a relative neglect of electromagnetic properties and processes in the Earth.

In his general publications we can trace the substantial evolution in Jeffreys's conception of both Earth structure and geophysical processes. For example, he initially argued that the solar system had been formed by tidal disruption of the primitive sun by a passing star. By 1952 he found defects in all theories of the origin of the solar system [*Jeffreys*, 1952]. At an early stage he inclined towards Darwin's resonance theory for the origin of the moon; later he wrote "the Moon was certainly not formed by fission from the Earth" [*loc. cit*., p. 290].

An illustration [*Bolt*, 1991] of how robust some of his early inferences were, however, comes from the growth of our knowledge of the terrestrial density distribution $\rho(r)$. In the first edition of *The Earth*, Jeffreys gave a critique of the primitive available knowledge on $\rho(r)$. The Earth model he preferred had been proposed by *Wiechert* [1897]. It consisted of two constant density shells and was based on abundances of rocky and iron meteorites. Jeffreys wrote (p. 198) that "The little knowledge that we have suggests that the interior is so dense as to be probably metallic while the outside is rocky. Such materials cannot mix freely and therefore we should expect a fairly strong boundary between them with a sudden discontinuity in density."

By 1929, the second edition reflected a radical change in the available observations. Jeffreys acknowledged the discovery by R. D. Oldham of a sharp decrease in P and S seismic velocities in the deep interior and the subsequent calculation for the depth of the mantle-core boundary of 2900 km by B. Gutenberg. Reliable distributions of velocity of both P and S waves in earthquakes had by then become available, allowing the compressional increase in density to be calculated using realistic elasticity assumptions. Jeffreys perceived that *Williamson and Adams* [1923] had provided "a method for recalculating the distribution of density." About two decades later, after a provisional solution in 1936, Jeffreys's student K. E. Bullen carried through this program [see *Bullen and Bolt*, 1985] using the much more representative V_p and V_s functions derived from the *Jeffreys-Bullen* [1940] *Seismological Tables* of travel times (the J-B tables). Succeeding editions of *The Earth* tabulated the density results of Bullen based on the Adams-Williamson differential equation.

With the great Chilean earthquake of 1960, terrestrial spectroscopy was born. While at Cambridge in 1961, I had reduced a long time series of ground motions from this earthquake written in the Grotta Gigante by a long pendulum devised by the late Professor A. Marussi. The Fourier analysis schemes available at the time provided significance tests only for the amplitudes of the largest eigen modes. *Jeffreys* [1964] noticed that uncertainties of the eigen frequencies themselves were not being estimated and published a note on the subject. He later [1977] remarked "What is often done is to compute for intermediate values [of the eigen frequency] by integration. This is pointless because the resulting values are simply interpolates and their errors are highly correlated .... I do not know of a single case where an uncertainty has been properly determined." Perhaps even more important, the use of the eigen frequency measurements to make inversions to properties of the Earth was criticized by *Jeffreys* [1967] on the grounds that no allowance had been made for anelasticity in the construction of Earth models. Work by *Randall* [1976] and others, "confirmed Jeffreys's objection and has shown that discrepancies between models based on free oscillation data and models based on travel-time data can be removed, and quantitative assessment of anelasticity made" [*Lapwood and Usami*, 1981, p. 17].

Despite Jeffreys's concerns regarding the measurements, the sixth edition of *The Earth* [1976] contained a description of the new spectral method of making density estimations. Jeffreys described (p. 288) the recording of the free oscillations of the Earth following the 1960 Chilean earthquake and noted their great usefulness in principle in revising interior elastic parameters, such as density and rigidity of the inner core. Such parameters "can be taken as an unknown and adjusted to fit the periods." Now, rather than proceeding by integration of the Adams-Williamson equations, with substitution of assumed bulk composition parameters, it was more promising to invert the problem using the terrestrial eigen frequencies [*Bullen*, 1975]. The critical strength was to allow strong *independent* inferences to be made on the materials in the deep interior.

As stated by Jeffreys as early as the second edition of *The Earth* [1929], "We may look at the matter from the other end. The mean pressures and the bulk moduli [inside the Earth] can .... be estimated, and hence so can the reduction in density if these pressures are taken off." In an extension of this idea, *Jeffreys* [1937] used this scaling principle to interpolate from the Earth's density distribution to those of other terrestrial planets and hence to infer their structure and composition [*Bullen*, 1975].

Fluid Dynamics and Nutations of the Earth and Moon

Jeffreys had an early interest in fluid dynamics, no doubt fortified by the environment at the Meteorological Office. His discussion of the eddy viscosity and convective processes in the atmosphere was influential and included the idea that cyclonic storms are essential to the general circulation. When the wind blows over the surface of a body of water, surface waves are formed. For well over a century, applied mathematicians have attempted to model the mechanism or mechanisms — with only partial success. An alternative to the unsatisfactory model of Helmholtz and

Kelvin was introduced by *Jeffreys* [1925]; it was based on "sheltering" on the leeward side of crests. Little progress was made subsequently until the mid-1950's. In other seminal work, beginning in 1926, *Jeffreys* [1926*b*] used stability arguments to establish realistic conditions for the onset of convection in a viscous fluid heated from below [*Jeffreys and Jeffreys*, 1971-77, *Collected Papers*, *4*, Section XII].

Jeffreys was deeply interested in the nutations of the Earth — both free and forced by the Moon. Oldham in 1906 and Gutenberg in 1914 had given seismological evidence that the major part of the core of the Earth was much less rigid than that of the mantle, in that no S waves were detected at relevant distances [see *Bolt*, 1982]. Jeffreys, using rough estimates of density and elastic parameters based on Wiechert's Earth model, calculated in 1915 that the resulting theoretical free period of the monthly nutation was too short when compared with observations. By 1926, after including the effective compression of density, he demonstrated that the seismological radius of the core obtained by Gutenberg was compatible with the observed free period if the core had negligible rigidity [*Jeffreys*, 1926*a*]. In celebration of this accomplishment, Jeffreys sent a postcard to Stoneley, his closest colleague and friend: "Die Wiechertsche und Oldham-Gutenberg Kerne Identisch sind" [see *Jeffreys*, 1976].

Jeffreys then went further. He concluded [1926*a*, p. 371] that "There seems to be no reason to deny that the Earth's metallic core is truly fluid." I take this demonstration of the core's fluidity as Jeffreys's most significant geophysical discovery, although other judgements can be strongly argued. The sequence of events which sustain this view is well described with detailed references by *Brush* [1980]. Brush writes: "In this context, 'discovery' simply means presenting a claim with arguments sufficient to convince the scientific community at the time; originality lies in the successful persuasion not in the conception of the claim." The majority view among geophysicists and geologists in the mid-20's was that the entire Earth was solid, a view apparently accepted by Jeffreys as late as 1925. About that time, Jeffreys had reworked some of the basic material. Brush argues that the credit for the discovery of the fluid core goes to Jeffreys because he was able to convince Gutenberg and others and refute certain arguments against the hypothesis. While Gutenberg appeared to accept Jeffreys's arguments of the non-rigidity of the core, he was reluctant to describe the core as liquid until as late as 1957 when one finds in print, "the core is liquid" [*Gutenberg*, 1957]. Much earlier, Gutenberg himself had considered a liquid core but rejected it. Although he had not observed S waves passing through it, his explanation was that they would be rapidly absorbed within the core by frictional attenuation.

Later, with Vicente, Jeffreys further investigated nutation of the Earth's axis [*Jeffreys and Vicente*, 1957]. This heavy computational work allowed for elastic properties of the mantle and fluid core and yielded fair agreement with the observed amplitude of the 19 yearly nutation. *Munk* and *McDonald* [1960] judged that "the subject [of rotation of the Earth] was reopened in the light of modern geophysical knowledge by Jeffreys. His contributions dominate the subject." He worked on the variation of latitude from at least 1916 almost to the end of his life.

Jeffreys's great interest in dissipative processes in the Earth is nowhere better illustrated than in his study of tidal friction in shallow seas [*Jeffreys*, 1920]. The work grew out of a demonstration by *Taylor* [1919] of the astronomical importance of energy loss by tidal action in the Irish Sea. Jeffreys considered two possible mechanisms for dissipation of the Earth's rotational energy: internal friction and torques due to tides raised by the Moon. He rejected the first and supported correctly the energy loss by frictional forces produced by currents in shallow seas around the world and along the coasts. *Knopoff* [1991] remarks "All of this work demanded extraordinary attention to the measurement of the Earth's gravity field, of the astronomical observations, of the rates of change of the length-of-day, the currents in the oceans, etc. Some of the results were done on mechanical calculating machines. The analysis could not have been done without Jeffreys's brilliance as an applied mathematician."

In a related field, Jeffreys did key work on the determination of the Earth's gravitational field [*e.g.*, *Jeffreys*, 1941, 1959]. He produced a plot of the deviation of the Earth's figure from hydrostatic shape, which showed gravity highs over the North Atlantic and the Pacific and gravity lows over the Caribbean and India [see *e.g.*, *Jeffreys*, *The Earth*, fourth edition, 1962, p. 187]. Although, as in other studies of his, these results were at first unfashionable and met with skepticism, they were eventually vindicated by observations of perturbations of artificial satellite orbits.

Seismological Travel-Times

One of Jeffreys's most lasting contributions to geophysics was his work in improving tables of travel times of seismic waves through the Earth. Such tables permit not only the precise location of remote earthquake sources, but allow mathematical inversions to estimate the seismic velocity structure. In the mid-1930's, Jeffreys set about correcting the generally used tables and was soon joined by Bullen as a Ph.D. student. By 1939 this research program, which produced the Jeffreys-Bullen tables and a corresponding seismic velocity model for the Earth, was complete. Contemporaneously, Gutenberg, with help from Richter, derived at the California Institute of Technology similar tables using rather different methods. Jeffreys excelled in statistical reduction of published times measured by others; Gutenberg, in personally correlating wave recordings. Jeffreys remarked that the closeness of the final results was most satisfactory given the independence of the analyses [see *Jeffreys*, 1939*b*, and *The Earth* , sixth edition, p.121].

The Jeffreys-Bullen travel times remain in 1991 in

standard use for most teleseismic locations today even though the corresponding derived velocity structure has been modified in important ways. Attempts have been made to refine them drastically, so far without much change, differences in mean times generally being within the standard errors estimated by Jeffreys. A recent revision, the 1991 IASPEI Tables [*Kennett*, 1991], that combine ISC data from various regions, gives a mean offset from J-B of about 1.85 seconds for P times. The variations are mostly within the J-B standard errors.

This result is in accord with *Jeffreys's* [1977] claim that "no other global tables could give better agreement with observation for all earthquakes." In fact, he was critical of efforts to revise global tables from mixed data that ignore regional differences. After World War II, Jeffreys spent much effort developing travel times for various regions of the Earth, but as yet this type of modification remains unused in routine global hypocenter programs.

Jeffreys's interest in geological processes [*Jeffreys*, 1935] made him well aware that, in the Earth, seismic waves do not propagate through a homogeneous, perfectly elastic medium. (His last research publication [*Jeffreys and Shimshoni*, 1987] dealt with regional differences in seismology.) He certainly foresaw the need to examine the three-dimensional structure of the Earth [see *Masters*, 1989]. Because he knew that rock discontinuities and strong velocity gradients would lead to serious inadequacies in the application of ray theory, his attention turned to numerical approximations of waveforms near cusps, caustics, and shadow zones. He also incorporated viscosity and scattering. In his wave analysis, he made incisive use of the Airy integral and independently rediscovered the Green-Liouville solution of the harmonic equation with a variable coefficient. This solution is now known as the JWKB approximation (named after Jeffreys, G. Wentzel, H. A. Kramers, and L. Brillouin).

Jeffreys stressed that one of the major questions remaining in seismology was the persistence of the coda in seismograms, a problem of current interest in understanding crustal processes [*The Earth*, fourth edition, 1962, p. 399]. Much modern seismological research incorporates the application of Rayleigh's Principle by Jeffreys for computing approximate wave eigen functions, as well as his results on viscous effects on waves and wave dispersion. Perhaps the only major parts of mathematical seismology that he did not visit in depth [*Lapwood*, 1982] were the inverse fault-plane method developed by P. Byerly and the related, but more general, theory and modelling of earthquake sources. Even so, Jeffreys addressed the process of rock fracture and melting in fault rupture [e.g., *The Earth,* 1962, p. 358].

Jeffreys had the insight to see that a fresh, strict analysis of the newly accumulated measurements from many widely distributed earthquakes could provide robust estimates of the unknown velocity structure of the Earth. In the course of the development of the J-B tables, he considered thoroughly the convergent nature of the process undertaken. He initiated methods of dealing with outlying observations by developing an early form of biweights ("the method of uniform reduction"). Because for travel-time curves there is no theoretical function available that can be fitted for estimating the adjustable parameters by least squares, he developed a "method of summary values" and connected the various branches of the curves using the chi-squared test in a masterly way.

My own introduction to the stature of Jeffreys came through work on travel-times suggested to me by Bullen, who from 1945 was Professor of Applied Mathematics at the University of Sydney, Australia. *Bullen* [1957] believed that Jeffreys's research approach which incorporated Bayes's theorem and probabilistic scientific inference was a quintessential example of "the spirit of applied mathematics."

SEISMIC VELOCITY STRUCTURE IN THE EARTH

By a nonlinear inversion, the seismic travel times lead to P and S velocity distributions of the Earth. This process is the fundamental seismological contribution to knowledge of deep Earth structure. In Figure 2, a comparison is given between prominent Earth models of Jeffreys and Bullen and the recent, widely-used PREM [*Dziewonski and Anderson*, 1981]. The comparison indicates that the velocity curves closely agree generally. The main exceptions are that the low-velocity zone outside the inner core (Region F) has now disappeared and, in the upper mantle, more discontinuities have been introduced. There are also small but geophysically important differences in detail in the narrow boundary layer D″ around the Earth's outer core.

Many of Jeffreys's other seismological contributions also use ingenious mathematical arguments. For example, a major problem in observational seismology in the early days was whether earthquakes were generated by deep sources. In 1922, *Turner* [1922] pointed out some discrepancies in travel times in earthquakes around the world, in particular from Japanese earthquakes. For some earthquakes, arrival times were later at antipodal stations than normal and earlier for others. Turner put "normal" focal depths at 200 km.

This surprising hypothesis sparked Jeffreys's interest, and he proposed as a crucial test a theorem from eigen theory which states that a mode of a vibrating system cannot be generated by a source at a node of that mode. It follows that because seismic surface waves have their motions restricted to surficial depths, they cannot be generated by deep sources. Later, *Jeffreys* [1976, p. 559] recalled that "I pointed out to Turner that ... a deep focus earthquake should excite very small surface waves or none. This could be tested by a simple inspection of the seismograms. But Turner was so satisfied by his own arguments that he would not look." The matter was settled decisively by *Wadati* [1928] in Japan, who published strong direct arguments

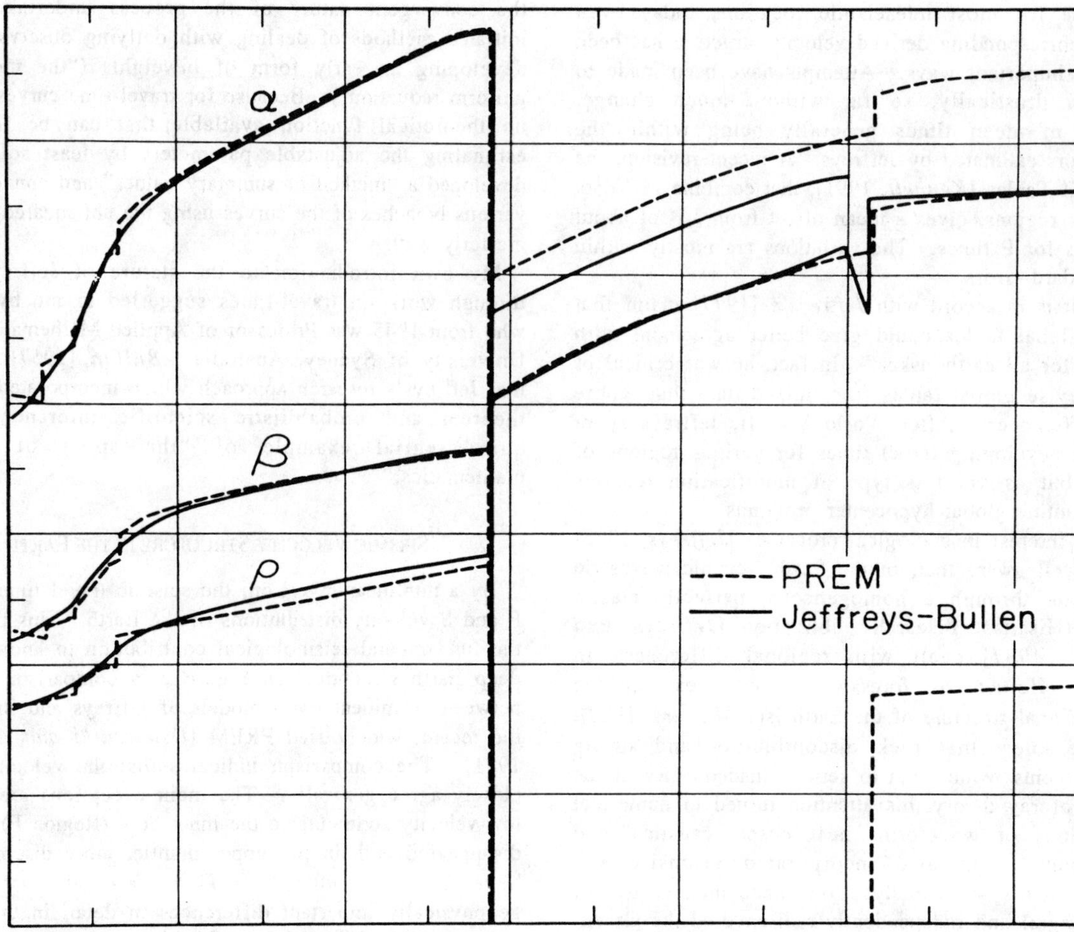

Fig. 2. Comparison of seismic P velocity (a) and S velocity (b) and density (r) distributions Model A and PREM in the Earth estimated by Jeffreys and Bullen (see *Bullen* and *Bolt*, 1985) and *Dziewonski* and *Anderson* (1981).

using local observations that showed that local earthquake sources ranged between tens and hundreds of kilometers in depth. The correct assessment of focal depth led ultimately to the mapping of subduction structure.

Geodynamics

There is, of course, much more that could be said about Jeffreys's voracious curiosity about geophysics. Given the spectacular accumulation of geophysical observations of diverse types from 1924 when the first edition of *The Earth* appeared, through the era of space satellites and massive oceanographic observations, it was inevitable that many hypotheses on structure and processes of the Earth waxed and waned during his working life. Through his major contributions in geodynamics and Earth structure he became involved in highly consequential controversies concerning global tectonic processes.

The best known was Jeffreys's rejection of continental drift and doubts on the geomagnetic evidence for plate tectonics. He had worked out models which he argued demonstrated that the cooling of the interior of the Earth would strain the crust sufficiently to account for mountain structures. Jeffreys considered convection of the Earth and he saw a place for it, but his studies of yielding led him to reject mantle convection as a prime mechanism.

The reasons for the rejection are not simple and can be only summarized here. It is generally agreed that there is a basic indeterminancy problem to overcome in selecting one Earth model or process from an infinite number. Jeffreys invoked the basic principle of simplicity that he first discussed as early as 1921 [*Wrinch and Jeffreys*, 1921; see also 1931, p. 38]. A modern view of the simplicity principle and Jeffreys's contribution to its evolution in statistical inference is discussed by *Jefferys and Burger* [1992]. Given this principle, Jeffreys went on to examine the consequences of a single law to describe imperfections of elasticity — one that was valid over all wave lengths: from S waves to free oscillations, to tidal yielding, to mountain building [*Jeffreys*, 1965]. In 1977, he stated that this problem was "his chief interest at present" [p. 11].

The model process that he studied most was the "modified Lomnitz law of damping" [*Jeffreys and Crampin*, 1970]; this entails that thermal movement once started will die down. The problem with the application of the Lomnitz law throughout the Earth is that the rheological coefficients are strongly temperature dependent. Dominant terms near the surface become small at the high temperatures of the mobile mantle. Most recent study on rheological properties of the Earth also makes it probable that there is no single simple law that applies universally whatever the time scale or whatever the strain.

For many of the present generation of geophysicists, Jeffreys's name is associated with his skepticism of the hypothesis of plate tectonics. Although notions of continents moving relative to each other can be found well before, a systematic account of continental displacement was presented in lectures by Alfred Wegener in Germany in 1912. But not for about ten years, probably because of the first World War, was the hypothesis taken seriously in the geological profession. In England on 16 February 1922 a short unsigned review of a book describing Wegener's theory appeared in *Nature* [p. 202-203], stating "This book makes an immediate appeal to physicists but is meeting with strong opposition from a good many geologists." Wegener postulated the breakup of a Carboniferous supercontinent, the lateral displacement of the fragments and the matching of geological structures and of biological and climatic zones. Both support for and objections to Wegener's theory became common in the ensuing decade, with major attacks and debates in England by such geologists as *Lake* [1922] of the Sedgwick Museum in Cambridge [see *Marvin*, 1991]. In a letter to *Nature*, *Jeffreys* [1923] said that Wegener's theory depended on two assumptions: first, that any force, however, small, can deform the Earth to any sizable extent if only it acts long enough. This he did not take to be incorrect, although there was some evidence against it. The second assumption, that a small force can overcome a larger force acting for the same time in the opposite direction was, he held, inconsistent with current physical knowledge. The objection to a mechanism for the rafting of continents through an upper layer of basaltic material with definite strength was, of course, correct. Indeed, as time went on, the seismological evidence grew that the structure of continents is deeper than earlier believed. On the other hand, paleomagnetic evidence during the 1950's and 60's became strong that drift has been occurring between North America and Europe since the Paleozoic era. Further history of the construction in the late 1960's of the plate tectonic model can be found in *Cox* [1973] and *Glen* [1982].

As has been noted, the first edition of *The Earth* appeared in 1924. In this treatise Jeffreys argued for a cooling, contracting Earth which had outer layers of finite strength down to hundreds of kilometers. This hypothesis had competition from several others, notable among which was the upper mantle convection current theory of *Holmes* [1931]. Later, *Holmes* [1965, p. 1202] wrote, "Following the lead of many influential geophysicists, most geologists were reluctant to admit the possibility of continental drift because no recognized natural process seemed to have the remotest chance of bringing it about." According to *Marvin* [1991], "Not until the 1960's was the existence established of a low-velocity zone in the upper mantle, nor did it become definite that earthquakes are only generated at depths of hundreds of kilometers at very special sites. Before that there was no reason to believe that the uppermost 700 kilometers of the interior were not equally stiff everywhere."

Oliver [1991] argues that Jeffreys was clearly on the track of plate tectonics before 1952 but because the globe-encircling rift systems were only partly mapped out at that time, he believed volcanism to be always intermittent. In analyzing melting at depth in the third edition of *The Earth* [1952, pp. 289-290], Jeffreys writes, ".... we should ordinarily have in this state a solid crust resting on a less dense liquid .... such a state would be stable if the crust was thick enough, and there is no reason why it should not become permanent. If it was too thin, or if it broke anywhere under some local disturbance, instability would arise and would lead to wholesale fractures of the outer crust. Solid blocks would be continually foundering and melting on the way down, while the fluid would actually come to the surface in places." As Oliver points out, this physical reasoning led to an Earth much like the one visualized in the theory of plate tectonics today. As always, Jeffreys looked for a crucial test. His argument continues: "This would be a tempting explanation of surface igneous activity, but unfortunately it requires a continuous connection among the liquid parts at the surface, with the solid blocks separated. Actual igneous activity is always local and the crust has remained connected throughout geological time." If this Bayesian assessment could have been conditioned on the observations of the midoceanic volcanic system, Jeffreys's views may well have changed. Even so, his argument [*Jeffreys*, 1974] was clearly set out for others to modify when the critical observations on magmatic flow rates were made. His final paper on the subject was in 1978 [*Jeffreys*, 1978].

Geophysical Inverse Problems

The modern literature of geophysical inverse theory [see *Backus and Gilbert*, 1970] was created towards the end of the most productive part of Jeffreys's career and was stated in terms of an algebraic formalism not found in Jeffreys's papers. Nevertheless, from the beginning of his research he was deeply involved in inverse problems which are nowadays called estimation, resolution, uniqueness, and so on. In the first edition of *The Earth* [p. 1] we find the statement: "Thus, the problem of the physics of the Earth's interior is to make physical inferences over a range of depth of over 6,000 km from data determined only for a range of 2 km at the outside." As the earlier discussion makes evident, although Jeffreys may not have used the term, mathematical modelling of physical processes in order to compare observations was his special mark.

As is illustrated in earlier sections, Jeffreys stressed the Bayesian approach to inference that uses previous knowledge and observations to help a scientist decide which hypothesis is most probable [*Zellner*, 1980]. He was brilliant in his treatment of uncertainty — particularly the differentiation between random errors and systematic errors. His methods are well illustrated in his analyses of gravity data [see *Jeffreys and Jeffreys*, *Collected Papers*, 1971-77, *3*, Section IV] and, as already mentioned, regional differences in seismological travel times.

It can be strongly argued that a clear understanding of Jeffreys's mode of thought on the structure and evolution of the Earth is impossible without consideration of his views on inference and probability. Many geophysicists do not realize that, like *The Earth*, Jeffreys's other opus, *Theory of Probability* [1939*a*] has also had great influence. Typically, Jeffreys not only developed therein a personal theory but included a wealth of applications to geophysics.

As described by *Lindley* [1991], Jeffreys was a contemporary at Cambridge of another giant in probability and statistics, the geneticist, Sir Ronald Fisher. Not only on the nature of scientific inference, but also on continental movement, were they in opposition. *Lady Jeffreys* [1991] has written, however, that "the apparent differences between Harold and R. A. Fisher have been much exaggerated." Sir Edward Bullard related the story of the day that the feud between them evaporated. After both hearing a lecture by Sir Arthur Eddington on scientific inference, they were both so horrified that they shook hands and promised not to write any more rude things about each other [*Box*, 1978, p. 422]. As statistics has evolved, the prevailing fashions have followed Fisher with his formulation of maximum likelihood, and J. Neyman with his use of confidence intervals. The essential difference is that *Jeffreys* [1977] considered probability to be the appropriate description for all uncertainty, whereas the majority of practicing statisticians restrict its use to the uncertainty associated with data.

It is of interest that *Runcorn* [1956] in his work on paleomagnetism in rock samples, turned to Fisher for a statistical way to test homogeneity in a cluster of directions of magnetization. *Fisher* [1953] subsequently produced his famous paper on statistical dispersion on a sphere, and the method was widely applied in geomagnetic investigations that became decisive evidence for tectonic plate movements.

An eminent geophysical theorist on the subject of inference of geophysical processes, *G. E. Backus* [1991] has lately reanalyzed the inverse problem of uniqueness very much along the lines explored by Jeffreys decades earlier. When prediction functionals are not linear combinations of the data functionals, prior information must be found. In cases where quadratic bounds exist, such as in modelling of the geomagnetic field at the core-mantle boundary [see *Bloxham, et al.*, 1989], Backus favors a confidence set inference (CSI) scheme derived from the developments of J. Neyman, rather than Bayesian prior probability distributions (BI). We cannot know what Jeffreys's answer would be to Backus's suggestions, but he generally took exception to such Fisher and Neyman constructs. *Jeffreys* [1977] did insist that the "actual scientific method consists of successive approximations to probability distributions." Indeed, he derived a useful form for the prior distribution of a parameter based on the log likelihood function. He says that he "once remarked to Fisher, that in nearly all practical applications we should agree, and that when we differed, we should both be doubtful" [*Jeffreys*, 1977]. Backus is, in fact, not far away from Jeffreys's position even today when he writes [1991]: "I advocate replacing BI by CSI only in certain circumstances. If prior information really is a probability distribution and the observer has confidence in it, then BI is preferable to CSI."

A quotation from Jeffreys's many stimulating philosophical writings helps us understand his insistence in geophysics (and indeed in all scientific argument) on the need for a universally consistent system of inference. "The theory of probability thus enables us to solve many of the major problems of scientific method ... it makes it possible to modify a law that has stood criticism for centuries without the need to suppose that its originator and his followers were useless blunderers." [*Jeffreys*, 1938]

Evidence for Jeffreys's continuing influence in modern times is easily found by consulting advanced texts [see *Lapwood*, 1979, 1982; *Jacobs*, 1987]. For example, in his discussion of the properties of terrestrial planets, *Kaula* [1968] writes, "The outstanding treatise of the mathematical treatment of the physics of the Earth's interior has long been *The Earth* by Jeffreys. Much of the material discussed in this chapter [On the Mechanical and Thermal Aspects of a Planetary Interior] is based on work of Jeffreys." A famous and seminal paper on the elasticity and constitution of the Earth's interior was published by Francis Birch in 1952. In his references, Birch cites the work of Jeffreys and Bridgman most frequently after his own. He states that Jeffreys in 1937 gave what he still regarded as the most reasonable values for the radii of the cores of the inner planets. I had occasion myself in rewriting Bullen's classical textbook, *An Introduction to the Theory of Seismology*, to realize how much of that book [*Bullen and Bolt*, 1985] was based upon the researches of Jeffreys in seismology. The most widely used advanced text in that subject today [*Aki and Richards*, 1980] gives a substantial list of references to Jeffreys's pioneering publications.

Acknowledgments. The helpful comments and interest of Lady Jeffreys are much appreciated. I benefited from suggestions on the paper from D. R. Brillinger, J. A. Hudson, J. A. Jacobs, L. Knopoff, and J. Oliver. Part of the work was done at the Institute for Theoretical Geophysics, Cambridge, England.

REFERENCES

Aki, K. and Richards, P. G., *Quantitative Seismology, Theory and Methods*, W. H. Freeman, New York, 1980.

Anonymous, Review of Wegener's theory of continental displacement, *Nature* (16 February 1922), 202-203, 1922.

Backus, G. E., Confidence set inference with a priori quadratic bound, *Geophys. J.*, *97*, 119, 1991.

Backus, G. E. and Gilbert, F., Uniqueness in the inversion of inaccurate gross Earth data, *Phil. Trans. Roy. Soc. London A*, *266*, 123-192, 1970.

Bates, C. C., Gaskell, T. F. and Rice, R. B., *Geophysics in the Affairs of Man*, Pergamon, Oxford, 1982.

Birch, F., Elasticity and constitution of the Earth's interior, *J. Geophys. Res.*, *69*, 227-286, 1952.

Bloxham, J., Gubbins, D. and Jackson, A., Geomagnetic secular variation, *Phil. Trans. Roy. Soc. London A*, *329*, 415-502, 1989.

Bolt, B. A., *Inside the Earth*, W. H. Freeman, New York, 1982.

Bolt, B. A., Memorial essay: Sir Harold Jeffreys (1891-1989), *Bull. Seismol. Soc. Am.*, *79*, 2006-2011, 1989.

Bolt, B. A., The precision of density estimation deep in the Earth, *Quart. J. Roy. Astron. Soc.*, *32*, 367-388, 1991.

Box, J. F., *R. A. Fisher: The Life of a Scientist*, Wiley and Sons, New York, 1978.

Brush, S. F., Discovery of the Earth's core, *Am. J. Phys.*, *48*, 705-724, 1980.

Bullen, K. E., The spirit of applied mathematics, *Austral. Math. Teacher*, *13*, 25-33, 1957.

Bullen, K. E., *The Earth's Density*, Chapman and Hall, London, 1975.

Bullen, K. E. and Bolt, B. A., *An Introduction to the Theory of Seismology*, 4th ed., Cambridge University Press, Cambridge, 1985.

Cook, A. H., Sir Harold Jeffreys, *Biographical Memoirs, Royal Soc. London*, *36*, 303-333, 1990.

Cox, A., ed., *Plate Tectonics and Geomagnetic Reversals*, W. H. Freeman, New York, 1973.

Darwin, G. H., *Scientific Papers, 1-5*, Cambridge University Press, Cambridge, 1907-16.

Dziewonski, A. M. and Anderson, D. L., Preliminary reference Earth model, *Phys. Earth Planet. Int.*, *25*, 297-356, 1981.

Fisher, R. A., Dispersion on a sphere, *Proc. Roy. Soc. London A*, *217*, 295-305, 1953.

Glen, W., *The Road to Jaramillo*, Stanford University Press, Stanford, 1982.

Gutenberg, B., The boundary of the Earth's inner core, *Trans. Am. Geophys. Union*, *38*, 570-753, 1957.

Holmes, A., Radioactivity and Earth movements, *Trans. Geol. Soc. Glascow*, *18*, 559-606, 1931.

Holmes, A., *Principles of Physical Geology*, 2nd ed., Ronald Press, New York, 1965.

Jacobs, J. A., *The Earth's Core*, 2nd ed., Academic Press, International Geophysics Series 37, London, 1987.

Jefferys, W. H. and Berger, J. O., Ockham's razor and Bayesian analysis, *Am. Scientist*, *80*, 64-72, 1992.

Jeffreys, B. S., Harold Jeffreys: some reminiscences, *Chance*, *4*, 22-23 & 26, 1991.

Jeffreys, H., Tidal friction in shallow seas, *Phil. Trans. Roy. Soc. London A*, *221*, 239-264, 1920.

Jeffreys, H., Letter on hypotheses of continental drift, *Nature*, *111*, 495-496, 1923.

Jeffreys, H., *The Earth. Its Origin, History and Physical Constitution*, Cambridge University Press, Cambridge, 1924 (subsequent editions in 1929, 1952, 1959, 1962 (reprinted with additions), 1970, 1976).

Jeffreys, H., On the formation of water waves by wind, *Proc. Roy. Soc. London A*, *107*, 189-206, 1925.

Jeffreys, H., The rigidity of the Earth's central core, *Mon. Not. Roy. Astro. Soc. Geophys. Suppl.*, *1*, 371-383, 1926*a*.

Jeffreys, H., The stability of a layer of fluid heated below, *Phil. Mag.*, *2* (7), 833-844, 1926*b*.

Jeffreys, H., *Scientific Inference*, Cambridge University Press, Cambridge, 1931 (subsequent editions 1957, 1973).

Jeffreys, H., *Earthquakes and Mountains*, Methuen, London, 1935.

Jeffreys, H., The density distributions of the inner planets, *Mon. Not. Roy. Astro. Soc., Geophys. Suppl.*, *4*, 62-71, 1937.

Jeffreys, H., Science, logic, and philosophy, *Nature*, *141*, 672-676, 716-719, 1938.

Jeffreys, H., *Theory of Probability*, Oxford University Press, 1939*a* (also 1948, 1961, paper-back 1983).

Jeffreys, H., The times of P, S, and SKS and the velocities of P and S, *Mon. Not. Roy. Astro. Soc., Geophys. Suppl.*, *4*, 498-533, 1939*b*.

Jeffreys, H., The determination of the Earth's gravitational field, *Mon. Not. Roy. Astro. Soc.*, *101*, 1-22, 1941.

Jeffreys, H., The origin of the solar system. Bakerian lecture, *Proc. Roy. Soc. London A*, *214*, 281-291, 1952.

Jeffreys, H., The reduction of gravity observations, *Geophys. J. Roy. Astron. Soc.*, *2*, 42-44, 1959.

Jeffreys, H., Note on Fourier analysis, *Bull. Seismol. Soc. Am.*, *54*, 1441-1444, 1964.

Jeffreys, H., The damping of S waves, *Nature*, *208*, 675, 1965.

Jeffreys, H., Radius of the Earth's core, *Nature*, *215*, 1365-66, 1967.

Jeffreys, H., Theoretical aspects of continental drift, *American Association of Petroleum Geologists, Memoir 23*, 395-405, 1974.

Jeffreys, H., Robert Stoneley 1894-1976, *Biograph. Mem. Roy. Soc. London*, *22*, 555-564, 1976.

Jeffreys, H., Probability theory in geophysics, *J. Inst. Maths. and its Applics.*, *19*, 87-96, 1977.

Jeffreys, H., On imperfection of elasticity in the Earth's interior, *Geophys. J. Roy. Astron. Soc.*, *55*, 273-281, 1978.

Jeffreys, H. and Bullen, K. E., *Seismological Tables,*, 1940, 1958, 1967, 1970; 4th reprinting, J. A. Hudson, ed., British Association Seismological Investigations Committee, 1988.

Jeffreys, H. and Crampin, S., On the modified Lomnitz law of creep, *Mon. Not. Roy. Astro. Soc.*, *147*, 295-301, 1970.

Jeffreys, H. and Jeffreys, B. S., ed., *Collected Papers of Sir Harold Jeffreys on Geophysics and Other Sciences (six volumes)*, Gordon and Breach, London, 1971-77.

Jeffreys, H. and Jeffreys, B. S., *Methods of Mathematical Physics*, Cambridge University Press, Cambridge, 1972 (originally published in 1946, with subsequent editions in 1950, 1956, paperback 1972; reprinted with corrections, 1980).

Jeffreys, H. and Shimshoni, M., On regional differences in seismology, *Geophys. J. Roy. Astron. Soc.*, *88*, 305-309, 1987.

Jeffreys, H. and Vicente, R., The theory of nutation and the variation of latitude, *Mon. Not. Roy. Astro. Soc.*, *117*, 142-161, 1957.

Kaula, W. M., *An Introduction to Planetary Physics*, Wiley and Sons, New York, 1968.

Kennett, B. L. N., *1991 IASPEI Seismological Tables*, Research School of Earth Sciences, Australian National University, Canberra, 1991.

Knopoff, L., Sir Harold Jeffreys: The Earth, its origin, history, and physical constitution, *Chance*, *4*, 24-26, 1991.

Lake, P., Wegener's displacement theory, *Geol. Mag.*, *59*, 338-346, 1922.

Lapwood, E. R., Citation, Fourth Award of the Medal of the Seismological Society of America, *Bull. Seismol. Soc. Am.*, *69*, 1305-1308, 1979.

Lapwood, E. R., Contributions of Sir Harold Jeffreys in theoretical geophysics, *Mathematical Scientist*, *7*, 69-84, 1982.

Lapwood, E. R. and Usami, T., *Free Oscillations of the Earth*, Cambridge University Press, Cambridge, 1981.

Lindley, D. V., Sir Harold Jeffreys, *Chance*, *4*, 10-14 and 21, 1991.

Marvin, U. B., The British reception of Alfred Wegener's continental drift hypothesis, *Earth Sci. Hist.*, *4*, 138-159, 1991.

Masters, T. G., Low Frequency seismology and the three-dimensional structure of the Earth, *Phil. Trans. Roy. Soc. London A*, *328*, 329-349, 1989.

Munk, W. H. and MacDonald, G. J. F., *The Rotation of the Earth*, Cambridge University Press, Cambridge, 1960.

Oliver, J., *The Incomplete Guide to the Art of Discovery*, Columbia University Press, New York, 1991.

Randall, M. J., Attenuative dispersion and frequency shifts of the Earth's free oscillations, *Phys. Earth Planet. Int.*, *12*, 1-4, 1976.

Runcorn, S. K., Paleomagnetic comparisons between Europe and North America, *Geol. Ass. Canada, Proc.*, *8*, 77-85, 1956.

Taylor, G. I., Tidal friction in the Irish Sea, *Phil. Trans. Roy. Soc. London A*, *220*, 133, 1919.

Turner, H. H., On the arrival of earthquake waves at the antipodes and on the measurement of the focal depth of an earthquake, *Mon. Not. Roy. Astro. Soc., Geophys. Suppl.*, *1*, 1-13, 1922.

Wadati, K., On shallow and deep earthquakes, *Geophys. Magazine*, *1*, 162-202, 1928 (Tokyo).

Wiechert, E., Ueber die Massenverteilung im Innern der Erde, *Nachr. Ges. Wiss. Göttingen Math, physik. Kl.*, 221-243, 1897.

Williamson, E. D. and Adams, L. H., Density distribution in the Earth, *J. Washington Acad. Sci.*, *13*, 413-428, 1923.

Wrinch, D. and Jeffreys, H., On certain fundamental principles of scientific inquiry, *Phil. Magazine*, *42*, 369-390, 1921.

Zellner, A., ed., *Bayesian Analysis in Econometrics and Statistics: Essays in Honor of Harold Jeffreys*, North-Holland, Amsterdam, 1980 (reprinted by R. E. Krieger Pub. Co., Florida, 1989).

Bruce A. Bolt, Department of Geology and Geophysics, University of California, Berkeley, CA 94720-4767

The Ionosphere

DAVID R. BATES

Department of Applied Mathematics & Theoretical Physics,
Queen's University, Belfast, Northern Ireland BT7 lNN.

The history of some aspects of ionospheric research is outlined. Seriously mistaken beliefs, for example that negative ions are the main carriers of negative charge in the E layer, have been widely held. Ionograms led to a perception on the diurnal variation of N, the number density of free electrons that forced theorists to propose that electrons and molecular positive ions disappear by a process that is many orders of magnitude faster than had earlier been inferred for any other loss process envisaged. The proposal regarding recombination is correct. However incoherent scatter radar studies have given that N in the E layer has an asymmetry with respect to noon that is in the opposite sense to that deduced from ionograms; and this behaviour has been satisfactorily reproduced by numerical simulation. The harmony is not complete good reason not having been given for rejecting the results of the ionogram research.

Just as remote sensing is used to obtain information on the interior of the Earth so it is used to obtain information on the ionosphere. In the case of the ionosphere remote sensing did not prevent seriously mistaken beliefs. Common causes of these were first the difficulty of not overlooking some facet of a complex phenomenon; second a tendency for physicists to presume that geophysical problems have simple solutions; and third a readiness to attach undue significance to order of magnitude agreement between theory and observation.

There are two main forms of remote sensing. They interact. I will begin with the older, the geomagnetic variations and as background information will recall some of the key advances. An historical account has been written by Chapman and Bartels [1940].

In 1722 the great London horologist, Graham, whose invention of the dead-beat escapement made really accurate mechanical clocks possible, discovered that a compass needle undergoes a small daily oscillation in direction. Over a hundred years later in 1850 Kreil found that the magnetic declination at Prague has a semi-lunar day variation. This was a remarkable feat the mean amplitude of the variation being only some 20 seconds of arc. Shortly after the recognition of the sunspot cycle Sabine in 1857 noticed that the solar variation increased with sunspot number. His work led Balfour Stewart to reason that since neither the Earth itself nor the lower atmosphere changes appreciably during the sunspot cycle the cause of the geomagnetic variations must lie in the upper atmosphere (as Gauss had speculated in 1839 that they do. See Kaiser, 1962). In 1882 he proposed the dynamo theory: that currents are induced by the motion of conducting air across the Earth's magnetic lines of force. He attributed the motion giving rise to the solar variation S to the heating action of the Sun on the upper atmosphere and that giving rise to the lunar variation L to the gravitational tide of the Moon. At the time it was thought that the air at great altitudes would be electrically conducting simply by virtue of being tenuous [cf Thomson, 1896]. Support for Balfour Stewart's thesis was provided in 1889 by Schuster who, using the spherical harmonics of Gauss proved that the main part of S does indeed originate above the Earth's surface. Taking the speed of the tidal wind to be the same as at the ground he calculated that the integrated electrical conductivity of the region concerned is 1.1×10^{-5} emu which is around the same as that of a layer of iron 10 meters thick.

In December 1901 Marconi confounded informed scientific opinion by succeeding in transmitting radio signals across the Atlantic from Cornwall to Newfoundland. Could diffraction of radio waves around the Earth be responsible? Macdonald of Cambridge University was privileged to be uniquely qualified to

Relating Geophysical Structures and Processes: The Jeffreys Volume
Geophysical Monograph 76, IUGG Volume 16
Published in 1993 by the International Union of Geodesy and Geophysics and the American Geophysical Union.

answer the question. An outstanding pure mathematician he had recently been awarded the prestigious Adams Prize for a treatise on Radio Waves in which he had devised the first exact solution to the problem of their diffraction by a prism. By January 1903 he was able to present a paper to the Royal Society giving the exact solution to the problem of their diffraction by a sphere. The solution was in the form of an infinite series involving products of Bessel and Legendre functions. [Macdonald 1903] was unable to sum the series analytically but after making an approximation to the Legendre functions he summed it numerically to obtain qualitative agreement with the results of Marconi Company operators.

What a year! The seemingly impossible had been achieved by Marconi and then explained by Macdonald using mathematics that few could follow.

Marconi's triumph endured but Macdonald's was fleeting. Within several months Rayleigh [1903] pointed out that the approximation that Macdonald made is invalid when the order of the Legendre function is large. He also showed that the phenomenon scaled so that if the trans Atlantic transmission were due the radio waves being diffracted then visible light waves would be diffracted through an angle of 45° by a sphere one inch in diameter which they certainly are not.

Few scientists of distinction have been as humiliated as badly as Macdonald. How was his career affected by his blunder? Not as it would have been in some occupations. Already a Fellow of the Royal Society he was awarded the Society's Royal Medal, elected President of the London Mathematical Society and appointed to a Chair in his old University, Aberdeen [Whittaker 1935]. Scientists happily, do not expect perfection in each other.

Even before the diffraction notion [Heaviside 1902] and [Kennelly 1902] had speculated that long distance radio transmission is in some way rendered possible by the upper air forming an electrically conducting layer (because of it being tenuous). Speculating similarly [Lodge 1902] suggested that the air is rendered conducting by electrons freed by solar ultra-violet radiation. Each of these three scientists was unaware of the ideas of the other two and of the central role of a conducting layer in the context of the geomagntic variations.

With one exception geomagneticians were slow to appreciate the relevance to them of the radio research. The exception was Chapman. He happened to be born in the town of Eccles and noticing a paper by a man called [Eccles 1912] in the Proceedings of the Royal Society he was curious enough to glance through it and recognized that radio scientists and geomagneticians were concerned with the same conducting layer. In the paper Eccles discussed how radio waves could bend around the Earth and sought to explain what then seemed to be a key problem: Marconi's observation that there is little difference between the day and night signals up to 500 miles from the sending station but that the day signals were unreadable at distances of 800 miles or more whereas the night signals were readable up to distances of 2000 miles. Eccles assumed firstly that there is a permanently conducting layer which is sharply defined and which reflects waves of all frequencies; secondly that the atmosphere below this reflecting layer is ionized in nearly horizontal strata the ionization increasing upwards so that rays are bent because they experience refraction. Eccles assumed that no free electrons exist. He supposed that the refraction is accompanied by absorption and that this is the origin of the day-night difference. Perhaps to atone for having a letter in which Heaviside put forward his speculation rejected by The Electrician Eccles tried to have the conducting layer called the Heaviside layer [see Ratcliffe 1974]. Before Appleton's nomenclature, the E-layer, became standard the name Kennelly-Heaviside layer was generally favoured. Lodge fared badly in this regard in view of being the only one of the three 1902 pioneers to recognize the origin of the conductivity.

During World War I Appleton, who came to dominate ionospheric research, was in the Army Signal Service. As part of his duties he was drawn into the development of radio, in particular the measurement of radio signals (see Clark 1971). When he returned to the Cavendish Laboratory in 1919 and had a radio scientist Barnett as a research student he had him record accurately the strength of the signals from the BBC's Bournemouth station. Barnett found that while the strength was constant during the day it varied periodically during the night. Appleton perceptively recognized that two waves were probably being received, a direct wave and a wave reflected from the E layer; and that the periodical variations were caused by interference effects associated with the E layer rising during the night and thus changing the path length. He also showed that if the frequency of the transmitter were changed steadily the interference pattern would change in a manner that would enable the height of the layer to be determined. He gave an account of his scheme to Larmor for his opinion. Larmor's letter of reply was characteristically brief. It read "If you can do this it will help much J.L." [Appleton 1961].

Appleton had an attribute often important in research: the ability to cajole. He persuaded the BBC to co-operate by arranging for their Bournemouth station to transmit special signals after normal close-down. He and [Barnett 1925] were thence able to fix the height of the E-layer as about 100 km. Two years later [Appleton 1927] found that the E-layer disappeared before dawn, presumably by recombination and that the interference pattern then indicated a layer at around 300 km. This was his discovery of the F-layer.

Appleton, overtly a very competitive scientist [Clark 1971] succeeded in being the first in almost every development of the exploration of the ionosphere by radio. He missed one important first because of his heavy involvement in exploiting the use of the BBC transmitter. It was Breit and [Tuve 1925] who introduced what was to become the usual method of sounding the ionosphere: transmitting a radio pulse vertically and recording the time interval before the echo was received as a function of the frequency of the carrier wave.

Meanwhile the interest of [Larmor 1924] in radio had been stimulated by the Appleton-Barnett work and he transformed our understanding of the action of the ionized layer. He took into account that the electromagnetic wave makes the electrons and ions in the layer oscillate which in turn makes them emit radiation of the same frequency as the primary wave. He showed that the effect of this emission is to reduce the phase refractive index to

$$\mu = (1 - \sum_{\text{all species}} 4\pi Ne^2/m\omega^2)^{1/2} \qquad (1)$$

where e is the electronic charge, N is the number density of species of mass m and ω is the angular frequency of the radio wave. It is evident from this formula that because of their great mass ions are far less effective than are electrons. We may disregard them in formula (1). For vertical incidence reflection occurs if N attains the value that makes

$$\mu = 0 \qquad (2)$$

If N_o is the maximum electron number density a vertically propagated radio wave having an angular frequency greater than the critical value

$$\omega_o = (4\pi N_o e^2/m)^{1/2} \qquad (3)$$

would be transmitted through the layer A measurement of ω_o thus enables N_o to be determined.

Having heard Larmor lecture on his research Appleton enquired if allowance should not be made for the influence of the Earth's magnetic field on the motion of the electrons. Larmor replied that it should but that he had overlooked the matter and he generously left it to Appleton to amend the theory appropriately [Appleton 1961]. This lead to the magneto-ionic theory. According to the magneto-ionic theory an ionized medium becomes doubly refracting in the presence of a magnetic field. For propagation normal to the field equation (1) remains valid for one of the two waves the, so-called ordinary wave. The corresponding equation for the other wave, the so-called extraordinary wave, is not as simple but need not concern us. [Gillmor 1982] has uncovered a painful facet of the history of the magneto-ionic theory by archival research.

Radio scientists having shown that the peak electron number density in the E-layer is about 1.5 x 10^5 cm^{-3} [Chapman 1926] calculated the conductivity of the layer. The value he obtained was in satisfying agreement with the value inferred from the geomagnetic variations. However Chapman, like Larmor, had overlooked the effect of the Earth's magnetic field on the motion of the electrons. [Pedersen 1927], who was familiar with the magneto-ionic theory pointed out that this made Chapman's calculated conductivity far too high. He postulated that the geomagnetic variations were evidence of an abundance of negative ions whose great mass prevents the Earth's magnetic field from significantly reducing the conductivity due to them (just as it makes them unobservable in radio sounding). Pedersen [1927] also gave the theory of the formation of an ionized layer by solar ultraviolet radiation. He has unjustly been seldom referenced in this context.

In his Bakerian Lecture to the Royal Society [Chapman 1931] acknowledged the correctness of Pedersen's criticism of his calculation of the conductivity of the E-layer and making allowance for the Earth's magnetic field he estimated that

$$\lambda \equiv N(\text{negative ions})/N(\text{electrons}) = 6 \times 10^2 \text{ to } 6 \times 10^3. \qquad (4)$$

Pekeris [1937] drew attention to a new feature of crucial importance: that contrary to what had hitherto been assumed the amplitude of a tidal wind in the upper atmosphere may differ greatly from the amplitude at ground level. In particular he showed that the solar semidiurnal tide at 100 km is amplified by a factor of up to 200 owing to resonance.

During the course of his Bakerian Lecture Appleton (1937) revised Chapman's estimate (4) to

$$\lambda = 10 \text{ to } 10^3 \qquad (5)$$

largely because of the work of Pekeris. The model of the ionosphere that he presented is of historical interest and is shown in Fig 1. The stratification into distinct E and F layers is much more pronounced than in profiles obtained from rockets and satellites (Fig 2). Information on the electron distribution in the valley between the layers has also been acquired by radio scientists with the aid of data from the extraordinary wave [cf Titheridge 1959]. The stratification depicted in figure 1 misguided theorists into fruitless research.

Appleton [1937] initiated consideration of how the electron number density N varies with time t at a fixed altitude. He took variation to be governed by the differential equation

$$dN/dt = q_o \cos\chi - \alpha N^2 \qquad (6)$$

in which χ is the solar zenith angle, q_o is the photoionization rate when χ is zero and α is the recombination rate coefficient. We will confine our attention to the E layer. Noting that the peak electron density in it does not differ appreciably from the equilibrium value $(q_o \cos\chi/\alpha)^{1/2}$ for much of the day Appleton deduced that

$$dN/dt << \alpha N^2 \qquad (7)$$

and from the measured dN/dt and N he further deduced that

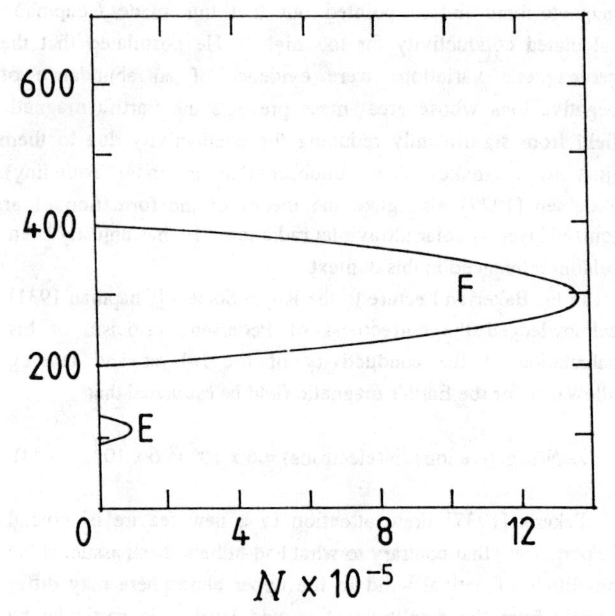

Figure 1. Idealized electron number density distribution in E and F layers during the day (after Appleton 1937).

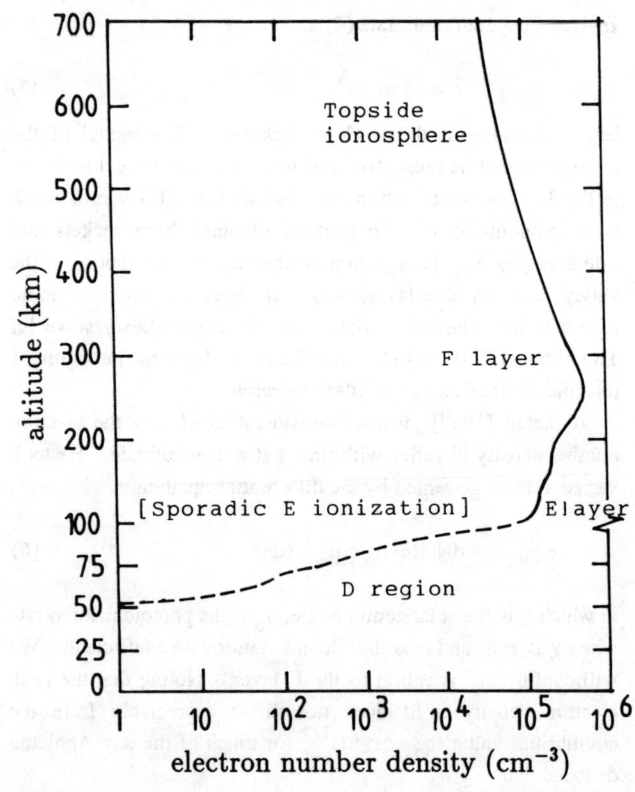

Figure 2. Representative electron number density distribution in E and F layers during a day near sunspot minimum. For clarity the vertical scale in the region below 100 km is double that in the region above (after Bates 1970).

$$\alpha > 10^{-8} \text{ cm}^3\text{s}^{-1}. \quad (8)$$

Such a large recombination coefficient was difficult to explain. In order to ease the problem theorists initially made a practice of replacing the inequality sign in equation (8) by an equality sign. Apparent justification of this was provided by [Appleton 1953] who pointed out that the finiteness of α makes the E layer sluggish in its response to the change in the photoionization rate through the day. He showed that the peak N is at the time interval

$$\Delta t = 1/2\alpha N_q \quad (9)$$

past noon N_q being its value at noon. He hence found that α is about 1×10^{-8} cm^{3}s^{-1}. In an investigation primarily concerned with solar tidal effects [Appleton et al 1955] refined the result to

$$\alpha = 1.0 \times 10^{-8} \text{ cm}^3\text{s}^{-1} \quad (10)$$

The N versus t curve has a very broad maximum and Δt is only about 5 min so it cannot simply be found by inspection of the curve. One of several methods used [see Appleton and Lyon 1957] was to use the best parabolic fit to five or seven N values spaced at hourly intervals around noon.

To see if he could explain the geomagnetic and radio results [Massey 1937] in an exploratory paper listed the atomic collision processes likely to be important and estimated the rate coefficient of each. He reckoned that the balance between the formation of negative ions by attachment and their destruction by collisional detachment gave that λ is approximately 100 and after estimating the rate coefficient for radiative recombination

$$O^+ + e \rightarrow O + h\nu \quad (11)$$

the only recombination process then thought to be important, he further reckoned that when allowance is made for positive ions (like negative ions) being much more abundant than electrons Appleton's lower limit to the rate coefficient α could be understood qualitatively.

The results of geomagnetic studies, radio research and atomic collision physics had meshed together quite convincingly. Was this the effective end of the problem? In 1937 many thought it was although they recognized that much fine tuning had to be done. Fine tuning can be a dull part of research but it is of course essential because the signal that emerges from the noise is sometimes not the signal expected.

On the atomic collision physics side a fairly lengthy investigation [Bates et al 1939, Bates and Massey 1943] was necessary. The main results of our quantal calculations were published in The Philosophical Transitions of the Royal Society where as we hoped they remained safely hidden until World War II ended and we were in a position to consider their implications properly. We found that they proved that λ of equation (4) is in fact very small compared to unity one reason being that photodetachment

$$O^- + h\nu \rightarrow O + e \quad , \qquad (12)$$

which [Massey 1937] had overlooked, is rapid. Contemporaneously with this the geomagnetic evidence that λ is large disappeared as understanding of the winds in the upper atmosphere and the calculation of the conductivity deepened [See Chapman 1956].

The problem of recombination in the E layer was now acute. By eliminating every other possibility we could conceive Massey and I [Bates and Massey 1947] judged it likely that dissociative recombination

$$O_2^+ + e \rightarrow O + O \qquad (13)$$

is rapid even though [Massey 1937] had argued persuasively that it must be extremely slow. We expressed ourselves cautiously because Nature's facility to surprise makes proceeding by a course of systematic elimination a high risk strategy. Nor did we attempt to explain how dissociative recombination could have a rate coefficient of at least 10^{-8} cm^3 s^{-1}. Even though our scheme [Bates and Massey 1947] explained the lower recombination rate that is observed in the F layer I was not sufficiently sure that we were correct to venture investing much of my time thinking about dissociative recombination. My attitude changed totally when Biondi and Brown (1949) reported measurements on the decay of the electron number density in a helium afterglow that appeared to give a recombination coefficient of 1.7 x 10^{-8} cm^3 s^{-1}. Having convinced myself that the ions involved were molecular I concentrated on the problem until I had devised a model that explains why dissociative recombination is very rapid (Bates 1950).

Experiments on the atmospheric gases are difficult but by the 1960's it had been shown by Biondi and his associates and others [cf Massey 1982a] that at 300K:

$$\begin{aligned} a(N_2^+) &= 1.8 \times 10^{-7}\ cm^3\ s^{-1} \\ a(O_2^+) &= 1.6 \times 10^{-7}\ cm^3\ s^{-1} \\ a(NO^+) &= 4.2 \times 10^{-7}\ cm^3\ s^{-1} \end{aligned} \qquad (14)$$

The photoionization rate in the E layer, q_o of equation (6) may be calculated from satellite data on the solar flux [cf Massey 1982b]. It provides a check on the correctness of the recombination coefficients measured in the laboratory. These are much greater than recombination coefficient (10) of [Appleton et al 1956]. Little attention was paid to the discrepancy because the tidal motion that Appleton and his associates were investigating made it necessary to rewrite equation (6) as

$$dN/dt = q_o \cos\chi - \alpha N^2 - \text{div}(N\ \underline{v}) \qquad (15)$$

where $\underline{v}$ is the velocity of the electron transport; and the true value of α might well be masked by the perturbation arising from the transport term. Appleton et al [1955] reasoned that transport takes the form of vertical drift and that this is least near the latitude of the S current focus (say between latitude 25° and 40°). Here an α that is independent of χ is obtained by use of an equivalent alternative to formula (9)

$$\alpha = \left| \frac{dN_m}{dt} \right| / N_m\ \Delta N_m \qquad (16)$$

where $\left| \frac{dN_m}{dt} \right|$ and N_m are the numerical averages of the am and pm values and ΔN_m is the amount by which the pm value of N_m exceeds the corresponding am value. Result (10) pertains to this current focus latitude. The natural inference was that the perturbation due to the transport term in equation (15) is large even here.

A dramatic change in perception took place when the electron number density N in the E layer was measured by incoherent scatter radar. Since the scattering cross section of an electron is only 1 x 10^{-24} cm^2 a powerful transmitter and a sensitive receiver are required. The measurement is akin to finding the size of a pea a hundred miles away.

Radar results [Monro et al 1976, Senior et al 1981, Gerard and Taieb 1986] show an asymmetry with respect to noon that is in the opposite sense to that obtained from the critical frequency ionograms. Recalling that satellite measurements by Stewart and Cravens [1978] have revealed that in the equatorial region the abundance of nitric oxide in the E layer doubles between sunrise and early afternoon, which behaviour has since been modelled by Rusch et al [1991], Senior et al [1981] suggested that as the day progresses charge transfer

$$O_2^+ + NO \rightarrow O_2 + NO^+ \qquad (17)$$

increases the $[NO^+]$ to $[O_2^+]$ ratio which decreases N because of the difference in the recombination coefficients. They reckoned that the decrease in N, should be by about 10 per cent. From their radar studies Gerard and Taieb [1986] found diurnal asymmetries in N of form 3 to 20 per cent. Numerical simulations that took into account both the growth in the nitric oxide abundance and the asymmetry in the diurnal variation of the atmospheric density are generally in good agreement with the measurements.

The results of radar studies, satellite measurements and numerical simulations have meshed together convincingly. Is this the effective end of the problem? The consensus appears to be that it is. After all the conflicting results of Appleton and his associates were obtained more than thirty years ago by a technique that is unsophisticated compared with incoherent scatter radar. Against them it has been argued that the derivation of the peak N in the E layer from critical frequency ionograms is complicated by the common presence of additional cusps corresponding to sporadic E stratification which makes it difficult either to determine the critical frequency unambiguously or to attach a clear physical significance to any value obtained. However Appleton and Lyon (1961) themselves refer to this point and state that "the greatest care has therefore been required to ensure that any conclusions are not invalidated" by the uncertainties. They averaged many results so that random variations should be largely eliminated; and they gave consideration to whether any regularities found could reasonably be attributed to some trend in the mode of measurement. In assessing the work Appleton's unrivaled experience in ionospheric physics cannot properly be ignored.

If the transport term is omitted and equation (6) is taken to be the continuity equation then α is not the true recombination coefficient but an apparent recombination coefficient which will be denoted by α(app) henceforth. Fig 3 shows how α(app), as determined by Appleton and Lyon [1961] by the use of formula (16), varies with cosχ during the day Their data points fit a simple curve very closely. The positive sign of α(app) signals that according to their critical frequency measurements the peak

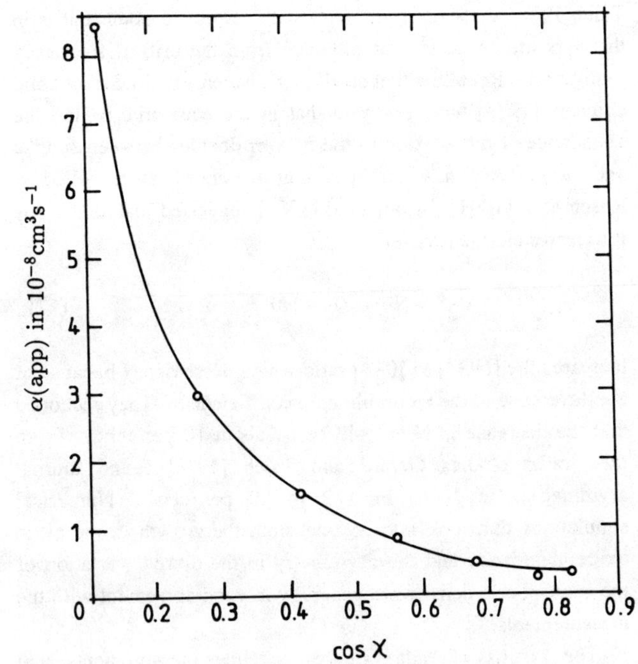

Figure 3. Diurnal variation of α(app) at Slough (51°N) for summer 1952 (after Appleton and Lyon 1961).

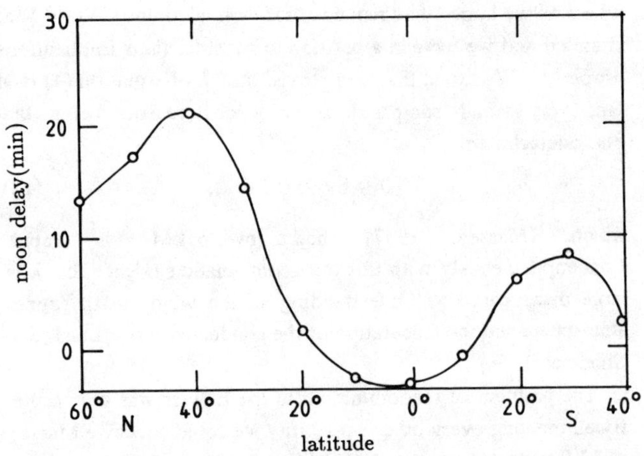

Figure 4. Variation with geographical latitude of delay Δt past noon of maximum electron density in E layer for June (mean of values for 1949 and 1953) (after Appleton et al 1955).

electron density in the E-layer is greater in the pm than in the am period. If the asymmetry with respect to noon had in fact the opposite sense it would be necessary to postulate that compared with those in the pm period the critical frequencies in the am period were too low by an amount that is a smooth function of cosχ.

As a test of the accuracy and consistency of their data Appleton and Lyon [1961] plotted Δt against the month of the year without correcting for the equation of time. On examining how Δt correlated with the equation of time they concluded that their computed values are significant to within a few minutes. Fig 4 shows the variation of the connected Δt of Appleton et al [1955] with latitude. The data points delineate a smooth curve remarkably well. In accord with figure 3 Δt is positive (except in the vicinity of the equator).

The results in figures 3 and 4 were once regarded as completely trustworthy. With the advent of radar studies they now present an enigma. If they are incorrect the reason may never be found since there is little appeal in puzzling over old measurements.

Acknowledgement

I thank Professor Henry Rishbeth for reading the first draft of this article and making helpful comments and the US Air Force for support under Grant No AFOSR 091 0261.

REFERENCES

Appleton, E.V. The existence of more than one ionized layer in the upper atmosphere. *Nature, 120*, 333, 1927.

Appleton, E.V. Regularities and irregularities in the ionosphere. *Proc. Roy. Soc. A, 162*, 451-479, 1937.

Appleton, E.V. A note on the "sluggishness" of the ionosphere. *J Atmos. Terr. Phys.*, *3*, 282-284, 1953.

Appleton, E.V. Sir Joseph Larmor and the ionosphere. *Proc. Roy. Irish Acad.*, *61A*, 55-66, 1961.

Appleton, E.V. and Barnett, M.A.F. On some direct evidence for downward reflection of radio waves. *Proc. Roy. Soc. A* , *109*, 621-641, 1925.

Appleton, E.V., Lyon, A.J. and Pritchard, A.G. The detection of the S_q current system by ionospheric radio sounding. *J. Atmos. Terr. Phys.*, *7*, 292-295, 1955.

Appleton, E.V. and Lyon, A.J. Studies of the E-layer of the ionosphere. I. Some relevant theoretical relationships. *J. Atmos. Terr. Phys.*, *10*, 1-11, 1957.

Appleton, E.V. and Lyon, A.J. Studies of the E-layer of the ionosphere. II. Electromagnetic pertubations and other anomalies. *J. Atmos. Terr. Phys.*, *21*, 73-99, 1961.

Bates, D.R. Dissociative recombination. *Phys. Rev.* *78*, 492, 1950.

Bates, D.R. Reactions in the ionosphere. *Contemp. Phys.* , *11*, 105-124, 1970.

Bates, D.R., Buckingham, R.A., Massey, H.S.W. and Unwin, J.J. Dissociative, recombination and attachment processes in the upper atmosphere. II. The role of recombination. *Proc. Roy. Soc. A*, *170*, 322-340, 1939.

Bates, D.R. and Massey, H.S.W. The negative ions of atomic and molecular oxygen. *Phil. Trans. Roy. Soc. A*, *239*, 269-304, 1943.

Bates, D.R. and Massey, H.S.W. The basic reactions in the upper atmosphere-I. *Proc. Roy. Soc. A*, *187*, 261-296, 1946.

Bates, D.R. and Massey, H.S.W. The basic reactions in the upper atmosphere-II. *Proc. Roy. Soc. A*, *192*, 1-16, 1947.

Biondi, M.A. and Brown, S.C. Measurement of electron ion recombination. *Phys Rev.* *76*, 1697, 1949.

Breit, G. and Tuve, M.A. A recent method for estimating the height of the conducting layer. *Nature*, *116*, 357, 1925.

Chapman, S. Ionization in the upper atmosphere. *Quart. J. Roy. Met. Soc.*, *51*, 225-236, 1926.

Chapman, S. Some phenonema of the upper atmosphere. *Proc. Roy. Soc. A*, *132*, 353-374, 1931.

Chapman, S. The electrical conductivity of the ionosphere: a review. *Il Nuovo Cimento Ser. X*, *4* *Suppl. 1*, 1385-1412, 1956.

Chapman, S. and Bartels, J. *Geomagnetism*, pp. 1049, Clarendon Press, Oxford, 1940.

Clark, R. *Sir Edward Appleton*, pp. 240, Pergamon Press, London, 1971.

Eccles, W.H. On the diurnal variations of the electrical waves round the bend of the earth. *Proc. Roy. Soc.*, *87*, 79-99, 1912.

Gérard, J-C. and Taieb, C. The E region electron density diurnal asymmetry at Saint Santin: observation and role of nitric oxide. *J Atmos. Terr. Phys.*, *48*, 471-483, 1986.

Gillmor, C.Stewart. "Wilhelm Altar, Edward Appleton and the magnetic ionic theory". *Proc. Amer. Philos. Soc.* *126*, 395-440, 1982.

Heaviside, O. *Telegraphy* in *Encyclopaedia Britannica*, *33*, 213-218, 1902.

Kaiser, T.R. The first suggestion of an ionosphere. *J. Atmos. Terr. Phys.*, *24*, 865, 1962.

Kennelly, A.E. On the elevation of the electrically conducting strata of the Earth's atmosphere. *Electrical World and Engineer*, *39*, 473, 1902.

Larmor, J. Why wireless electric waves can bend around the earth. *Phil. Mag.*, *48*, 1025-1036, 1924.

Lodge, O. Mr. Marconi's results in day and night wireless telegraphy. *Nature*, *66*, 222, 1902.

Macdonald, H.M. The bending of electric waves around a conducting obstacle. *Proc. Roy. Soc.*, *71*, 251-258, 1903.

Massey, H.S.W. Dissociation, recombination and attachment *Processes in the upper atmosphere. I. Proc. Roy. Soc. A*, *163*, 542-553, 1937.

Massey, H.S.W. Photochemistry of the midlatitude ionosphere in *Applied Atomic Collision Physics* (Edited by Massey, H.S.W., McDaniel, E.W. and Bederson, B.) I. pp. 22-75, Academic Press, New York, 1982a.

Massey, H.S.W. Atomic collisions and the lower ionosphere at midlatitudes. *ibid*, pp. 105-148, 1982b.

Monroe, P.E., Nisbet, J.S. and Stick, T.L. Effects of tidal oscillations in the neutral atmosphere on electron densities in the E region. *J. Atmos. Terr. Phys.*, *38*, 523-529, 1976.

Pedersen, P.O. The propagation of radio waves along the surface of the Earth and in the atmosphere. Danmarks Naturvidenskabelige Samfund A. Nr. 15 a/b, pp. 244 Copenhagen, 1927.

Pekeris, C.L. Atmospheric oscillations. *Proc. Roy. Soc. A*, *158*, 650-671, 1937.

Ratcliffe, J.A. Scientist's reactions to Marconi's transatlantic radio experiment. *Proc. Inst. Elect. Eng.*, *121*, 1033-1038, 1974.

Rayleigh, Lord On the bending of waves around a spherical obstacle. *Proc. Roy. Soc.*, *72*, 40-41, 1903.

Rusch, D.W., Gérard, J-C. and Fesen, C.G. The diurnal variation of NO, $N(^2D)$, and the ions in the thermosphere: a comparison of satellite measurements to a model. *J. Geophys. Res., 96*, 11331-11339, 1991.

Senior, C., Bauer, P., Taieb, C. and Petit, M. The effect of nitric acid on the variation of the electron density in the E region with solar zenith angle. *C.r. Acad. Sci. Paris, 292*,(Serie II), 1195-1202, 1981.

Stewart, A.I. and Cravens, T.E. Diurnal and seasonal effects in E region low latitude nitric oxide. *J. Geophys. Res., 83*, 2453-2456, 1978.

Thomson, J.J. *Recent researches in electricity and magnetism*, pp. 578. Cambridge Press, 1896.

Titheridge, J.E. The use of the extraordinary ray in the analysis of ionospheric records. *J. Atmos. Terr. Phys., 17*, 110-125, 1959.

Whittaker, E.T. *Hector Munro Macdonald 1865-1935 Obit. Not. Fellows Roy. Soc., 1*, 551-558, 1935.

D.R. Bates, Department of Applied Mathematics and Theoretical Physics, Queen's University, Belfast, Northern Ireland BT7 1NN.

Hurricanes and Atmospheric Processes

YOSHIO KURIHARA

Geophysical Fluid Dynamics Laboratory/NOAA
Princeton University, Princeton, New Jersey

Hurricanes are strong cyclonic vortices in which the gradient wind relationship holds during their evolution. The development of hurricanes proceeds with a persistent thermal forcing and the continual adjustment of fields to a new state of gradient wind balance. The forcing is largely due to the release of the latent heat received at the ocean surface. The adjustment is achieved by the development of a transverse circulation and the generation of inertia gravity waves. The behavior of the vortex also strongly depends on how it responds and adjusts itself to the environmental forcing. Thus, the spatial and temporal variability of the tropical cyclone climatology is related to regional and seasonal changes in the conditions of the larger scale environment. As exemplified by the evolution of hurricanes, the dynamics of an atmospheric system which undergoes slow structural change is controlled by the processes of forcing and adjustment. An atmospheric process may play a dual role: it contributes to the adjustment of one system while it provides forcing to another system, thus linking atmospheric systems of distinctly different scales.

1. INTRODUCTION

A dynamical system in the atmosphere can be classified according to a time scale characterizing the evolution of the individual system. For the purpose of discussion in this article, a system is considered to be slow when a long time scale, typically a day or longer, makes the time tendency term of the governing equations smaller than the other major terms by an order of magnitude. The structure of a slow system represents a nearly balanced state between all the major dynamical effects. The structural change of the system is a succession of nearly balanced states. In contrast, the short time scale of a fast system, on the order of hour and less, makes the time tendency term one of the primary terms in the governing equations. The structure of a fast system represents a state of dynamical imbalance.

A theme of this paper is to point out some important common features of atmospheric processes which are involved in the development of slow systems. The evolution of hurricanes will be used as a specific example. Hurricanes are strong vortices which develop over the tropical region of the Atlantic in summer and fall. (Vortices having similar structure to hurricanes also develop in different ocean basins and are called typhoons or cyclones, depending on the region. Arguments made in this paper can be applied to all of these tropical cyclones.) Although the hurricane is just one special slow atmospheric phenomenon and each hurricane exhibits its own unique life cycle, the meteorological processes pertaining to its evolution possess dynamical features in common with other slow systems as well.

Relating Geophysical Structures and Processes: The Jeffreys Volume
Geophysical Monograph 76, IUGG Volume 16

Tropical cyclones have favorable regions and periods for their development. This suggests that, while the internal dynamics is essential for the formation of a hurricane, its evolution is significantly influenced by the environmental conditions. The understanding of the life cycle of a hurricane requires the consideration on the effects of interactions between the vortex and both the smaller and the larger scale systems. These interactions may be called the internal forcing and the environmental forcing, respectively.

In the present paper, the problem concerning the structure of a balanced vortex in a calm environment is first discussed. Then, how a balanced vortex changes in time in response to the internal forcing is considered in Section 3. By treating an idealized quasi-balanced change followed by a nearly balanced change, two important motions associated with the adjustment process of the vortex structure are discussed. Some interesting issues in the dynamics of hurricane development are addressed in Section 4. The scope is extended in Section 5 to briefly discuss the effect of the environmental forcing on the life of a hurricane and the implications for the climatology of tropical cyclones. Finally, in Section 6, the results obtained with respect to hurricanes are used to characterize dynamical features which are common to slow atmospheric systems in general.

2. A BALANCED VORTEX

As indicated by observational analysis [e.g., Willoughby, 1990] as well as by results of numerical simulations [e.g., Kurihara, 1975], the gradient wind relationship holds fairly well for hurricane vortices above the boundary layer. Namely, the sum of the centrifugal and Coriolis forces acting on the mean tangential or azimuthal component of wind is nearly balanced by

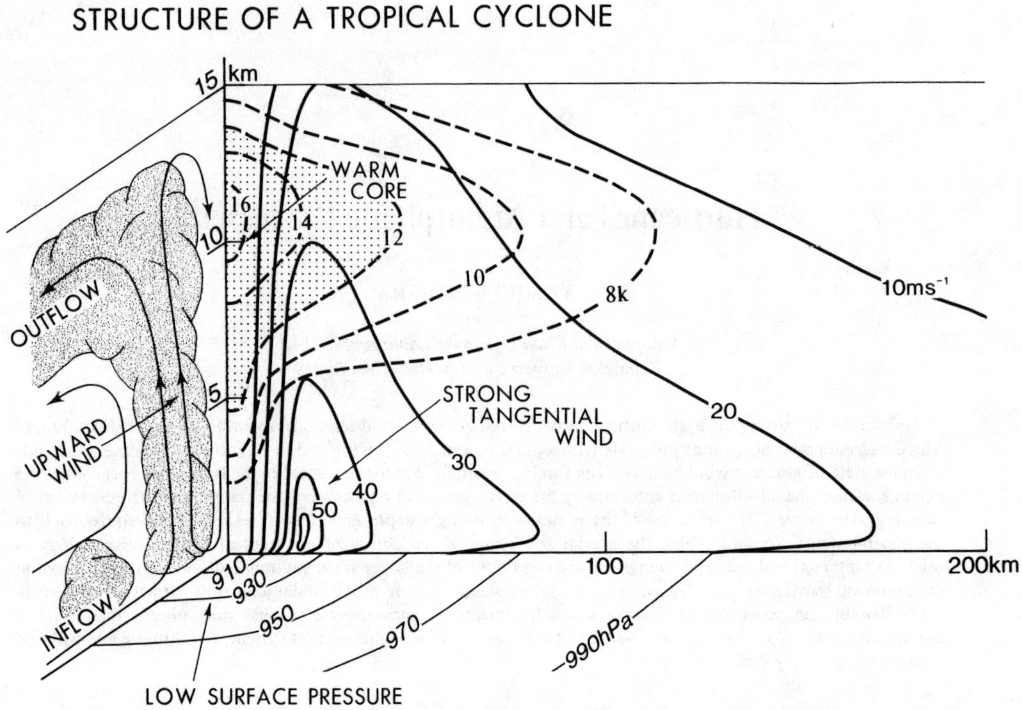

Fig.1 Radius-height cross-sections showing a typical axisymmetric structure of hurricanes. In one section, distributions of tangential or azimuthal winds (m s^{-1}) and the temperature deviation from the far distant ambient atmosphere (K) are shown. In another section, transverse circulation and cumulus clouds are schematically illustrated. Surface pressure field (hPa) is also drawn.

the pressure gradient force. A typical axisymmetric hurricane structure is schematically shown in Fig.1. Specific features include low central pressure at the surface, strong low level wind rotating cyclonically (counterclockwise in the Northern Hemisphere and clockwise in the Southern Hemisphere) and very warm temperature in the core column or eye. At upper levels in the outer region (not shown), the pressure gradually decreases outward and the wind is anticyclonic. Also important is the transverse circulation consisting of radial inflow in the planetary boundary layer, strong upward motion causing a deep eyewall cloud and outflow at upper levels.

In this section, a determination of the balanced vortex structure is considered. The vortex treated here is nonviscous, stands on an insulated frictionless surface and has no transverse circulation. The gradient wind relation must be exactly satisfied within this vortex. Now suppose that an axisymmetric distribution of temperature, T, is specified on the radial(r)-vertical(z) plane; it has a realistic warm core and the stratification is statically stable everywhere, i.e.,

$$\frac{\partial \theta}{\partial z} > 0 \quad \text{or} \quad \frac{\partial T}{\partial z} > -\frac{g}{c_p} \tag{1}$$

where θ is the potential temperature (the temperature which a parcel of dry air would attain if brought to the standard pressure of 1000hPa), g the acceleration of gravity and c_p the specific heat at constant pressure. Also, the vortex is assumed to be in hydrostatic balance,

$$\frac{\partial p}{\partial z} = -\frac{g}{\alpha} \tag{2}$$

where p is the pressure and α the specific volume of the air. Combining the above with the equation of state

$$p\alpha = RT \tag{3}$$

where R is the gas constant of the air, one obtains the formula for the pressure field:

$$\frac{p}{p_s} = \exp\left(-\frac{g}{R}\int_0^z \frac{1}{T}\,dz\right) \tag{4}$$

where p_s is the surface pressure.

Hereafter, the pressure coordinate system in which the pressure is taken as the vertical coordinate will be used. Horizontal derivatives in this coordinate system are taken along isobaric surfaces. The gradient wind relationship is then expressed by

$$\frac{v^2}{r} + fv = \frac{\partial \phi}{\partial r} \tag{5}$$

where f is the Coriolis parameter (the component of the earth's vorticity about the local vertical) and it is assumed to be constant; ϕ denotes the geopotential ($\phi = gz$) of an isobaric surface and can be easily obtained from (2). The left hand side of (5) expresses the sum of the centrifugal force and the deflecting force acting on the air in cyclonic flow v. The right hand side represents the pressure gradient force with its sign reversed. It is easy to compute a balanced cyclonic flow from (5) for a known pressure or geopotential field. By expressing it in terms of the absolute angular momentum, M, which is defined in reference to the vortex center on the constant-f plane as

$$M = r\left(v+\frac{f}{2}r\right) \tag{6}$$

the balance relation (5) takes the form

$$\frac{M^2}{r^3} - \frac{1}{4}f^2 r = \frac{\partial\phi}{\partial r} \tag{7}$$

There exists just one balanced cyclonic wind for a given pressure field. On the other hand, given a temperature field, the number of pressure fields satisfying hydrostatic balance is limitless, because the surface pressure $p_s(r)$ in (4) can be arbitrarily specified. More generally, a pressure value can be prescribed at an arbitrary height h (h=h(r)). It is due to this arbitrariness that an infinite number of tangential flow fields satisfy the gradient wind balance with the same temperature field $T(r,z)$. A similar argument was made by Jeffreys [1933] for the case of midlatitude large scale flow. He stated that an infinite number of geostrophic flows exist for a given steady distribution of temperature.

Apparently, some constraints in the real atmosphere prevent the realization of a large number of possible balanced vortices. If the actual pressure value is specified at the tropopause level, i.e., at the bottom of the stratosphere, the pressure fields obtained from (4) for a given realistic profile of temperature will also be fairly realistic. In this respect, the presence of a very stable stratosphere can be such a constraint. On the other hand, as will be discussed in Section 4, the energetical constraint alone can also limit the minimum surface pressure under certain circumstances. Another possible constraint is the requirement that a balanced vortex should be dynamically stable. A state of unstable balance cannot be sustained when it is perturbed. The dynamical stability of an axisymmetric vortex is measured by the following three factors;

$$S = -\frac{\alpha}{\theta}\frac{\partial\theta}{\partial p} \qquad \text{Static stability} \tag{8}$$

$$I = \frac{2M}{r^3}\frac{\partial M}{\partial r} = \xi\eta \qquad \text{Inertial stability}$$

$$\text{where} \quad \xi = \frac{2M}{r^2} = f+\frac{2v}{r} \tag{9}$$

$$\eta = \frac{1}{r}\frac{\partial M}{\partial r} = f+\frac{\partial rv}{r\partial r}$$

$$B = -\frac{\alpha}{\theta}\frac{\partial\theta}{\partial r} = -\frac{\partial\alpha}{\partial r} = \frac{\partial}{\partial r}\frac{\partial\phi}{\partial p} \qquad \text{Baroclinicity} \tag{10}$$

A balanced vortex is stable if it is statically stable ($S > 0$), inertially stable ($I > 0$) and baroclinically stable for symmetric overturning ($SI - B^2 > 0$) [Ooyama, 1966]. In the balanced vortex, the vertical variation of M is related to B through the pressure derivative of (5). In this case, the potential vorticity, PV, of the axisymmetric vortex

$$PV = g\left(\frac{\partial\theta}{\partial r}\frac{\partial M}{\partial p} - \frac{\partial\theta}{\partial p}\frac{\partial M}{\partial r}\right) \tag{11}$$

becomes

$$PV = \frac{g}{r}\frac{\theta}{\alpha}\frac{r^3}{2M}(SI-B^2) \tag{12}$$

Therefore, the condition for baroclinic symmetric stability of the vortex is equivalent to requiring that the product of M and the potential vorticity is positive everywhere. Probably, the stability condition mentioned above is a strong constraint and restricts the structure of balanced vortices to a reasonable range.

3. Evolution of a Vortex

Suppose that the structure of a balanced vortex is disturbed by forcing due to atmospheric processes of one kind or another. Subsequent changes in the vortex structure will depend on how the mass and wind fields are readjusted. When a new state of the gradient wind balance is established in the vortex by an ideal transverse circulation, the change is called a quasi-balanced change by analogy with the quasi-geostrophic change of a geostrophically balanced system. In more general cases, a nearly balanced change involves high frequency waves in addition to a transverse circulation. The evolution of a hurricane is considered in this paper as a succession of almost balanced states.

The equations governing the time change of absolute angular momentum and the thermodynamic energy are respectively written,

$$\frac{\partial M}{\partial t} = -u\frac{\partial M}{\partial r} - \omega\frac{\partial M}{\partial p} + rF_t \tag{13}$$

and

$$\frac{\partial}{\partial t}\left(-\frac{\partial\phi}{\partial p}\right) = -u\frac{\alpha}{\theta}\frac{\partial\theta}{\partial r} - \omega\frac{\alpha}{\theta}\frac{\partial\theta}{\partial p} + \frac{R}{c_p p}Q \tag{14}$$

In the above equations, $-\partial\phi/\partial p$ is the thickness representing the temperature; u denotes the radial component of the wind and ω its vertical component ($\omega = dp/dt$). The continuity equation holds with respect to u and ω

$$\frac{\partial ru}{r\partial r} + \frac{\partial\omega}{\partial p} = 0 \tag{15}$$

Use of the stream function ψ yields

$$u = \frac{1}{r}\frac{\partial\psi}{\partial p} \tag{16}$$

and

$$\omega = -\frac{1}{r}\frac{\partial\psi}{\partial r} \tag{17}$$

In (13) and (14), F_t and Q represent the internal forcing which affects the tangential wind and temperature of the vortex, respectively. These effects are derived from various kinds of small scale processes. Processes such as the turbulence, the cumulus convection and associated gravity waves redistribute the effects of air-sea interaction, i.e., the frictional stress and evaporation, and, hence, provide sources or sinks of momentum and internal energy for the vortex dynamics.

First, a very special case is considered in which a vortex undergoes an ideal quasi-balanced change. The maintenance of the gradient wind balance relation requires that the time derivative of (7) be valid everywhere in a vortex:

$$\frac{\partial}{\partial t}\left(\frac{M^2}{r^3}-\frac{1}{4}f^2r\right)=\frac{\partial}{\partial t}\left(\frac{\partial\phi}{\partial r}\right) \tag{18}$$

Taking the pressure derivative of the above equation yields a requirement which the time tendencies of the absolute angular momentum and the thickness have to satisfy whenever a vortex transforms from one balanced state to another:

$$\frac{\partial}{\partial p}\left(\frac{2M}{r^3}\frac{\partial M}{\partial t}\right)=\frac{\partial}{\partial r}\left(\frac{\partial}{\partial t}\frac{\partial\phi}{\partial p}\right) \tag{19}$$

Using (13), (14), (16) and (17), the above condition becomes

$$\begin{aligned}\frac{\partial}{\partial r}\left(\frac{S}{r}\frac{\partial\psi}{\partial r}+\frac{B}{r}\frac{\partial\psi}{\partial p}\right)+\frac{\partial}{\partial p}\left(\frac{B}{r}\frac{\partial\psi}{\partial r}+\frac{I}{r}\frac{\partial\psi}{\partial p}\right)\\=\frac{\partial}{\partial r}\left(\frac{R}{c_p p}Q\right)+\frac{\partial}{\partial p}\left(\frac{2M}{r^2}F_t\right)\end{aligned} \tag{20}$$

The coefficients S, I and B appearing in (20) are the stability parameters previously defined by (8), (9) and (10). If $SI-B^2>0$, i.e., if PV (see (12)) and M of the balanced vortex are positive, the streamfunction and, hence, the transverse circulation can be obtained by solving (20) with properly specified boundary conditions. As mentioned before, positive $PV \cdot M$ implies the symmetric stability of a balanced vortex. Therefore, a stably balanced vortex can undergo a quasi-balanced change through the adjustment by the transverse circulation obtained above. Circulation patterns responding to diabatic heating and frictional forcing have been studied [e.g., Shapiro and Willoughby, 1982; Holland and Merrill, 1984] and are schematically illustrated in Fig.2. The circulation with such patterns effectively spreads the influence of forcing to a larger domain, thereby reducing the direct local impact of the forcing. For example, the upward vertical motion cools the diabatically heated region, while the subsidence warms the surrounding area. The transverse circulation derived from (20) will establish a new balanced state, if the time tendencies of the angular momentum and the temperature actually satisfy the ideal condition (19). Is a balanced change unrealizable otherwise?

In general, the radial flow u is not diagnostically constrained by the streamfunction of (20) but governed by the equation of motion (radial component),

$$\begin{aligned}\frac{\partial u}{\partial t}&=-u\frac{\partial u}{\partial r}-\omega\frac{\partial u}{\partial p}\\&\quad+\left(\frac{v^2}{r}+fv-\frac{\partial\phi}{\partial r}\right)+F_r\end{aligned} \tag{21}$$

The quantity in the parentheses on the right hand side of (21) represents the imbalance between the wind and mass fields. The dynamics of the above free vortex requires the presence of high frequency mode in addition to the slow mode. Namely, the inertia gravity waves are immediately generated once a balanced state is disturbed. These mesoscale waves make an important contribution to the adjustment of the fields to a state of balance by transporting mass and momentum. In transitions of a balanced flow in which the available potential energy is consumed, some amount of the total energy of the vortex is lost by wave radiation. In a recent observational study [Sato et al., 1991], signals of gravity waves which were apparently emitted from a typhoon were identified in the radar data taken during storm passage.

4. DEVELOPMENT OF HURRICANES

Whether a hurricane will evolve from a weak tropical depression depends on how forcing and adjustment processes function. Of two types of forcing, internal and environmental, only the former is considered in this section and the effect of the latter will be discussed in the next section. It is obvious that, in order for the vortex to intensify, the distribution of forcing has to be favorable and persistent. The distribution is, of course, most favorable if only a slow mode having hurricane-like characteristics is amplified. It is not advantageous if the forcing contributes largely to the generation of the vortex scale fast modes [Syono and Yamasaki, 1966].

Basically, the energy of hurricanes originates from the latent and sensible heat supply from the warm tropical ocean. Furthermore, when the energy flux at the surface is combined with the winds of a hurricane, a continual positive feedback cycle is established. Therefore, the sea surface temperature, the air-sea interaction processes and the boundary layer behavior are important factors controlling the evolution of tropical cyclones.

The latent energy supplied at the ocean surface becomes a source of internal forcing when it is released in cumulus clouds. The distribution of latent heat release is related to the form of convection. For example, whether convection is penetrative or slants along the transverse flow makes a difference. Convection in real storms seems to be a combination of the two types [Jorgensen, 1984]. The heating distribution also depends on the thermodynamical behavior of the clouds, e.g., whether the convection process is pseudo-adiabatic or reversible. Fig.3 illustrates some of these different types of convection. The cloud behavior is also affected by the entrainment of the surrounding air as well as by the microphysics of the cloud. It should be stressed here that the released heat in each individual convection element has to spread over a reasonable area to produce an internal forcing for the vortex system, i.e., Q of (14). This may be achieved rapidly by the circulation and inertia gravity waves of the cloud scale. Understanding of the scale linkage between the cloud and vortex should rely on the knowledge of the cumulus

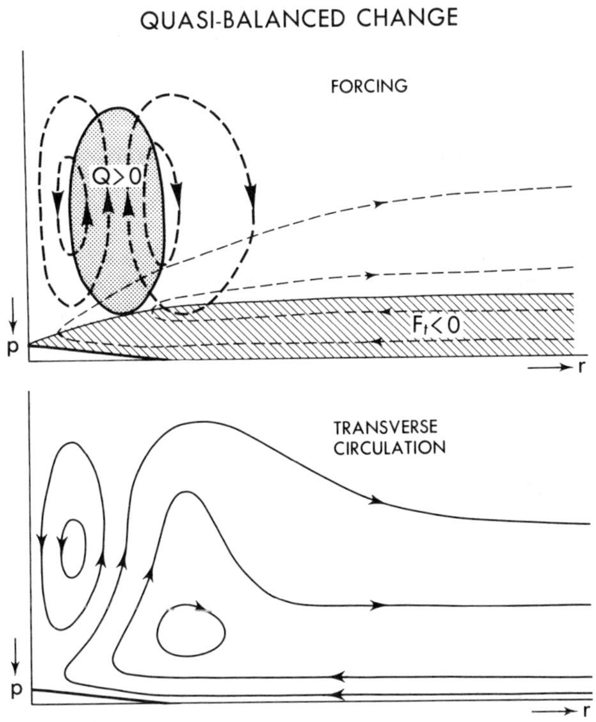

Fig.2 Schematic picture of the transverse circulation associated with quasi-balanced change of a vortex. Upper panel shows the flow (thick dashed line) due to heating in a shaded region and that (thin dashed) due to frictional effect in a hatched region. Sum of the above two flows yields the transverse circulation shown in the lower panel.

dynamics with cloud physical processes taken into account [e.g., Yamasaki, 1983].

Problems concerning the limit of the surface pressure drop or potential intensity of hurricanes are of interest. The pressure at the surface can be determined by the hydrostatic relation (see (4)) if the pressure at a certain level and the vertical profile of temperature are known. Malkus and Riehl [1960] proposed an empirical model in which an undisturbed level was assumed at about the tropopause height. Also, the warmest possible temperature profile in the eyewall cloud was assumed for a chosen equivalent potential temperature, Θ_E, a quantity representing the thermodynamical property of the surface air. The specification range of Θ_E is limited because of the assumed presence of an undisturbed upper level. The obtained formula for the minimum possible surface pressure below the eyewall cloud is, for a range of Θ_E between 350K and 365K,

$$-\delta p_s = 2.5\ \delta\Theta_E \tag{22}$$

where δp_s denotes the amount of surface pressure drop below 1000hPa and $\delta\Theta_E$ is the deviation of Θ_E above 350K. To estimate the minimum pressure at the center of a hurricane, Miller [1958] computed the temperature and moisture sounding of the eye on the basis of an eye model consisting of the subsidence of the air and the mixing with the eyewall. The obtained maximum intensities for given sea surface temperatures range from 987hPa for 26C to 892hPa for 31.1C. In a different approach, Emanuel [1988] discussed the potential intensity of hurricanes from an energetical viewpoint, in which a tropical cyclone is regarded as a Carnot heat engine. Energy is supplied from the sea surface, lost at the cold upper outflow levels and the converted work is ultimately dissipated at the surface. Let T_s be the sea surface temperature, T_o the temperature at the outflow level and η the maximum efficiency of the heat engine. A surface air parcel travelling from an ambient location to a central location with its temperature maintained at T_s changes its pressure from p_a to p_c and its mixing ratio of water vapor from the unsaturated value r_a to the saturation value r_c. Total heat received by the air parcel is the sensible heat needed for the isothermal expansion plus the latent heat, i.e.,

$$\text{heat} = RT_s \ln\left(\frac{p_a}{p_c}\right) + L(r_c - r_a) \tag{23}$$

where L denotes the latent heat of evaporation. As the energy dissipation at the surface is balanced by the work done by the pressure gradient force, it is given by

$$\text{dissipation} = RT_s \ln\left(\frac{p_a}{p_c}\right) \tag{24}$$

Hurricanes will intensify if

$$\frac{\text{heat} \times \eta}{\text{dissipation}} > 1 \tag{25}$$

where the efficiency η is dependent on T_s and T_o

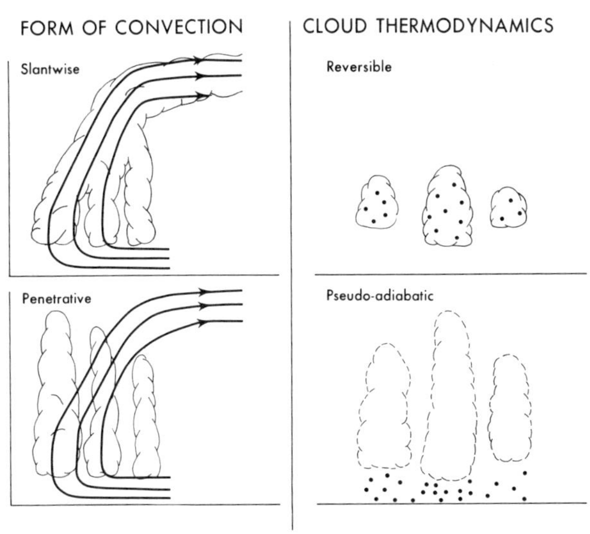

Fig.3 Schematic picture of clouds with different dynamics and thermodynamics. The left part shows clouds developed slantwise along the mean flow (upper panel) and vertically penetrating clouds (lower). The right part indicates the reversible clouds which retain the condensed droplets (upper panel) and the pseudo-adiabatic clouds which do not possess droplets (lower).

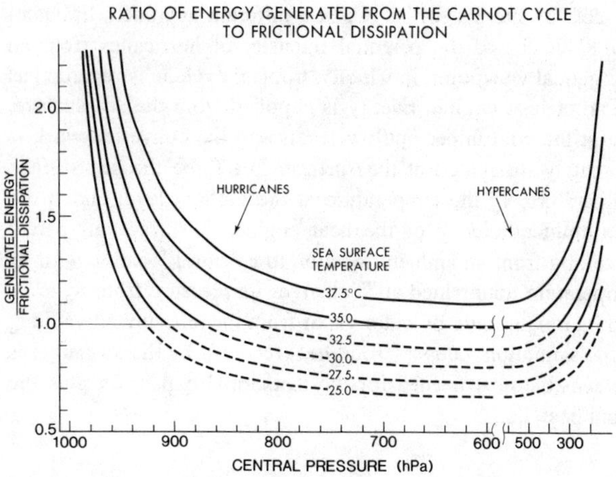

Fig.4 Diagram showing the ratio of the generated energy to the frictional dissipation at the surface (in ordinate) in the Carnot cycle model of hurricanes for different values of central surface pressure (hPa, in abscissa) and sea surface temperature (25 to 37.5C at 2.5C interval, solid and dashed lines). The surface pressure and relative humidity of the ambient air are assumed to be 1010hPa and 80%, respectively. The temperature of the upper level outflow is set to -75C. When the ratio is above unity (above the thin horizontal line), a system develops until the ratio becomes unity. Therefore, only a solid line portion is realizable for a given sea surface temperature.

$$\eta = \frac{T_S - T_O}{T_S} \tag{26}$$

It is clearly indicated from (25) that the evaporation at the surface is a necessary ingredient of hurricanes. The ratio of the generated energy, i.e., that part of supplied heat which becomes available for work, against dissipation of energy is shown in Fig.4 as a function of p_c for different values of T_s. In the calculation, T_o is set to 198.2K and, for the ambient surface air, p_a to 1010hPa and the relative humidity to 80%. It is shown that the potential minimum surface pressure decreases as the sea surface temperature increases up to about 35C. Note that the potential minimum pressure can be determined with some certainty when T_s is relatively low. For higher T_s, the potential intensity becomes very sensitive to a small change in the parameter values. If the scheme is valid for $T_s > \sim 35$C, a vortex would continue to develop and become what is called a hypercane. However, as the vortex intensifies, the energy dissipation may drastically increase with the development of turbulence or multi-cell structure, or fast modes may remove large amount of energy from the system. This may prevent the storm from reaching the hypercane range or stop the run away intensification.

If a vortex reaches a quasi-equilibrium mature stage, its structure in the free atmosphere is in an almost balanced state. In addition to a balanced gradient wind, the vortex includes the transverse circulation (see Fig. 1) which counteracts the heating and frictional forces. For example, the strong heating effect caused by cumulus convection in an inner region is balanced by the adiabatic cooling due to upward motion. As tropical cyclones intensify, the eye formation tends to result from the boundary layer dynamics [Ooyama, 1969b; Eliassen, 1971]. In the eye, adiabatic warming due to mean sinking motion can be balanced with cooling due to the outward heat flux by the evolved eddies [Kurihara and Bender, 1982].

5. INTERACTION WITH LARGE SCALE SYSTEMS

Small scale systems contained in a hurricane, such as cumulus convection and turbulence, provide internal forcing to a vortex. On the other hand, a hurricane is embedded in a large scale environment, and its evolution can be affected also by the condition of the environment. Actually, a triangular relation of scale interactions is formed by large scale systems, the vortex system and small scale systems.

Forcing due to the interaction with large scale systems tends to induce a secondary circulation or asymmetric structure in a vortex which eventually affects the symmetric structure. The vertical shear of the environmental wind can alter the vertical coherence in the momentum and temperature fields and also influence the heating pattern. The impacts of such effects on hurricane development have been emphasized [e.g., Gray, 1979; Tuleya and Kurihara, 1981; Tuleya, 1991]. Also, interaction between a vortex and the environmental flow can produce horizontal fluxes of momentum which force changes in the wind field [e.g., Pfeffer and Challa, 1981]. A symmetric vortex on the earth cannot remain symmetric because of the generation of a dipolar vorticity structure. Such an structure, known as the beta gyre, is forced by the planetary vorticity advection by the symmetric flow which provides negative tendency to the relative vorticity in the eastern half of the vortex and positive tendency in the western half. The resultant cross center flow can make a significant contribution to the motion of a tropical cyclone [e.g., Ross and Kurihara, 1992; M. A. Bender et al., unpublished manuscript, 1992].

Interaction with large scale flows generally causes imbalance between the wind and mass fields in a vortex. How the vortex responds to the imbalance can strongly influence the subsequent growth or decay of the system. If the adjustment to a new balanced vortex is quickly and smoothly made, a system may retain its main features and remain as a tropical cyclone. If the adjustment is slow, the structure of a system can be substantially modified and a tropical cyclone may decay. In extreme cases, a tropical cyclone may disintegrate.

As stated above, the environmental conditions are important factors for controlling the evolution of a tropical cyclone. This implies that only some tropical depressions reach the intensity of hurricanes since different environmental conditions affect their development. Another implication can be found in the spatial and temporal variability of the tropical cyclone climatology. It is probable that hurricane formation potential in a certain area and period is subject to the quasi-stationary flow conditions at the corresponding basin and season. If large scale flow conditions change, for instance by El Nino, tropical cyclone activity may become above normal in some regions and below normal in other areas [e.g., Dong, 1988]. It is also likely that intraseasonal variation of environmental conditions causes active and quiet periods of tropical disturbances. In addition to its direct influence, the large scale systems indirectly affect hurricanes by influencing the behavior of small scale systems.

The condition at the ocean and land surfaces is another kind of environmental condition influencing the life cycle of tropical cyclones. Various aspects of this issue have been studied, e.g., the structure and intensity changes at the landfall of hurricanes [e.g., Tuleya et al., 1984], the effect of topography [e.g., Bender et al., 1987], sensitivity to the sea surface temperature [e.g., Ooyama, 1969a], and the mutual response of oceans and tropical cyclones [e.g., Khain and Ginis, 1991]. Also, the importance of radiation on the development of hurricanes has been suggested [Kurihara and Tuleya, 1981].

6. Summary and Generalization

In the present article, the evolution of hurricanes is considered as the adjustment of forced vortices to a state of near balance. A energy source of hurricanes is the heat supplied at the ocean surface. Moist convection and turbulence are internal processes which eventually force a vortex to change its structure. The process of adjustment reestablishes a state of balance and is accomplished by the joint effects of the transverse circulation and inertia gravity waves. The life cycle of a tropical cyclone also seems to strongly depend on how a vortex responds to the forcing from large scale systems. The forcing should be favorable and persistent for steady development of hurricanes. The adjustment should be quick for a smooth evolution of hurricanes. The above aspect has important implications not only for the fate of an individual vortex but also for spatial and temporal variations of tropical cyclone activity.

Although hurricanes are a special phenomenon in the atmosphere, a number of their dynamical features pertain to many other slow mode systems as well. Structures of these systems are characterized by certain relationships expressing balance between the wind and mass fields. When a balance relation in a particular system is disturbed by a forcing process, adjustment processes try to reestablish a balanced state. Forcing of a system is generally provided by both smaller and larger systems. The adjustment process may involve the contribution of fast modes generated by forcing. Only with continual favorable forcing and quick adjustment, will the slow system smoothly develop. It has to be stressed that a single atmospheric process can play dual roles, both forcing and adjustment. Namely, the same process may provide forcing to one system and adjustment of another system at the same time. For example, the poleward heat transport process in the middle latitudes implies both forcing of the synoptic scale eddies and the adjustment of the basic zonal mean temperature field. In another instance, the momentum transport by inertia gravity waves indicates adjustment of a source system and forcing of sink systems. Sometimes, forcing and adjustment proceed in a feedback manner between two systems. In short, the dual function of an atmospheric process serves to link systems of distinctly different scales.

Acknowledgments. The author would like to express his thanks to Prof. Keiiti Aki, University of Southern California, for inviting him to prepare and present this paper at the Jeffreys Symposium in the Twentieth General Assembly of the International Union of Geodesy and Geophysics, Vienna, 1991. He is grateful to Ngar-Cheung Lau and Rebecca J. Ross for making a number of suggestions for clarification and improvement on many points in the original manuscript. His thanks are also extended to Kerry Emanuel for very valuable suggestions for improvement of the original manuscript. Thanks are also due to Philip Tunison, Catherine Raphael and Jeffrey Varanyak for the preparation of figures.

References

Bender, M. A., R. E. Tuleya, and Y. Kurihara, A numerical study of the effect of island terrain on tropical cyclones, *Mon. Wea. Rev., 115*, 130-155, 1987.

Dong, K., El Nino and tropical cyclone frequency in the Australian region and the northwest Pacific, *Aust. Met. Mag., 36*, 219-225, 1988.

Eliassen, A., On the Ekman layer in a circular vortex, *J. Meteor. Soc. Japan, 49*, Special Issue, 784-789,1971.

Emanuel, K. A., The maximum intensity of hurricanes, *J. Atmos. Sci., 45*, 1143-1155, 1988.

Gray, W., Hurricanes: their formation, structure and likely role in the tropical circulation, in *Meteorology over the Tropical Oceans*, edited by D. B. Shaw, 155-218, Royal Meteorological Society, 1979.

Holland, G., and R. T. Merrill, On the dynamics of tropical cyclone structural change, *Quart. J. Roy. Met. Soc., 110*, 723-745, 1984.

Jeffreys, H., The function of cyclones in the general circulation, *Proces-Verbaux de l'Association de Meteorologie*, UGGI (Lisbon), Part II (Memoires), 219-230, (1933), in *Theory of Thermal Convection*, edited by B. Saltzman, pp.200-211, Dover, New York, 1962.

Jorgensen, D. P., Mesoscale and convective scale characteristics of mature hurricanes. Part II: Inner core structure of Hurricane Allen (1980), *J. Atmos. Sci., 41*, 1287-1311, 1984.

Khain, A., and I. Ginis, The mutual response of a moving tropical cyclone and the ocean, *Beitr. Phys. Atmosph., 64*, 125-141, 1991.

Kurihara, Y., Budget analysis of a tropical cyclone simulated in an axisymmetric numerical model, *J. Atmos. Sci., 32*, 25-59, 1975.

Kurihara, Y., and R. E. Tuleya, A numerical simulation study on the genesis of a tropical storm, *Mon. Wea. Rev., 109*, 1629-1653, 1981.

Kurihara, Y., and M. A. Bender, structure and analysis of the eye of a numerically simulated tropical cyclone, *J. Meteor. Soc. Japan, 60*, 381-395, 1982.

Malkus, J. S., and H. Riehl, On the dynamics and energy transformations in steady-state hurricanes, *Tellus, 12*, 1-20, 1960.

Miller, B.I., On the maximum intensity of hurricanes, *J. Meteor., 15*, 184-195, 1958.

Ooyama, K., On the stability of the baroclinic circular vortex: a sufficient criterion for instability, *J. Atmos. Sci., 23*, 43-53,1966.

Ooyama, K., Numerical simulation of the life cycle of tropical cyclones, *J. Atmos. Sci., 26*, 3-40, 1969a.

Ooyama, K., Numerical simulation of tropical cyclones with an axisymmetric model, *Proceedings of the WMO/IUGG Symposium on numerical weather prediction in Tokyo, Nov.26-Dec.4, 1968*, III 81-88, Japan Meteorological Agency, Tokyo, 1969b.

Pfeffer, R. L., and M. Challa, A numerical study of the role of eddy fluxes of momentum in the development of Atlantic hurricanes, *J. Atmos. Sci., 38*, 2393-2398, 1981.

Ross, R. J., and Y. Kurihara, A simplified scheme to simulate asymmetries due to the beta effect in barotropic vortices, *J. Atmos. Sci., 49*, 1620-1628, 1992.

Sato, T., N. Ao, M. Yamamoto, S. Fukao, T. Tsuda, and S. Kato, A typhoon observed with the MU radar, *Mon. Wea. Rev., 119*, 755-768, 1991.

Shapiro, L. J., and H. E. Willoughby, The response of balanced hurricanes to local sources of heat and momentum, *J. Atmos. Sci., 39*, 378-394, 1982.

Syono, S., and M. Yamasaki, Stability of symmetrical motions driven by latent heat release by cumulus convection under the existence of surface friction, *J. Meteor. Soc. Japan, 44*, 353-375, 1966.

Tuleya, R. E., Sensitivity studies of tropical storm genesis using a numerical model, *Mon. Wea. Rev., 119*, 721-733, 1991.

Tuleya, R. E., and Y. Kurihara, A numerical study of the effects of environmental flow on tropical storm genesis, *Mon. Wea. Rev., 109*, 2487-2506, 1981.

Tuleya, R. E., M. A. Bender, and Y. Kurihara, A simulation study of the landfall of tropical cyclones using a movable nested-mesh model, *Mon. Wea. Rev., 112*, 124-136, 1984.

Willoughby, H. E., Gradient balance in tropical cyclones, *J. Atmos. Sci., 47*, 265-274, 1990.

Yamasaki, M., A further study of the tropical cyclone without parameterizing the effects of cumulus convection, *Papers Meteor. Geophysics, 34*, 221-260, 1983.

Y. Kurihara, Geophysical Fluid Dynamics Laboratory/NOAA, Princeton University, Princeton, NJ 08542.

Upper Mantle Density Anomalies, Tectonic Stress in the Lithosphere, and Plate Boundary Forces

MARTIN H. P. BOTT

Department of Geological Sciences, University of Durham, England

Large lateral variations of density occur in the upper mantle mainly as a result of hot spot activity, asthenospheric upwelling at ocean ridges and subduction. The effect of such sub-lithospheric loads on the isostatic state and stress in the lithosphere is modelled by finite element analysis. The anomalous pressure and shear stress caused by the loading produce a flexural isostatic response and give rise to loading stress which concentrates in the strong core of the upper lithosphere. This is the main source of tectonic stress in the lithosphere. Anomalously hot, low density upper mantle produces deviatoric tension and cool, dense upper mantle produces deviatoric compression. As an example, it is shown that a 500 km wide region with density reduced by only 10 kg/m^3 between 200 and 400 km depths would give rise to a 70 MPa horizontal deviatoric tension in a 20 km elastic layer near the top of the lithosphere. If such a stress system is cut through by a zone of weakness such as occurs beneath the crest of an ocean ridge, or by a subduction fault, then the stresses in both plates are redistributed and plate boundary forces are produced. Normal ridge push modelled in this way produces deviatoric compression of about 40 MPa in old ocean floor and a plate boundary force relative to old ocean floor of about 2.5×10^{12} N/m. Substantially greater compression and ridge push occurs if the ridge is underlain by a hot spot. Similarly, subducting lithosphere gives rise to compression above, but when this is intersected by a weak subduction fault, slab pull and trench suction occur as the sinking slab draws the plates towards the subduction plate boundary. It is also shown that dense sinking lithospheric mantle beneath mountain ranges is a major source of compression and may give rise to a collision pull effect.

INTRODUCTION

The strong upper part of the lithosphere acts as a reservoir for strain energy. This is being dissipated by ongoing tectonic activity at a rate which must be substantially greater than that of release of tectonic energy by earthquakes, that is about 3×10^{10} W. The strain energy must therefore be renewed at approximately the same rate as it is being dissipated. The process of lithospheric stress renewal must be sought in fluid motions driven by major regions of anomalous density within the underlying mantle. The well-established density anomalies of this type are (1) upper mantle hot spots beneath oceanic and continental swells, (2) hot low density upwelling asthenosphere beneath ocean ridges, and (3) cool dense subducting lithosphere. These all appear to be related to the escape of heat from the interior of the Earth by convection. This paper aims to show how these anomalous density structures in the upper mantle give rise to tectonic stresses in the lithosphere which drive most tectonic activity including the plate motions.

Loper [1985] has proposed an attractive hypothesis of mantle convection which directly relates the above three types of major density anomaly in the mantle to the pattern of convection. Heat escapes to the surface from the deep interior in two stages. In the *first stage*, narrow plumes carry hot material up from the thermal

Relating Geophysical Structures and Processes: The Jeffreys Volume
Geophysical Monograph 76, IUGG Volume 16
Published in 1993 by the International Union of Geodesy and Geophysics and the American Geophysical Union.

boundary layer just above the core-mantle boundary and discharge it into the upper mantle, producing upper mantle hot spots. Additionally, heat is carried up from the lower to the upper mantle by slow advection. In the *second stage*, material from the asthenosphere, not normally from a hot spot region, rises to the surface at ocean ridges where heat is lost by progressive cooling of the newly formed oceanic lithosphere. The return downflow of cooled material is essentially by the subducting slabs. Thus the two above types of low density sub-lithospheric loading are directly related to the two stages of heat escape and the single type of high density sub-lithospheric loading is related to the single-stage return flow of subducting material.

Our problem is to understand how major sub-lithospheric density anomalies give rise to renewable stress in the lithosphere and to the associated plate boundary forces which drive the plates. This paper draws together the main results from a series of papers by the author and colleagues dealing with different aspects of stressing the lithosphere and origin of plate boundary forces, using elastic-viscoelastic finite element analysis. Technical detail of the methods is avoided for the sake of clarity, but can be found in the relevant papers.

STRESSES CAUSED BY ANOMALOUS DENSITIES

The lithosphere responds by flexural isostasy to the subsurface loading associated with anomalous densities within it or below it. This gives rise to a secondary topographical load at the surface. The surface and subsurface loads are of opposite polarity but equal in total magnitude as required by isostasy. The combined effect of the surface and subsurface loading is to produce anomalous stresses which mainly concentrate into the strong upper part of the lithosphere. Two types of stress are produced, bending stress and simple loading stress. Bending stress is the response to asymmetrical surface and subsurface loading. Simple loading stress is the response to symmetrical surface and subsurface loading such as occurs in local isostasy. The vertical principal stress is modified by an amount equal to the additional normal stress exerted by the load, but the horizontal principal stress is effectively unchanged. As a result, anomalous deviatoric stresses approximately equal to half the normal stress produced by the equal and opposite loads occur in the region between them, provided that the strength is not exceeded. Loading stress concentrates upwards into the relative strong elastic core of the upper lithosphere and, in contrast to bending stress, its magnitude is inversely proportional to the thickness of this strong elastic core [*Kusznir and Bott*, 1977]. In flexural isostatic situations, both bending and simple loading stresses occur. In engineering applications, bending stress is of paramount importance. Within the earth, however, it is the local loading stress which is of greatest tectonic importance.

The tectonic force T_f is defined as the difference between vertical and horizontal normal stresses integrated with respect to depth z down to the bottom of the anomalous densities at depth D. Thus

$$T_f = \int_0^D (\sigma_{xx} - \sigma_{zz})\, dz \qquad (1)$$

In the ideal case of an infinitely wide uniform structure, this can be calculated from the density-moment function [*Parsons and Richter*, 1980; *Dahlen*, 1981], giving

$$T_f = \int_0^D gz\Delta\rho\, dz \qquad (2)$$

where g is gravity and $\Delta\rho$ is the anomalous density at depth z referenced to a standard density–depth distribution such as that beneath old ocean floor. The geoid anomaly is proportional to the density-moment function [*Haxby and Turcotte*, 1978], giving a useful relationship between geoid anomaly and lithospheric stress due to loading.

Tectonic force is a useful concept in discussing stress in the lithosphere and in estimating plate boundary forces. It can be estimated from the anomalous density–depth distribution for wide structures using Equation (2). Bending stress is approximately eliminated as its average value integrated across the bent elastic layer is zero. The tectonic force is independent of the rheological structure of the lithosphere provided that the lithospheric strength is not exceeded. After upward concentration of stress is complete, the deviatoric stress in an elastic layer of specified thickness can be calculated by dividing tectonic force by twice the layer thickness. Tectonic force needs to be referenced to a standard lithospheric density–depth distribution to be meaningful, as was earlier recognised for plate boundary force by *Lister* [1975].

SUB-LITHOSPHERIC LOADING – HOT SPOT EXAMPLE

The mechanism by which a small but deep sub-lithospheric negative density anomaly gives rise to isostatic uplift and deviatoric stressing of the strong upper part of the lithosphere is studied in this section using elastic–viscoelastic finite element analysis. The results are com-

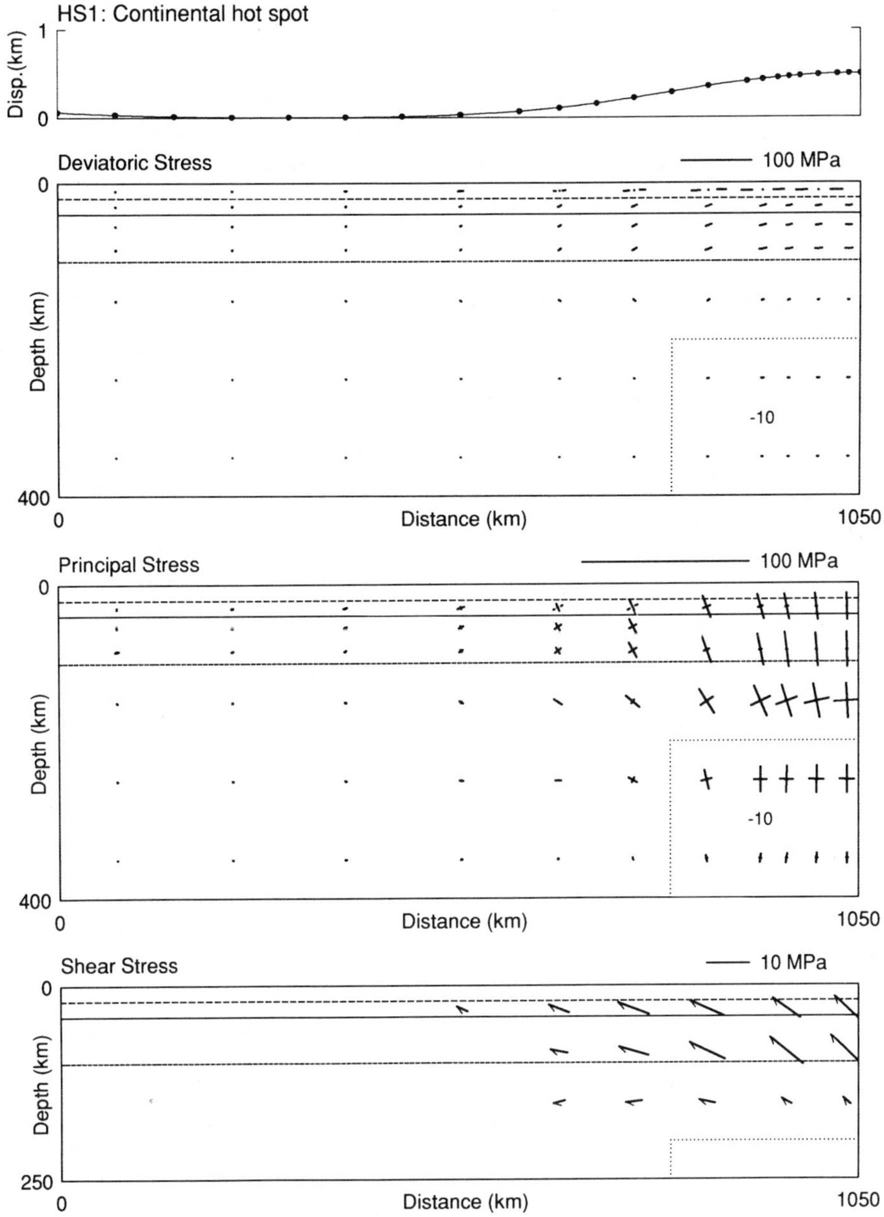

Fig. 1. The left half of the symmetrical Model HS1 of sub-lithospheric loading by a deep density anomaly of -10 kg/m^3 outlined by the dotted line, representing an incipient sub-continental hot spot in the upper mantle. The diagram shows, from top to bottom: vertical displacement of the surface in isostatic response to the deep load; deviatoric stress; principal stress; shear stress (note different stress scales). The base of the crust is marked by a solid line, and the base of the upper elastic crust and of the lithosphere are denoted by broken lines of different spacing. In this and succeeding figures, the stresses are shown at the centre of elements; a broken line with a dot at the centre denotes tension and a solid line denotes compression; anomalous densities are shown in kg/m^3 and mechanical properties and other specifications are as stated in the text.

pared to theoretical estimates obtained using the density-moment function. The model used gives insight into the origin of tension characteristic of uplifted continental rift systems.

Model HS1 (Figure 1) consists of a two-dimensional finite element grid of 77 isoparametric rectangular elements which extends 1050 km horizontally and down to 400 km depth. The crust is 40 km thick and the lithosphere extends to 100 km depth. Young's modulus is 0.90×10^{11} Pa for the crust and 1.75×10^{11} for the upper mantle. Poisson's ratio is 0.27 throughout. The uppermost 20 km of the crust is elastic. The underlying part of the lithosphere has been assigned a viscosity of 10^{23} Pa s and the asthenosphere a value of 10^{21} Pa s. The nodes at the right edge are constrained to zero horizontal displacement so that the grid represents the left half of a symmetrical structure 2100 km wide. The nodes along the bottom and left edge are free, and isostatic boundary conditions are applied at the surface and at the Moho assuming the density increases downwards by 2930 and 400 kg/m^3 respectively. In computing the stresses, it has been assumed that the out-of-plane deviatoric stress is zero, but the in-plane stress differences are almost identical for the plane strain assumption. With this assumption, only one stress needs to be displayed, as the other is equal and opposite.

A rectangular region with an anomalous low density of −10 kg/m^3 between 200 and 400 km depths occupies the central 500 km of the symmetrical model (that is 250 km wide adjacent to the right edge of the grid). This represents a deep upper mantle region with its temperature raised by about 100 K above the normal, such as might be produced by a plume. In reality, such a hot region would normally be in contact with the base of the lithosphere, but it has been placed well below to demonstrate convincingly the coupling between a deep low density region and the lithosphere above. Further models of this type have been presented by *Bott* [1991a, 1992] to which reference should be made for technical details and fuller discussion.

Figure 1 shows the result of running this viscoelastic model for 1000 time steps of 500 yr each, so that dynamic equilibrium is approached. The deep load causes flexural isostatic uplift at the surface which reaches a maximum of 490 m at the centre of the symmetrical model (only half shown in Figure 1). The result of the loading is to develop a sub-horizontal deviatoric tension which reaches a maximum of 68 MPa in the elastic layer. Upward concentration into the elastic layer is not yet fully complete. The model has also been run for 2000 time steps (i.e. 1 Myr) and the maximum stress has increased to 82 MPa. Numerical computation of the tectonic force shows that the final equilibrium stress level will be about 103 MPa, and thus that this value is 66% attained after 1000 time steps and 79% after 2000 time steps. Upward concentration of stress from the asthenosphere is effectively complete by 1000 iterations. The high viscosity of the lower lithosphere is the reason for the slow attainment of equilibrium. The tension is inversely proportional to the thickness of the elastic layer, so that where the layer is thinner the stress is proportionally larger. This feature is not shown in Figure 1, but see *Bott* [1992].

The process of attainment of dynamic viscoelastic equilibrium is not of geological significance, as it would be for loading on a shorter timescale such as postglacial loading and unloading. The timescale of development of major tectonic structures such as continental rift systems is much longer than that of attainment of viscoelastic equilibrium, so that we can assume that equilibrium is effectively maintained as the structure develops.

The distribution of anomalous principal stress relative to the standard lithostatic pressure–depth distribution at the left edge of the model is shown in the middle of Figure 1. This diagram shows that the vertical normal stress acting on the base of the lithosphere is anomalously high above the deep load. It is this excess normal stress, acting on the base of the upper elastic part of the lithosphere, which supports the flexural isostatic uplift. The anomalous vertical normal stress, which extends through the upper elastic lithosphere, also contributes to the loading stress. In this example this contributes almost 10 MPa to the deviatoric tension in the elastic layer. However, this is only a small contribution to the actual tension.

A further, and more substantial, contribution to the deviatoric tension comes from the advective flow driven by the region of anomalous low density. This exerts outward directed shear stress on the base of the upper elastic lithosphere. This is shown at the bottom of Figure 1 where shear stress in the advecting region between the load and the lithosphere is plotted at the centre of elements. The symmetrical outward drag produced in this way causes a contribution to the deviatoric tension in the elastic layer which eventually reaches about 90 MPa at the centre of the symmetrical model. In contrast to loading stress caused by normal stress acting on the base of the elastic layer which is independent of the depth of a wide load, the contribution from advective drag increases with the depth of the load as the advection affects a wider region.

It is of interest to compare the tectonic force computed from the stress distribution using Equation (1) with the theoretical value of 5.88×10^{12} N/m obtained for an infinitely wide structure using Equation (2). Integrating the stress difference at the centre of the symmetrical model yields a tectonic force of 4.15×10^{12} N/m for the 100 km thick lithosphere. The theoretical value is higher than the computed value because of the finite width of Model HS1.

Although Model HS1 has been presented mainly to demonstrate the mechanism of sub-lithospheric loading, nevertheless the results apply to uplifted plate interior regions supported by an anomalously hot, low density upper mantle, such as present-day uplifted continental rift systems (e.g. East Africa, Baikal). The modelling indicates that substantial local deviatoric tension should occur relative to the adjacent unaffected regions. This is borne out by stress observations, which indicate that such regions display tension in contrast to normal continental regions where compression dominates [*Zoback et al.*, 1989]. More detailed modelling of uplifted continental rift systems, including lithospheric thinning and faulting, is given by *Bott* [1992].

RIDGE PUSH

Plate boundary forces originate when a lithospheric stress system caused by sub-lithospheric loading is intersected by a zone of weakness. The weak region is unable to withstand large deviatoric stresses, with the result that the boundary forces acting on the edges of the adjacent strong plates are profoundly modified. This radically modifies the stress distribution not only locally but also within the adjacent plate interiors. If the deviatoric loading stresses intersected by the zone of weakness are inherently tensional as at an ocean ridge, then the supplementary stress system caused by the weakness is compressional and the boundary forces act to drive the plates away from each other. If they are inherently compressional as at a subduction plate margin, then the supplementary stress system is tensional and the boundary forces draw the plates towards the boundary. The way in which this works at a modelled ocean ridge is shown here, using the models described by *Bott* [1991b] to which reference is made for a detailed technical description.

Ocean ridges are underlain by a relatively shallow low density subsurface load in the form of hot upwelled asthenosphere and hot thinned lithosphere. It is this anomalous density distribution which supports the topography of the ridge relative to old ocean floor. If the elastic lithosphere continued uninterrupted across the crest of the ridge, then strong horizontal deviatoric tension would occur as in Figure 1. However, the lithosphere is extremely thinned and intersected by a zone of weakness extending to the surface at the ridge crest.

This situation has been modelled [*Bott*, 1991b] based on the cooling plate model of *Parsons and Sclater* [1977] and the associated elastic thickness model of *Watts* [1978 for an assumed spreading rate of 15 mm/yr. The resulting anomalous density distribution relative to that beneath the old ocean floor is shown in Figure 2 (Model RIG2). The model is a symmetrical one about the ridge crest and only the right half is shown. A passive margin is included on both sides of the ocean, with 35 km thick continental crust having a -400 kg/m^3 density contrast to the upper mantle. This supports the topography which is represented by a surface load in local isostatic equilibrium. The zone of weakness is modelled by including a pair of weak viscoelastic elements extending to the surface at the ridge crest. The mechanical properties are as in Model HS1 (Figure 1). The thin oceanic crust has an insignificant effect and is excluded for simplicity. The nodes at the distant edge of the elastic layer have been constrained to zero horizontal displacement to enable the ridge push force to be measured in absence of underside drag on the plate, and thus the gradual decrease in compression away from the ridge caused by underside drag is excluded from the displayed stresses. The nodes beneath the ridge crest are constrained to zero horizontal displacement to make the model symmetrical, but the remaining nodes along the base and distant edge are free.

The models have been run for 2000 time steps of 500 yr each. It should be noted, with the distant edge fixed to eliminate underside drag, that the plates are not actually moving apart significantly as equilibrium is approached. Thus the spreading rate of 15 mm/yr only refers to the geometry of the structure.

Figure 2 (upper part) shows the flexural uplift of the ocean ridge and the deviatoric stress in the upper elastic part of the lithosphere for Model RIG2 of a normal oceanic ridge as described above. An enlarged plot of the central region is shown below the main model. The deviatoric stress has been concentrated into the upper elastic part of the lithosphere. Small tensions occur near the ridge crest as a result of the ongoing extension, representing the ridge resistance. These give way laterally at about 40 km to horizontal deviatoric compressions which increase to about 40 MPa at about 200 km from the ridge crest, remaining at about this level out to the oldest ocean floor because of trade-off between increasing tectonic force and thickening elastic layer. Comparison with observations is complicated by the probable dominance of thermal stresses between ages of about 2.5 to 35 Ma [*Bratt et al.*, 1985], but there is good agreement with the observed tensional regime near the ridge crest [*Sykes*, 1967] and the compressional regime in old ocean floor [*Bergman*, 1986].

The deviatoric compressions are reduced to 9 MPa in the continental region as a result of the loading stress from the low density continental crust but would be

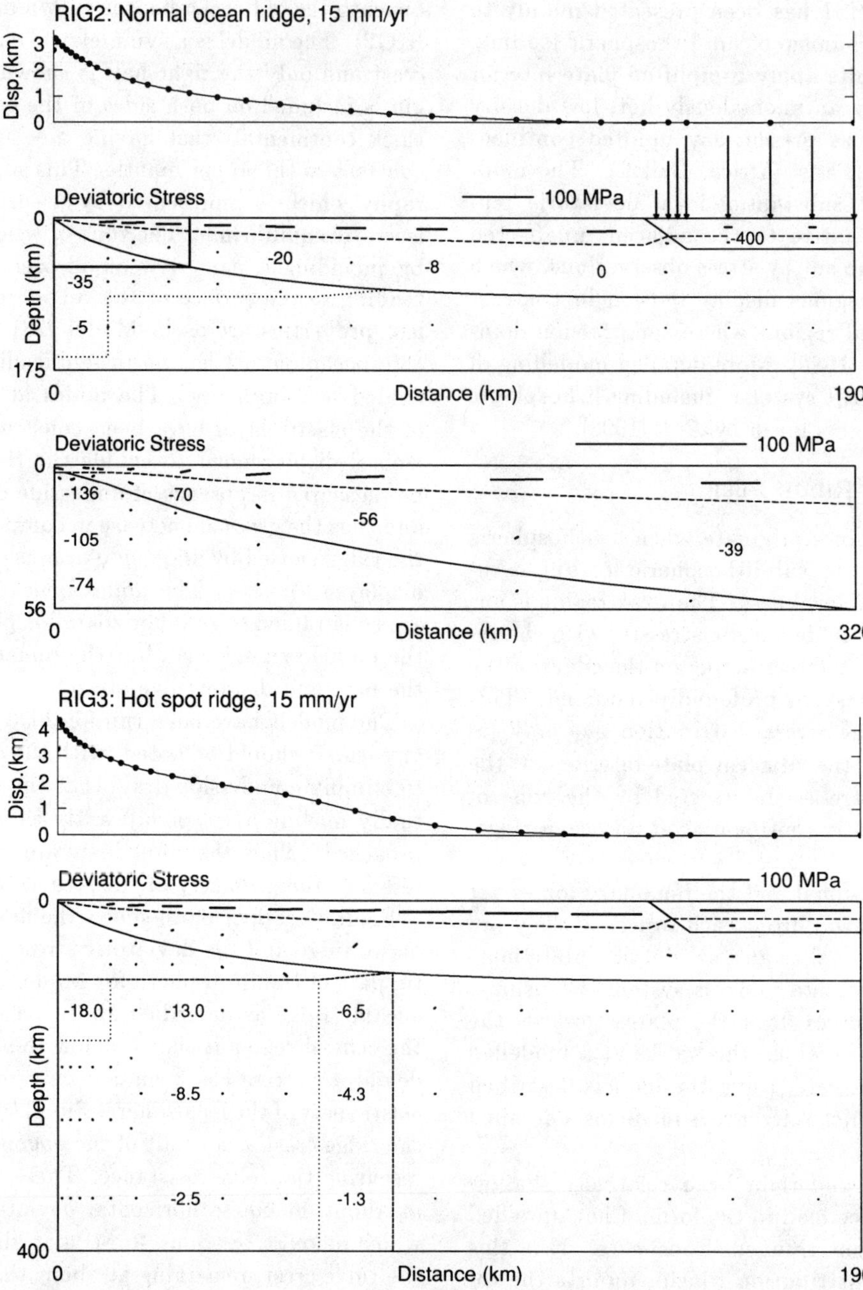

Fig. 2. The vertical isostatic displacement at the surface and the deviatoric stresses shown for the right half of symmetrical models of an ocean ridge spreading at 15 mm/yr. Model RIG2 (above) represents a normal ocean ridge and Model RIG3 (below) represents a ridge underlain by an anomalously low density upper mantle. The ocean is bordered by passive margins with continental crust outlined by the solid line in local Airy equilibrium with a surface load representing the topography (shown on the same scale as the stresses for Model RIG2 only). Representative values of the anomalous density field relative to old ocean floor are shown in kg/m^3; Model RIG3 densities are only shown where they differ from those of RIG2. The base of the lithosphere is marked by a solid line and the base of the upper elastic part of it is denoted by a broken line. The central region of Model RIG2, as outlined by the thick line, is shown enlarged beneath the main model.

larger if the underlying upper mantle is denser than at comparable depths beneath the old ocean floor [*Fleitout and Froidevaux*, 1983]. The plate boundary force, referenced to the old ocean floor, can be determined by integrating the stress difference across the elastic lithosphere, yielding a value of 2.5×10^{12} N/m which is in excellent agreement with other estimates such as that based on the density-moment function [*Dahlen*, 1981]. The stress distribution and ridge push force are almost identical for models based on a faster spreading rate [*Bott*, 1991b].

How does the ridge push originate? The weak region at the crest cannot withstand large stress differences. As a result the horizontal normal stress becomes nearly equal to the anomalous vertical normal stress caused by the surface and subsurface loading. It is this compressional horizontal normal stress acting on the edges of the elastic plates which gives rise to the ridge push and to the distant compressions. For a normal ridge this gravity wedging effect is dominant and the advective drag contribution is relatively small [*Bott*, 1991b].

Some oceanic regions are underlain by an anomalously hot upper mantle, possibly associated with a nearby mantle plume. An example is the North Atlantic region north of 30°N where the low density of the upper mantle is reflected in anomalously shallow bathymetry focussing on Iceland and the Azores [*Cochran and Talwani*, 1978]. It is of interest to examine the effect of such a hot spot beneath an ocean ridge on the associated plate interior stresses and the ridge push force.

Model RIG3 (Figure 2, lower part) represents an ocean ridge underlain by an anomalously hot, low density upper mantle. As well as the density anomaly associated with the normal ridge, there is additionally an anomalous low density grading stepwise from -14.5 kg/m^3 at 90 km depth to zero at 400 km depth. The additional isostatic uplift of the ocean floor is about 1 km at the ridge crest. The additional small but deep density anomaly has the effect of greatly increasing the compressional stress in old oceanic lithosphere by more than a factor of two, consistent with increase of stress with source depth indicated by Equation (2). Similarly, the plate boundary force referenced to the old ocean floor is 6.2×10^{12} N/m which is also more than twice that of a normal ridge. The much increased ridge push force and plate interior compression above those of the normal ridge is partly explained by increased pressure in the weak region but is mainly the result of shear drag exerted on the base of the elastic layer by the deep load, as demonstrated by *Bott* [1991b].

In contrast to the normal ridge, the high compressional stress in the adjacent continental crust is a spectacular feature of the hot spot model. It was suggested by *Bott* [1991b] that the North Atlantic hot spot may explain in this way the substantial compressional stresses characteristic of the North American and European continents [*Zoback et al.*, 1989] and may be an important factor in driving the mountain building of the Alpine belt.

SUBDUCTION PULL

The stressing associated with a subduction plate margin is of opposite polarity to that at an ocean ridge. The major feature giving rise to the stress in the surface plates is the anomalously cool and dense subducting material. This gives rise to isostatic downflexure at the surface and to compressional loading stress in the elastic upper lithosphere above the slab. When this is intersected by a weak subduction fault, then stress is redistributed within both plates giving rise to tension within the distant plate interiors and causing both plates to be pulled towards the subduction plate boundary. This situation has been modelled by *Bott et al.* [1989] and by *Whittaker et al.* [1992].

The stress situation associated with an ideal subduction plate margin is shown in Figure 3 using a replotted model from *Whittaker et al.* [1992]. The model is 2800 km wide and extends to 650 km depth, but is only shown down to 400 km depth in the figure. The lithosphere is 90 km thick and the uppermost 30 km is elastic. A 30 km thick continental crust in local Airy isostatic equilibrium is incorporated in the overriding plate, but oceanic crust is not included in the subducting plate for simplicity. The mechanical properties are the same as in Models HS1 and RIG2 except that the lower lithosphere has a viscosity of 5×10^{22} Pa s and the mantle transition zone below 400 km depth has a Young's modulus of 2.8×10^{11} Pa and a viscosity of 10^{22} Pa s. A 45° dipping slab with an anomalous density of $+50$ kg/m^3 extends to nearly 300 km depth. The nodes at the base at 650 km depth are fixed vertically and those at the distant edges of the lithosphere are constrained to zero horizontal displacement to measure the pull on the plates in absence of underside drag. Isostatic boundary forces are applied at the surface. The models have been allowed to relax for 250,000 years.

Model SLAB1 (top) shows the deviatoric stresses and the vertical surface displacement profile when the subduction fault is locked, preventing active subduction. Apart from ongoing rollback as the slab becomes progressively steeper, equilibrium is approached. The load-

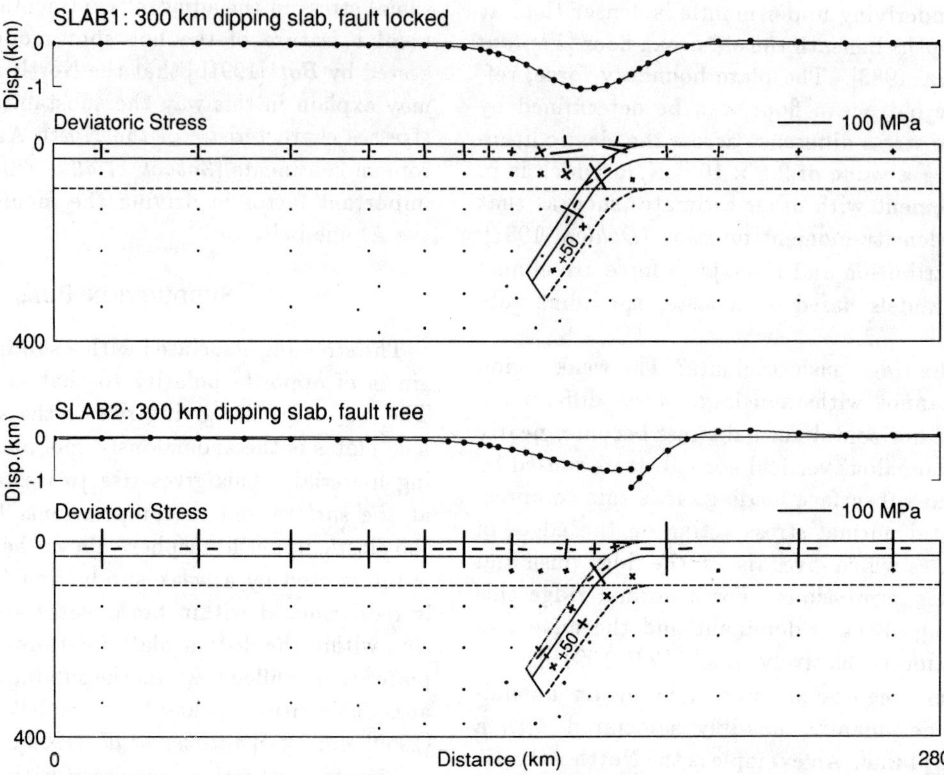

Fig. 3. Deviatoric stress distributions and surface displacement profiles produced by models of a subducting slab with excess density of 50 kg/m^3 dipping at 45° and extending to about 300 km depth. Model SLAB1 (above) has the subduction fault locked, and Model SLAB2 (below) has the subduction fault free. Both of the in-plane principal stresses are shown. Vertical exaggeration is ×1.5. The base of the lithosphere is marked by a broken line and the base of the upper elastic layer (which coincides with the continental Moho on the overriding plate) is marked by a solid line.

ing stresses produced by the downpull of the dense slab give rise to sub-horizontal deviatoric compressions of up to 70 MPa in the 30 km thick elastic layer above the slab, but these die away at distance from the slab in both directions. A small horizontal deviatoric tension of about 25 MPa dominates the more distant parts of the elastic layer of the overriding plate, representing the loading stress of the continental crust.

Model SLAB2 (bottom) differs from SLAB1 only in that the fault is unlocked, with zero friction. At the end of the 250,000 yr run the slab is sinking with a downdip velocity of 8 mm/yr and a realistic rollback of 10 mm/yr. The downdip velocity is on the slow side but the resulting change from downdip tension to downdip compression at about 200 km depth indicates that the resistance at the base of the slab, which has been increased by fixing the basal nodes vertically, is realistic. The effect of the free fault on the near surface stresses in both plates is profound. The result of annihilating the shear stress on the fault plane is that supplementary tractions are applied which cause local vertical flexural motion (not shown) and supplementary pervasive deviatoric tension which extends throughout both plates out to their fixed edges. The supplementary tension has the effect of approximately annulling the compression above the slab, and giving rise to deviatoric tension of 65 MPa near the fixed edge of the subducting plate and 70 MPa (including the continental loading stress of 25 MPa) near the fixed edge of the overriding plate. The result of the weak fault is to permit both plates to be pulled towards the subduction plate margin as the slab beneath sinks. The estimated plate boundary forces referenced to old ocean floor are -4.8×10^{11} N/m for the slab pull on the subducting plate and -4.2×10^{11} N/m for the trench suction acting on the overriding plate. They are of comparable magnitudes, as might be anticipated from the overall equilibrium of the model.

The results show good general agreement with focal

mechanism results for slabs [*Isacks and Molnar*, 1971; *Zhou*, 1990], with one principal stress approximately downdip, and downdip tension at shallower depths giving way to downdip compression towards the bottom of the slab. Within the two models (Figure 3), the fault is either fully locked or fully free. However, a fault with some friction would produce an intermediate situation with trench–arc compression giving way to back-arc tension in agreement with the observed situation in some arc regions [*Nakamura and Uyeda*, 1980].

Some further models, not shown here, with vertical and 45° dipping slabs extending to about 300 and 400 km depths were presented by *Whittaker et al.* [1992]. As is to be expected, the stresses and plate boundary forces increase with increasing depth extent of the slab, but surprisingly the slab dip has a relatively insignificant effect on the stresses. The effect of dense subducting material in the lower mantle has not been modelled and may have a significant additional influence on the stresses and plate boundary forces. However, the inferred increase in viscosity with depth at the mantle transition zone may subdue the effect of very deep, dense material on the lithosphere.

COLLISION PULL

The locally originating stress associated with an ideal collision mountain range was modelled by *Bott* [1990]. The stress arises from two main factors. *First*, the low density crustal root, and the mountain topography which it supports isostatically, together give rise to

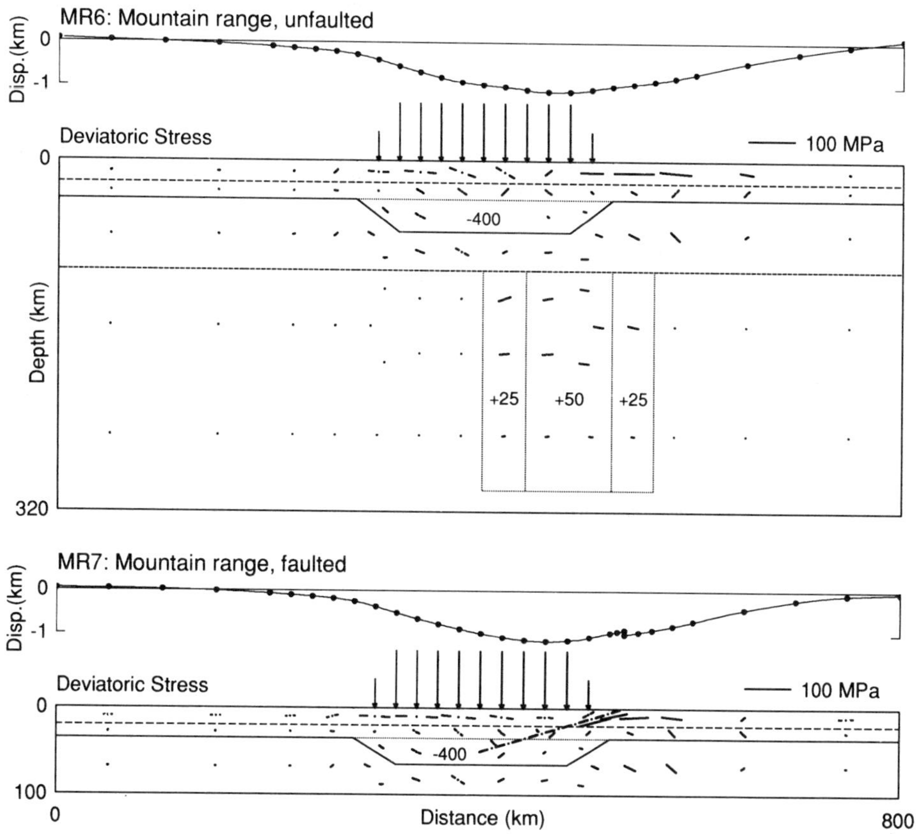

Fig. 4. Models of a collision mountain range incorporating a crustal root extending to 60 km depth and a 200 km thick slab of downbulging lithospheric mantle asymmetrically located beneath one flank of the crustal root. A solid line marks the Moho and dotted lines outline the regions of anomalous density. The base of the upper elastic layer and of the lithosphere are marked by broken lines. The surface loads represent the mountain topography supported by the crustal root in local Airy equilibrium (shown on the same scale as the stresses). The displacement profiles show the flexural downpull caused by the lithospheric slab but exclude the mountain topography. Model MR6 (above) is unfaulted. Model MR7 (below) incorporates an unlocked thrust fault cutting across the elastic layer with the edges of the lithosphere constrained to zero horizontal displacement.

deviatoric tension in the relatively strong near-surface layer. This tends to cause gravitational collapse of the mountain range and opposes the stress which is driving the mountain building process. *Second*, the presence of a dense downbulge of cool lithospheric mantle has recently been inferred from seismological and gravity observations across the Alps [*Mueller and Panza*, 1986]; such a downbulge would be expected to occur beneath collison ranges as a result of convergence of the plates and may have been present at the time of initial collision as a result of subduction at the margin of the closing ocean. The dense downbulging region causes isostatic flexural depression of the region above and gives rise to deviatoric compression of comparable magnitude to the crustal root-related tension.

The actual state of locally originating stress is the superimposition of these two stress systems, with the further addition of stress due to distant plate boundary forces which is not included here. If the crustal root occurs symmetrically above the lithospheric root, then these two stress systems tend to annul each other so that the resulting deviatoric stresses are small. However, if the lithospheric root is asymmetrically located beneath one of the flanks of the mountain range, as appears to apply for the Alps, then compression would dominate this flank whereas tension would affect the central and distant parts of the mountain range.

Figure 4 (top) shows the stress distribution for Model MR6 representing the asymmetrical situation envisaged above. The full model is 800 km wide and extends to 650 km depth, with mechanical properties as in Model SLAB1 (Figure 3), except for a viscosity of 10^{23} Pa s for the lithospheric mantle. It has been run for 1000 time steps of 500 yr each. The crustal root is 30 km in depth extent and has an anomalous low density of -400 kg/m^3 supporting a surface load representing the mountains in local equilibrium. The lithospheric root, which is comparable to that beneath the Alps, is 200 km in depth extent, with an anomalous density of $+50$ kg/m^3 in the central region and $+25$ kg/m^3 in the flanks, corresponding to a temperature of 250–500 K relative to the adjacent asthenosphere. For simplicity, the upper layer 20 km thick has been assumed to be elastic throughout. The results show that sub-horizontal deviatoric compressions of up to 94 MPa occur above the lithospheric root and tensions of up to 59 MPa dominate the parts of the mountain range distant from this flank.

Figure 4 (bottom) shows Model MR7 which includes a plane thrust fault with zero friction cutting across the elastic layer in the region of compression, with the edges of the plates constrained to zero horizontal displacement so that the inward pull can be measured. Small deviatoric tensions occur out to the edge of the model, as in model SLAB2, showing that the lithospheric downbulge has a similar (but smaller) effect to subduction in presence of a lithospheric zone of weakness. The resulting collision pull, referenced to normal continental lithosphere, is about 1.0×10^{12} N/m and acts on both plates. It is suggested that this may be a critical factor in driving the plates towards a collision plate boundary.

The modelling explains the observed occurrence of extensional collapse of part of a mountain range contemporaneous with compressional thrust tectonics on one flank [*Dewey*, 1988]. It also demonstrates that mountain building may not be just the passive response of a weak region to distant plate boundary forces, but that the dense downsinking lithospheric mantle may be an important factor in driving the compressional tectonics and in pulling the plates towards the orogenic belt.

CONCLUSIONS

It has been demonstrated how large density anomalies in the upper mantle below the lithosphere produce tectonic stress in the lithosphere and drive the plate motions. The three major types of density anomaly discussed are associated with plume-derived upper mantle hot spots, upwelling asthenosphere beneath ocean ridges and dense subducting material. All of these can be associated with the escape of heat from the interior of the Earth to the surface and are by-products of the mantle convection process.

The first model presented (Figure 1) is of quite a small density anomaly of -10 kg/m^3 of 500 km width extending between 200 and 400 km depths beneath continental lithosphere, such as might result from a 100 K temperature rise produced by a plume. This gives rise to updoming of the type observed in continental rift systems. It also produces a large horizontal deviatoric tension of about 100 MPa in the 20 km thick elastic layer, which could give rise to continental rifting. The relatively small but deep load couples to the lithosphere as a result of both anomalous normal stress above the load (which causes the isostatic uplift) and outward directed shear stress applied to the base of the elastic layer by advection driven by the density anomaly. Both the pressure and the shear stress contribute to the deviatoric tension, although the shear stress contribution is substantially greater for this deep load.

If a lithospheric stress system produced by sub-lithospheric loading is intersected by a weak zone which cuts right across the strong lithosphere, then the stresses within the plates are redistributed as a result of the near annihilation of the deviatoric stresses in the zone or plane of weakness. This has the effect of modifying the horizontal boundary forces acting on the adjacent plate edges, thus giving rise to the plate boundary force and also radically modifying the stresses within the adjacent plate interiors.

This is illustrated by modelling the stress system associated with a normal ocean ridge with a weak crest. In the model (Figure 2), a small tension occurs near the ridge crest representing the ridge resistance required to drive the ongoing viscoelastic deformation in the weak zone. This gives way laterally to a compression of about 40 MPa which dominates the flanks and the old ocean floor. The ridge push force, referenced to old ocean floor, is estimated to be 2.5×10^{12} N/m. The stress and ridge push force are effectively independent of spreading rate. An ocean ridge underlain by an anomalous hot spot upper mantle producing an additional seabed elevation of 1 km (e.g. North Atlantic hot spot) has also been modelled (Figure 2), demonstrating that the ocean floor compression and ridge push force are both more than doubled as a result of the small but deep additional density anomaly.

The opposite situation applies at a subduction plate boundary (Figure 3). The dense subducting lithosphere gives rise to a local deviatoric compression in the trench–arc region above when the subduction fault is locked, but the compression is effectively annihilated and tension pervades the distant plate interiors when the fault is unlocked. It is shown that slab pull and trench suction of over 4.0×10^{12} N/m, referenced to old ocean floor, are produced by a slab penetrating to about 300 km depth.

A downbulge of dense sinking lithospheric mantle, such as has been observed to underlie the Alps asymmetrically, may be a general feature of collision mountain belts as a result of the plate convergence. The root of thickened crust gives rise to extensional stress but the downbulge gives rise to compression. With an asymmetrically located downbulge, one flank of a mountain range may be subjected to compressional tectonics at the same time as the central region and opposite flank undergo extensional collapse as modelled (Figure 4). It is also shown that the downbulge may give rise to a small collision pull acting on both plates to pull them towards the orogenic belt.

It has thus been demonstrated that hot, low density regions in the upper mantle have an important role in producing local plate interior tension and the ridge push force. Cool and dense sinking lithospheric material beneath subduction and collision plate boundaries produces local compressional stress above and gives rise to plate boundary forces which pull the plates toward the boundary. In this way, the escape of heat from the deep interior produces mantle density anomalies which are the most important factor in stressing the lithosphere and driving the plates.

REFERENCES

Bergman, E. A., Intraplate earthquakes and the state of stress in oceanic lithosphere, *Tectonophysics*, *132*, 1-35, 1986.

Bott, M. H. P., Stress distribution and plate boundary force associated with collision mountain ranges, *Tectonophysics*, *182*, 193-209, 1990.

Bott, M. H. P., Sublithospheric loading and plate-boundary forces, *Philos. Trans. R. Soc. London, Ser. A*, *337*, 83-93, 1991a.

Bott, M. H. P., Ridge push and associated plate interior stress in normal and hot spot regions, *Tectonophysics*, *200*, 17-32, 1991b.

Bott, M. H. P., Modelling the loading stresses associated with active continental rift systems, *Tectonophysics*, *215*, in press, 1992.

Bott, M. H. P., G. D. Waghorn, and A. Whittaker, Plate boundary forces at subduction zones and trench–arc compression, *Tectonophysics*, *170*, 1-15, 1989.

Bratt, S. R., E. A. Bergman, and S. C. Solomon, Thermoelastic stress: how important as a cause of earthquakes in young oceanic lithosphere?, *J. Geophys. Res.*, *90*, 10,249-10,260, 1985.

Cochran, J. R., and M. Talwani, Gravity anomalies, regional elevation, and the deep structure of the North Atlantic, *J. Geophys. Res.*, *83*, 4907-4924, 1978.

Dahlen, F. A., Isostasy and the ambient state of stress in the oceanic lithosphere, *J. Geophys. Res.*, *86*, 7801-7807, 1981.

Dewey, J. F., Extensional collapse of orogens, *Tectonics*, *7*, 1123-1139, 1988.

Fleitout, L., and C. Froidevaux, Tectonic stresses in the lithosphere, *Tectonics*, *2*, 315-324, 1983.

Haxby, W. F., and D. L. Turcotte, On isostatic geoid anomalies, *J. Geophys. Res.*, *83*, 5473-5478, 1978.

Kusznir, N. J., and M. H. P. Bott, Stress concentration in the upper lithosphere caused by underlying visco-elastic creep, *Tectonophysics*, *43*, 247-256, 1977.

Isacks, B., and P. Molnar, Distribution of stresses in the descending lithosphere from a global survey of focal-mechanism solutions of mantle earthquakes, *Rev. Geophys.*, *9*, 103-174, 1971.

Lister, C. R. B., Gravitational drive on oceanic plates caused by thermal contraction, *Nature*, *257*, 663-665, 1975.

Loper, D. E., A simple model of whole-mantle convection, *J. Geophys. Res.*, *90*, 1809-1836, 1985.

Mueller, S., and G. F. Panza, Evidence of a deep-reaching lithospheric root under the Alpine arc, in *The origin of arcs*, edited by F.-C. Wezel, pp. 93-113, Elsevier, Amsterdam, 1986.

Nakamura, K., and S. Uyeda, Stress gradient in arc–back arc regions and plate subduction, *J. Geophys. Res.*, *85*, 6419-6428, 1980.

Parsons, B., and F. M. Richter, A relation between the driving force and geoid anomaly associated with mid-ocean ridges, *Earth Planet. Sci. Lett.*, *51*, 445-450, 1980.

Parsons, B., and J. G. Sclater, An analysis of the variation of ocean floor bathymetry and heat flow with age, *J. Geophys. Res.*, *82*, 803-827, 1977.

Sykes, L. R., Mechanism of earthquakes and nature of faulting on the mid-oceanic ridges, *J. Geophys. Res.*, *72*, 2131-2153, 1967.

Watts, A. B., An analysis of isostasy in the world's oceans 1. Hawaiian–Emperor seamount chain, *J. Geophys. Res.*, *83*, 5989-6004, 1978.

Whittaker, A., M. H. P. Bott, and G. D. Waghorn, Stresses and plate boundary forces associated with subduction plate margins, *J. Geophys. Res.*, *97*, 11,933-11,944, 1992.

Zhou, H., Observations on earthquake stress axes and seismic morphology of deep slabs, *Geophys. J. Int.*, *103*, 377-401, 1990.

Zoback, M. L., *et al.*, Global patterns of tectonic stress, *Nature*, *341*, 291-298, 1989.

M. H. P. Bott, Department of Geological Sciences, University of Durham, South Road, Durham DH1 3LE, England.

A Simple Rheological Framework for Comparative Subductology

TOSHIKO SHIMAMOTO AND TETSUZO SENO

Earthquake Research Institute, University of Tokyo, 1-1-1 Yayoi, Bunkyo-ku, Tokyo 113, Japan

SEIYO UYEDA

Department of Geophysics, Texas A&M University, College Station, Texas 77863-3114 U.S.A.; and Department of Marine Science and Technology, Tokai University, 3-20-1 Orido, Shimizu, Shizuoka 424, Japan

Abstract. The origin of the diversity of subduction zones, particularly with respect to the maximum size of inter-plate thrust-type earthquakes, is still controversial, and previous models fail to account for why only moderate earthquakes occur in some subduction zones (e.g., Mariana). This paper demonstrates that the diversity of subduction-zone seismicity and the existence of nearly aseismic subducting plate boundaries, free from large and great earthquakes, can be explained by a simple rheological model. The model subdivides a subducting plate boundary into three zones: i.e., (1) shallow decoupled zone, (2) intermediate seismogenic zone, and (3) lower aseismic interface that rebounds aseismically following large earthquakes. The first zone, assumed to be shallower than about 30 km, is weak and aseismic due primarily to the massive solution-transfer processes under the presence of abundant H_2O released through progressive metamorphism. Without this shallower cutoff, the seismogenic zone becomes too wide and the lack of large and great earthquakes in some subduction zones cannot be explained. The boundary between zones (2) and (3) should be associated with the shift in the frictional properties from velocity weakening to velocity strengthening. Although many factors could be involved in this shift, it is assumed here for simplicity that temperature is the primary factor. A crude examination, using temperature distributions in various subduction zones, reveals that the wider the seismogenic zone (w in km), the greater the expected earthquake magnitude (M_S or M_W) and that M_S or M_W = 1.2 (±0.4) log w + 6.5 (±1.6) holds.

Relating Geophysical Structures and Processes: The Jeffreys Volume
Geophysical Monograph 76, IUGG Volume 16

Introduction

The diverse nature of subduction zones has been extensively discussed [e.g., Karig, 1974; Molnar and Atwater, 1978; Chase, 1978; Uyeda and Kanamori, 1979; Uyeda, 1979, 1982, 1984; Uyeda and Nishiwaki, 1980; Dewey, 1980; Kelleher and McCann, 1976; Ruff and Kanamori, 1980; Peterson and Seno, 1984]. This diversity is particularly notable with respect to the maximum size of inter-plate thrust-type earthquakes. The extreme cases are central Chile where truly great earthquakes occur and the Mariana arc where only moderate earthquakes of at most magnitude-7 class occur.

It now seems generally accepted that the diversity with respect to the subduction-zone seismicity is due to the difference in the degree of plate coupling in subduction zones; i.e., the stronger the plate coupling, the greater the earthquakes. However, the primary source for the variation in the plate coupling is not resolved as yet. Uyeda and Dewey [papers cited above] argued that the movement of the overriding plate relative to the trench is the major controlling factor for the plate coupling, if the asthenosphere is inert or only slowly moving. That is, there would be strong plate interactions producing great earthquakes when the overriding plate is moving toward the trench; whereas, the plate interaction must be much less producing only moderate earthquakes when the overriding plate is retreating from trench.

Although such an explanation may appeal to intuition, it cannot quite account for why only moderate earthquakes occur in the Mariana-type subduction zones. Irrespective of the movement direction of the overriding plate at a subduction zone, the subducting plate is downthrusting under the overburden of the upper plate. Moreover, the downthrusting is taking place along a relatively cool subducting plate boundary due to the subduction of a cold slab. Thus, it is not immediately clear how great earthquakes could be eliminated by the retreating upper plate or by the extensional stress field in the overriding plate. There are examples of earthquakes of magnitudes up to 7.7 in association with normal faulting [Bonilla et al., 1984], so that large, if not great, earthquakes can indeed occur even in the extensional stress regimes.

Another model, proposed to account for the diversity of subduction zones, is Kanamori's [1977b] evolutionary model. In this model, subduction is assumed to start with low angle thrusting and the dip of the slab steepens as an arc evolves. The essence of this model is that the smaller the angle of subduction, the wider the contact zone between the overriding and subducting plates, and hence the greater the plate coupling and resulting earthquakes. Thus, the variation of subduction zones with respect to the size of thrust-type earthquakes can be predicted by this simple geometric model. The model, however, cannot explain why only moderate earthquakes occur in some subduction zones by the same token stated above.

Lay and Kanamori [1981] proposed an asperity model to account for the diversity in subduction-zone seismicity. However, the physical nature of the asperities is still unclear. They may be some geometrical irregularities, again not clearly specified, of the contact surfaces. Hence, the model cannot predict the diversity of subduction-zone seismicity at this time.

Ruff [1989] suggested that the subduction of a large amount of sediments increases the degree of plate coupling. In our opinion, however, the subduction of more sediments will bring more pore fluids and hydrous minerals such as clays into the subducting plate interface to make the coupling even weaker than in the case with less sediment subduction (for more discussions see Wang [1980] and Shimamoto [1985]). Moreover, consolidated pelitic sediments are similar in the composition of major constituent minerals to granitic rocks and are not expected to be stronger at greater depths than the mafic rocks constituting the oceanic crust or the ultramafic rocks in the wedge mantle. From the mechanical viewpoint, therefore, the subduction of thick sediments will not foster the occurrence of large earthquakes.

We shall examine the problem here with the view that the primary controlling factors are the rheological properties of rocks and the thermal state at the plate boundary. The physical mechanisms of earthquake phenomena have long been one of the the central topics in rock rheology [Bridgman, 1951; Griggs and Handin, 1960; Mogi, 1962; Brace and Byerlee, 1966; Dieterich, 1979; among many others]. With regard to the origin of the thrust-type earthquakes in subduction zones, the rheological model of subducting plate boundaries after Shimamoto [1985] is an outcome of such attempts. On the other hand, there have been a number of investigations during the past two to three decades on the thermal regimes of subduction zones. These investigations include both marine and land measurements of heat flow distribution, estimation of thermal structures of the crust and upper mantle based on seismic wave characteristics, and theoretical modelling to explain these results. As will be shown in a later section, the central problem here has been how to explain the high temperature under some arc and backarc regions where a cold slab is subducting.

This paper presents a first-step synthesis of these approaches and demonstrates that a simple rheological model [Shimamoto, 1985, 1991], combined with thermal structures of various subduction zones estimated from simple modeling, accounts for (1) the diversity of subduction zones with respect to the size of thrust-type earthquakes and, in particular, (2) the lack of large and great earthquakes in some subduction zones. As for the thermal structures of various subduction zones, we will use results from a comparative modeling study by Honda and Uyeda [1983]. This modeling is an extremely simple application of the corner flow to various subduction zones, but proved useful for our present purpose.

A Rheological Model For Subducting Plate Boundary

Shimamoto [1985] discussed the seismicity and deformation mechanisms in subduction zones and divided a subducting plate boundary into the following three zones.

(1) Shallow aseismic and decoupled interface

A common feature of many subduction zones is that there is an aseismic zone along the subducting plate boundary down to the depth of 20~30 km (TA in Figure 1). The existence of this zone is best exemplified by the distribution of microearthquake foci (see Fukao [1979], Seno and Kroeger [1983] and Hirata et al. [1985]). There are some depth distributions, particularly those from routine determinations, showing the microseismic activity shallower than 20 km [e.g., Zhang and

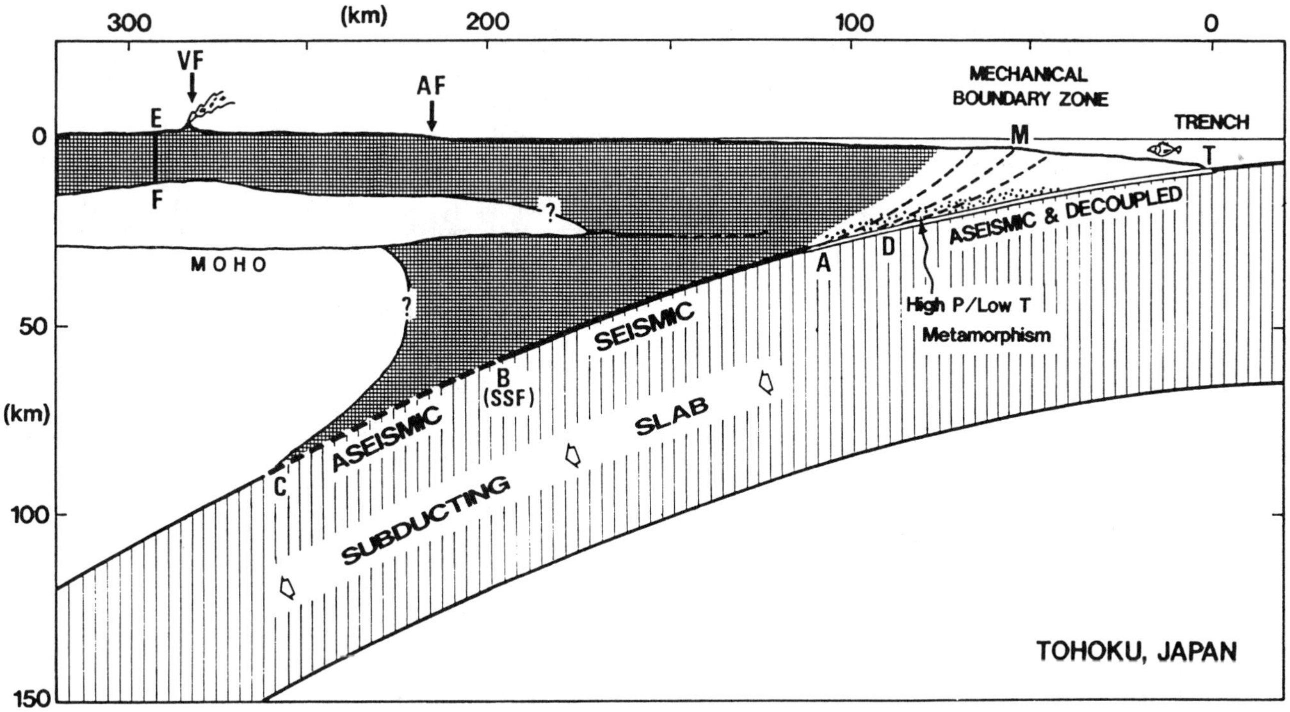

Fig. 1. A simple rheological model of a subducting plate boundary after Shimamoto [1985], schematically drawn for Tohoku region, northeast Japan, as a typical example. The portion with a fine checked pattern roughly indicate the lithosphere in the overriding arc. T, trench, VF, volcanic front, AF, aseismic front [Yoshii, 1979], SSF, seismic-slip front [Kawakatsu and Seno, 1983], M, mechanical boundary zone [Nakamura et al., 1984].

Schwarz, 1992]. However, it should be noted that the depth determination of shallow earthquakes near the trench is extremely difficult. Hence, the depths reported on the routine basis, such as ISC and Harvard CMT, are often different from those determined in individual studies [e.g., Honda et al., 1990; Zhang and Lay, 1992]. At present, microearthquake studies using ocean bottom seismometers [e.g., Hirata et al., 1985] and very careful depth determination using body waves [e.g., Seno and Kroeger, 1983] are the most reliable sources. They show that the microseismicity becomes scarce at depths shallower than 20 km and rapidly increases below the depth of about 30 km.

This shallow interface is nearly decoupled during much of the interseismic period [Shimazaki, 1974], and hence the interface coupling must be much weaker here than at the deeper portions where large or great earthquake initiates. This shallow interface is a unique place in the earth where high-pressure and low-temperature type progressive metamorphism takes place and a large amount of H_2O (more than 10 % in volume) is released constantly through metamorphic reactions down to the depth of 30~35 km (Figure 2). Such enormous amount of H_2O would raise the pore pressure and, perhaps more significantly, promote deformation via solution-transfer processes [Shimamoto, 1985]. When the processes are fully operative, rocks behave like viscous fluid with very low strength [e.g., Rutter and Mainprice, 1979; Rutter, 1983; Spiers et al., 1990]. The low seismicity and low strength of the shallow interface are perhaps due largely to the predominantly ductile deformation fostered by the solution-transfer processes (for other possible factors see Shimamoto [1985]).

(2) Intermediate seismogenic interface

The seismicity along subducting plate boundaries increases abruptly from the depth of about 30 km and diminishes at the depth of 60~70 km (AB in Figure 1; Kawakatsu and Seno, [1983]). Ruptures causing large or great thrust-type earthquakes typically initiate from the depth of 30~50 km, propagate upward to deviate off from the plate interface at around point D, and reach the mechanical boundary zone, M in Figure 1 [Nakamura et al., 1984]. Perhaps, the most important factor causing the shift from the

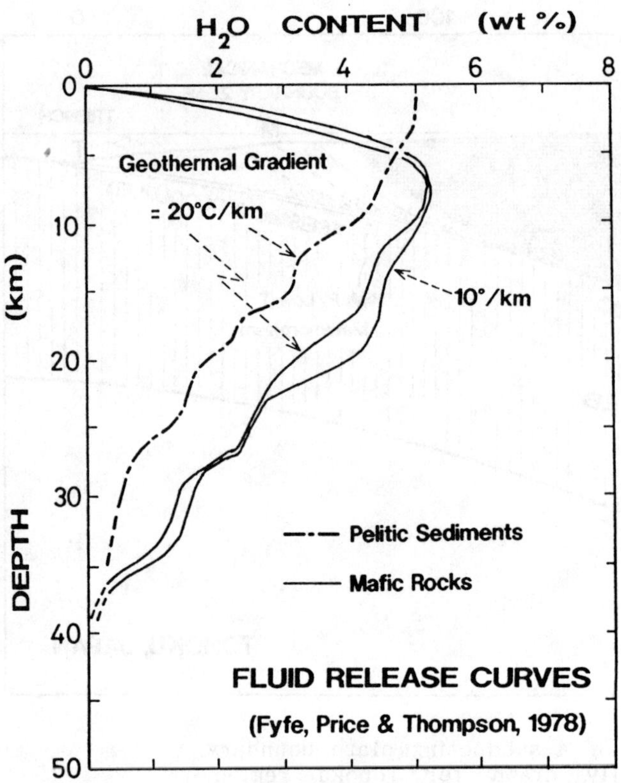

Fig. 2. Fluid-release curves for pelitic sediments and mafic rocks during progressive metamorphism, under linear geothermal gradient of 10 and 20 °C/km (after Fyfe et al. [1978]). The H_2O content is given in weight percents and indicates the amount of chemically-bound water in metamorphic rocks of various metamorphic grades.

shallow aseismic zone to the intermediate seismogenic zone is that the amount of fluid release becomes diminishingly small at the depth of 30~35 km (Figure 2). Cracks and pores will be cemented quickly at such great depths to suppress the circulation of H_2O in rocks. These processes both make the solution-transfer processes less effective and presumably enhance the seismicity. Unstable fault motion in the semiductile regime and the dehydration embrittlement under relatively dry environments are the likely candidates for the mechanisms of the thrust-type earthquakes [Shimamoto, 1985], although much work needs to be done in the future to understand these mechanisms more comprehensively.

(3) Deep aseismic interface

The deep portion of the subducting plate boundary has been shown to rebound aseismically for the time period of a few tens of years following a great thrust-type earthquake (BC in Figure 1) [Shimazaki, 1974; Thatcher and Rundle, 1979]. The shift from the seismogenic to aseismic interface must be associated with the change in the frictional property from velocity weakening to velocity strengthening that takes place at a depth slightly shallower than that for the onset of fully plastic deformation (Figure 3; compiled from Shimamoto [1985, 1986, 1989]). At the deep interface, the strength of rock is still high but seismogenic motion no longer occurs (B to C in Figures 1 and 4).

Thermal Regimes in Subduction Zones

The distribution of heat flow in trench-arc-backarc systems are characterized by low heat flow in the trench and forearc region and high heat flow in the arc's axial zone and backarc basin (for reference see Watanabe et al. [1977] and Yamano and Uyeda [1989]). This was first established in the Japanese subduction zones [e.g., Uyeda and Horai, 1964; Uyeda and Vacquier, 1968] and later extended to other subduction zones [Sclater et al., 1972; Uyeda and Watanabe, 1982; Henry and Pollack, 1988; and many others]. These surficial thermal manifestations were found to be generally consistent with the deeper thermal structures inferred from the distribution of seismic wave velocity and attenuation characteristics; namely the downgoing high V-high Q slab and low V-low Q upper mantle under arc and backarc regions [e.g., Utsu, 1971; Isacks et al., 1968]. Here again, there is a diversity; that is, high heat flow in the backarc region and high mantle temperature have not been observed in the Chilean-type subduction zones [Uyeda and Watanabe, 1982; Henry and Pollack, 1988]. This shows that the backarc spreading process is an important factor controlling the thermal regime of backarc regions.

The central problem has been how to explain the high heat flow and high upper-mantle temperature and associated backarc extentional tectonics in terms of the subduction of a relatively cold slab. It was shown that simple solid heat conduction regimes can not explain them [e.g., Toksöz et al., 1971] and that some massive heat transfer in the upper mantle is needed [Hasebe et al., 1971].

Another very important indication of the same enigma is the existence of arc volcanism. Arc volcanism is almost ubiquitous in all arcs, although it is interrupted where the asthenospheric wedge does not exist under the arc like in central Chile [e.g., Sacks, 1977]. This fact also suggests that the convective heat transfer in the mantle wedge, as advocated by many researchers [McKenzie, 1969; Andrews and Sleep, 1974; Toksöz and Hsui, 1978; Bodri and Bodri, 1978; Honda and Uyeda, 1983], is an important factor.

In this work, we will make use of the thermal structures of different arcs obtained by a comparative study based on a simple model by

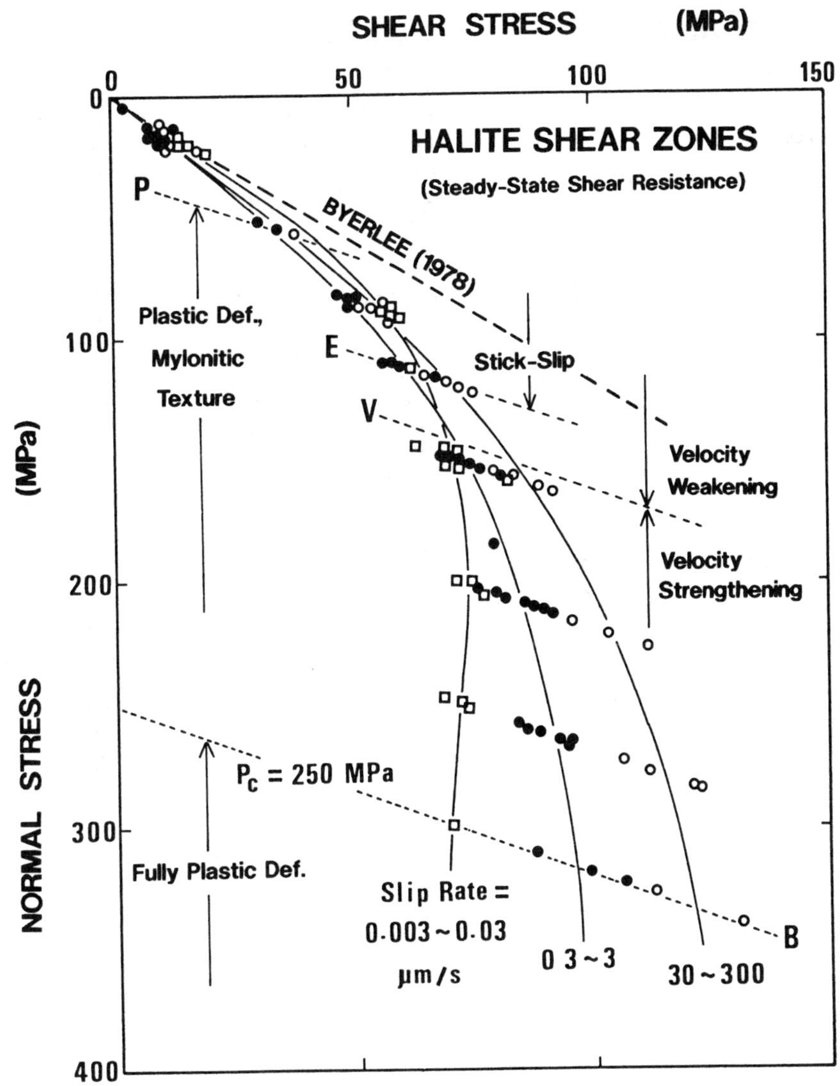

Fig. 3. Steady-state or nearly steady-state shear resistance of synthetic halite shear zones (dry and 0.3 mm thick) plotted against the normal stress, for different ranges of slip rates (compiled from Shimamoto [1986], Hiraga and Shimamoto [1987]). Heavy dashed line indicates the frictional strength of many brittle rocks [Byerlee, 1978]. P and B denotes, respectively, the onsets of plastic deformation and fully plastic deformation of low temperature type. Unstable fault motion ceases at E, slightly shallower than V at which the frictional property changes from velocity weakening to velocity strengthening.

Honda and Uyeda [1983]. Leaving the details to their paper, the model is a two-dimensional, constant viscosity (taken as $2x10^{20}$ poise) corner flow type [Batchelor, 1967] as shown in Figure 5a, in which an oceanic plate of age t1 subducts with a prescribed dip angle and velocity U under a landward plate with a horizontal velocity V. The landward plate can be either an oceanic plate of age t2 or a continental plate. Only viscous heating is considered as a heat source in the thermal modeling. When the parameters are taken to fit various subduction zones, one obtains the results such as those shown in Figure 5b-d as representative examples. From these diagrams, one can assess the temperature along the plate boundary, i.e., the interface between the subducting and overriding plates for various subduction zones as a function of distance from the trench axis (Figure 6).

For precise thermal modeling of subduction processes, one must take into account various factors such as frictional heating, radioactive heating, heat sink due to dehydration, and heat

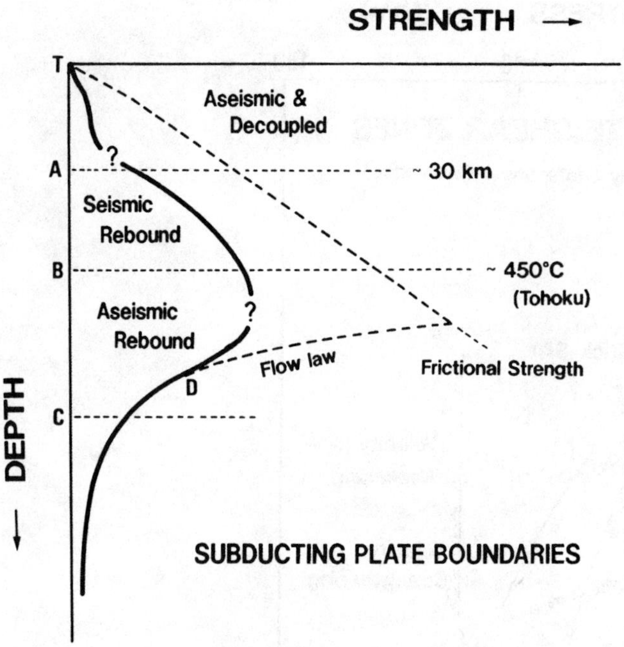

Fig. 4. A conceptual sketch for the strength profile along subducting plate boundaries. TA, shallow aseismic and decoupled zone, AB, intermediate seismogenic zone, and BC, deep aseismic zone (cf. Figure 1). The strength in the shallow zone is low owing to the massive solution-transfer processes. Heavy dashed line indicates the frictional strength of many brittle rocks and the flow stress in the fully plastic regime. The strength of rocks along subducting plate boundaries is perhaps much smaller than the frictional strength due to the weakening effects of abundant H_2O and chemical reactions on the strength of rocks. The deviation of the strength profile from the flow law at point D is due to the presumed localization of plastic deformation [Hobbs et al., 1986].

transfer due to fluid migration. However, critical information is too meager to perform sophisticated thermal modeling for different subduction zones in any impartial way. The simple model used here [Honda and Uyeda, 1983] was shown to be able to represent fairly well at least the Tohoku (NE Japan) arc where an observed heat flow profile across the arc is available [e.g., Honda, 1985].

Implications for Comparative Subductology

The diversity of subduction zones will now be examined based on the simple plate-boundary model, described above. From the rheological viewpoint, the size of thrust-type earthquakes must be controlled largely by the width of the seismogenic interface (AB in Figure 1), by the rheological properties of the same interface, and by the ambient conditions. Although faults from D to M in Figure 1 are parts of the focal regions, not much elastic strain energy will be stored in the wedge (TDM) because the shallow interface (TA) is nearly decoupled. Thus, these faults cannot be regarded as significant in controlling the size of earthquakes. By the same token, the degree of plate coupling across the subducting plate boundary should be controlled by the intermediate and deep portions of the boundary (A to C in Figure 1). However, the precise and quantitative estimation of the size of earthquakes and the degree of plate coupling is not possible at this time, because the rheological properties under shear have not yet been understood fully for rocks under relevant conditions.

Hence, we shall simplify the problem and assume that the greater the width of the seismogenic zone (AB in Figure 1), the larger the plate interaction and the greater the resulting earthquakes. The transition between the shallow decoupled zone and the seismogenic zone (point A in Figure 1) can be taken to be about 30 km in depth at which the dehydration is nearly completed (Figure 2) and the occurrence of microearthquakes sharply increases. Since the fluid-release curves are rather insensitive to the geothermal gradient (Figure 2), point A in Figure 1 was assumed to be the same for all subduction zones. Thirty-kilometers depth for various subduction zones is indicated by small arrows in Figure 6.

The controlling parameters for the shift from the seismogenic zone to the deep aseismic zone (point B in Figure 1) are not known at present. But the transition takes place at a depth between that for the onset of plastic deformation (P in Figure 3) and that for the onset of fully plastic deformation (B in the same figure; cf. Figure 4). In nature, the onsets of partially-operative plastic deformation and fully plastic deformation both depend primarily on temperature, and hence the transition depth to the aseismic interface was assumed, as a crude approximation, to occur at a given temperature. According to the data in the Tohoku region where seismicity and heat flow have been investigated closely [e.g., Honda, 1985], the transition seems to take place at a temperature of about 450° C (vertical broken line passing through SSF in Figure 6).

With these simplifications, the width of the seismogenic zone, w, as measured along the plate boundary, has been determined for various subduction zones, using the temperature profiles after Honda and Uyeda [1983] (for representative results see Figure 5). The results are plotted against typical magnitude, M_s or M_w, of thrust-type earthquakes in the subduction zones [Kanamori, 1978, Ruff and Kanamori, 1980] (Figure 7). Although data points are somewhat scattered, the earthquake magnitude increases in

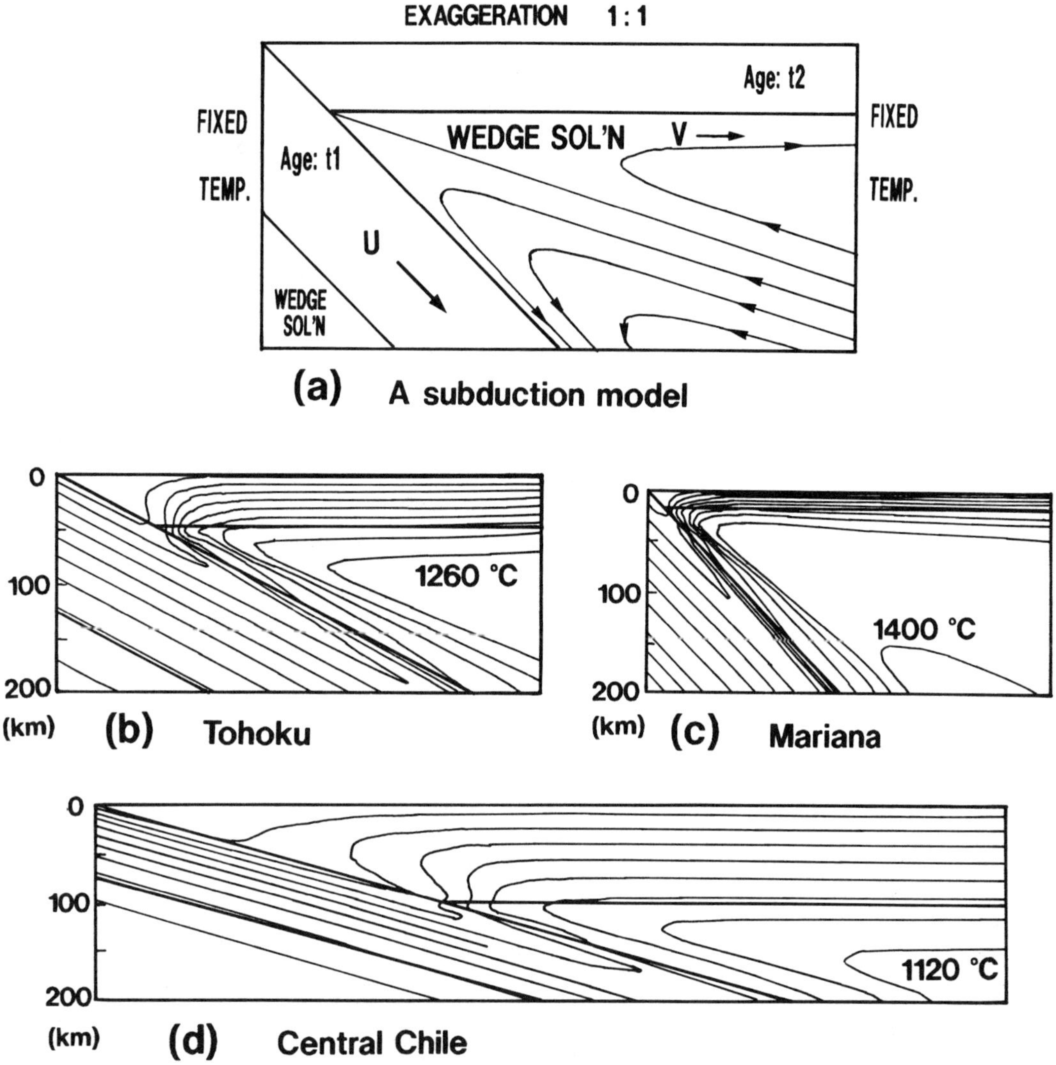

Fig. 5. (a) A two-dimensional subduction zone model in which an oceanic plate of age t1 subducts underneath an oceanic plate of age t2 or continental plate at a constant speed of V and with a given dip angle [Honda and Uyeda, 1983]. Constant-viscosity corner flow in the wedge mantle is incorporated in the computation of temperature distribution using the finite difference method. Representative results for temperature distribution are shown for (b) Tohoku, Japan, (c) Mariana, and (d) central Chile. The contour interval is 140° C.

proportion to the logarithm of the width of the seismogenic zone; that is,

$$M_s,\ M_w = 1.1(\pm 0.4)\ \log\ (w) + 6.1(\pm 1.6) \qquad (1)$$

holds with the correlation coefficient of 0.7. The present, first-order rheological model thus accounts for the diversity of subduction-zone seismicity.

A unique feature of subducting plate boundaries, that has never been taken into account in the past arguments, is the shallow decoupled interface (TA in Figure 1). This zone is much wider than the weak zone conventionally interpretted as due to the presence of unconsolidated sediments in the accretionary prism. Indeed, the zone is several times as wide as the vertical extent of the upper-crustal seismogenic zone behind the volcanic front (cf. EF and TA in Figure 1). Furthermore, the lower plate interface is not seismogenic. Thus, for some subduction zones such as Mariana and New Hebrides, the shallow decoupled zone and the deep aseismic zone occupy almost the entire plate interface and the seismogenic interface is only on the order of 10 km or even less. This

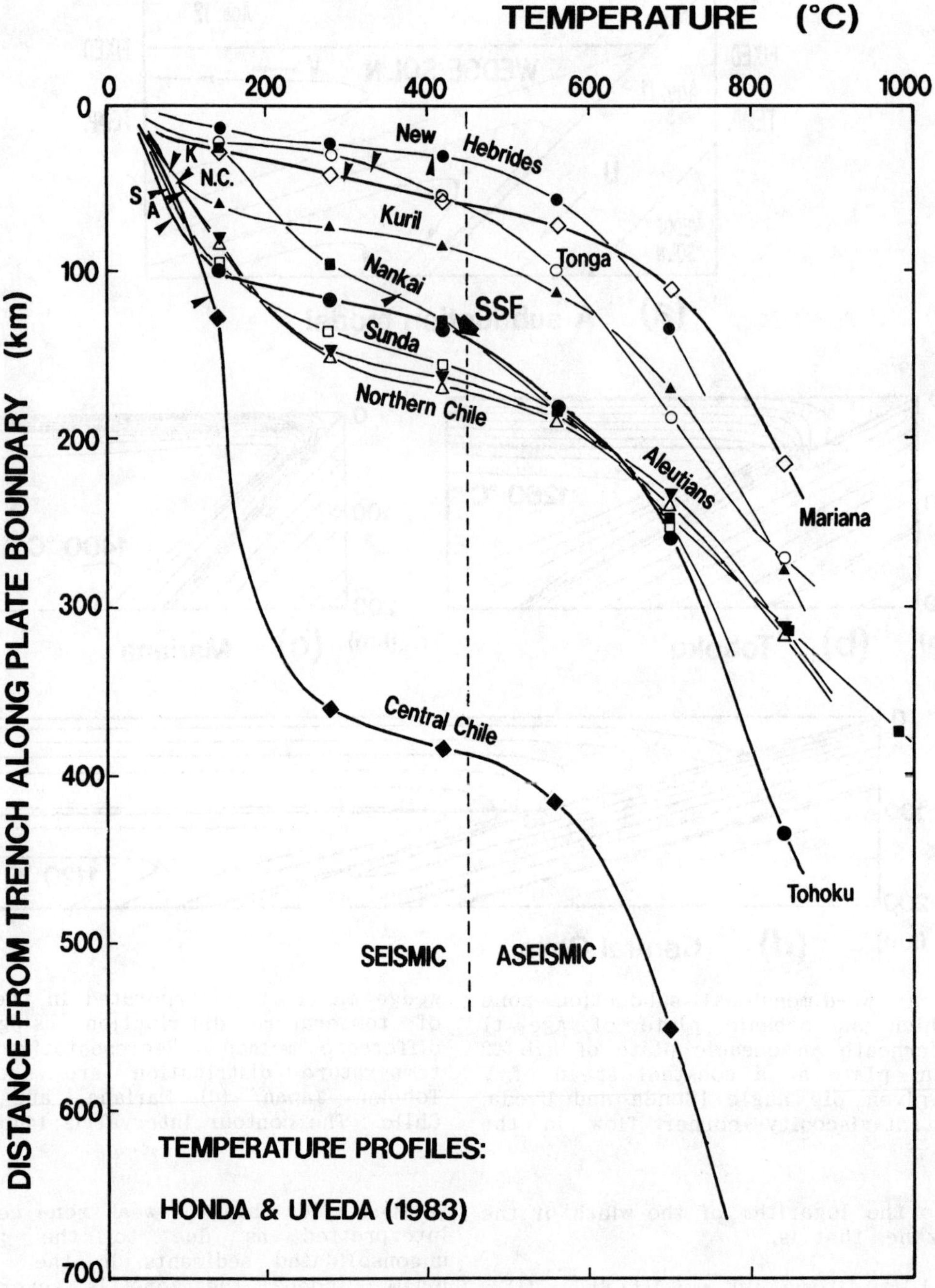

Fig. 6. Temperature along plate boundaries for various subduction zones (data from Honda and Uyeda [1983]). Small arrows correspond to the depth of 30 km which is assumed to be the lower bound of the shallow decoupled zone. SSF denotes the seismic slip front for the subduction zone in Tohoku, northeast Japan [Kawakatsu and Seno, 1983]. The capital symbols on the upper-left corner are the initials of the names of the subduction zones.

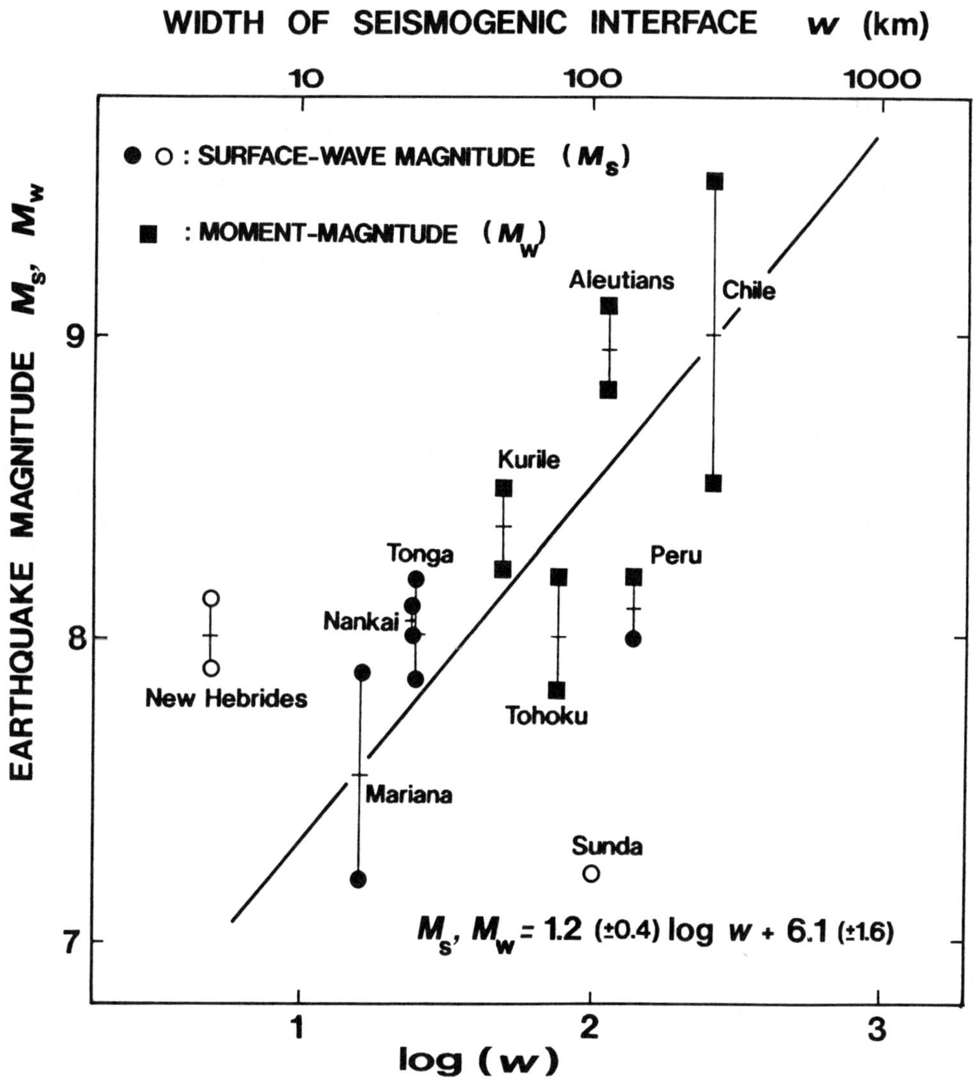

Fig. 7. Magnitude of typical large or great earthquakes for various subduction zones [Kanamori, 1977a, 1978; Ruff and Kanamori, 1980; Sykes and Quittmeyer, 1981; Seno and Eguchi, 1983; Peterson and Seno, 1984; Newcomb and McCann, 1987], plotted against the width (*w*) of the seismogenic zone (A to B in Figure 1). The width was measured along a subducting plate boundary in the direction normal to the trench. M_s: surface-wave magnitude, M_w: moment magnitude for great earthquakes.

is perhaps the primary reason why only moderate earthquakes occur in some subduction zones.

Discussions and Conclusions

The great diversity in the subduction zone seismicity and its origin [Kanamori, 1978] is one of the most important issues addressed in "comparative subductology" [e.g., Uyeda, 1982, 1984]. Present work has demonstrated that a simple rheological and thermal model accounts for the great diversity of subduction zones in the size of maximum earthquakes and for the existence of nearly aseismic behavior of some subducting plate boundaries. It is indeed remarkable that such a crude model can explain the gross behavior of subducting plate boundaries. This implies that our basic premise is generally right and that the diversity of subduction zones is due primarily to the diversity in the thermal state of the subducting plate boundaries, rather than due to other factors. Thus the source of the diversity of the thermal state along subducting plate boundaries emerges as the most fundamental problem in comparative subductology.

Previous work [McKenzie, 1969; Andrews and Sleep, 1974; Toksöz and Hsui, 1978; Honda and Uyeda, 1983] indicates that a smaller dip angle of the subducting slab, a lower temperature in the upper plate, and a higher velocity of plate subduction all increase the width of the seismogenic zone. Thus, shallow-dip subduction under a continental plate provides the largest seismogenic zone, and steep-dip subduction under a young backarc basin plate the smallest. Needless to say, these two cases are the Chilean and Mariana type subduction zones in the comparative subductology.

The dip angle of a slab has been argued to be determined by the balance between the torque exerted by its own weight and that exerted by the induced mantle flow [e.g., Stevenson and Turner, 1977; Hsui et al., 1990]. The torque due to the slab's weight, which is larger for older slabs, tends to steepen the dip; while that due to the induced flow tends to make it shallower. In addition to these factors, Uyeda and Kanamori [1979] suggested that the westward net rotation of the lithosphere relative to the asthenosphere [Minster and Jordan, 1978] would make the west dipping Mariana slab steeper and the east dipping Chilean slab shallower. Although still somewhat speculative, this is potentially an important possibility and worth further investigation as it has recently been demonstrated that such a net lithospheric rotation is theoretically possible if the mantle viscosity is laterally non-uniform [Ricard et al., 1991; O'Connell et al., 1991].

Despite the apparent success of our model, it must be kept in mind that the model does not predict the mechanical interaction of subducting and overriding plates to any quantitative degree because the rheological properties of a subducting plate boundary have not been established as of yet. In particular, future studies need to address (1) rheological properties of rocks in the strength-peak regime and (2) rheology of rocks undergoing solution-transfer processes, chemical reactions, and phase changes.

The significance of the first problem is of no question, since the degree of mechanical interaction of the plates should largely depend on the rheological properties of rocks in the strength-peak regime between brittle and fully plastic regimes. Moreover, the mechanisms of large thrust-type earthquakes cannot be solved unless the rheology in this regime is established [Shimamoto, 1985, 1989]. And yet, a series of experiments on halite, on which we based our arguments in this paper, are about the only systematic studies so far in this regime. Even the form of constitutive laws has not yet been established in the strength-peak regime. A high-temperature biaxial frictional testing machine and a high-temperature and high-pressure rotary frictional testing machine have been installed at the laboratory of the first author to perform a thorough experimental study on this in the near future.

A paradox in plate tectonics in the light of rock rheology is how subduction can take place so freely along a cold and wide plate boundary. Indeed, the analysis of the driving forces of plates [Forsyth and Uyeda, 1975] has revealed only weak interactions along subducting plate boundaries. Hence there must be some weakening mechanisms operative there. The primary candidate for this is the solution-transfer processes, which we have argued in the present paper to be effective at the plate interface shallower than about 20~30 km. Such processes may be partially operative at deeper parts of the subducting plate boundaries to lower the strength, but the degree of their involvement cannot be evaluated at this time. Thus partially-operative solution-transfer processes in very impermeable crystalline rocks at depths have to be incorporated into the constitutive properties of plate boundaries. Moreover, the effects of chemical reactions and phase changes, that no doubt occur in subduction zones, on rock deformation still remain largely unexplored. The mechanical behavior of subducting plate boundaries cannot be modeled unless those processes are fully studied in the future.

After the completion of this paper, we encountered recent work by Tichelaar and Ruff [in review], addressing to the same type of problem. They have determined earthquake depths of about 90 thrust-type earthquakes with magnitudes greater than 5.5 and have examined thermal structures of individual subduction zones, using thermal modeling by Molnar and England [1990]. Based on these results, they arrived at two critical temperatures (400° C and 550° C) corresponding to the lower bounds of seismic slip at subducting plate boundaries.

This result may warrant some comments, although we feel that it is still a tentative conclusion at least for the following three reasons. (1) Ninety earthquakes are not enough to determine the depth range for seismic slip for most major subduction zones in the world, since it is unlikely that the entire seismogenic zone ruptures during the medium to large earthquakes analyzed by Tichelaar and Ruff. (2) One must be careful in determining the intrinsically seismogenic zone using focal regions, because earthquake ruptures during large and great earthquakes may propagate into aseismic zones [e.g., Strehlau, 1986]. (3) They claim that the lower bound for the seismic slip is about 40 km in Northeast Honshu, Japan, but this is not consistent with the previous result of 60km from a detailed study on the region by Kawakatsu and Seno [1983].

Many complicating factors, such as the effects of lithology of the plate boundary, the velocity of the subducting plate, and irregularities of the plate interface, are disregarded in our rheological model, since their effects cannot be evaluated at this time. It is not surprising therefore that the temperature at the base of seismogenic interface varies from one subduction zone to the other. However, we assumed a single critical temperature for the cutoff of seismicity (of course as a crude assumption) for the following reasons.

If the cutoff of seismicity is indeed associated with the shift in the frictional property from velocity weakening to velocity strengthening (Figure 3), the cutoff depth is bounded by the depths corresponding to the onsets of partially-operative plastic deformation (semibrittle or semiplastic deformation) and fully plastic deformation. Although little experimental data on such a shift in the frictional property of rocks at high temperatures are available at present, one can get some insight into the problem from experimental data on the fully plastic deformation of rocks. The temperature for the onset of easy plastic deformation depends mainly on rock type and strain rate.

Flow laws of rocks [e.g., Ranalli and Murphy, 1987] suggest that this temperature could vary by several tens of degrees in centigrade for a change in the strain rate by one order of magnitude. Thus, the variation in the velocity of subduction (normally less than one order of magnitude) may cause that much of variation in the critical temperature for the cutoff of seismicity, if the thicknesses of the zones of concentrated shearing deformation at subducting plate boundaries are about the same.

There are more uncertainty in the effects of rock types on the critical temperature. However, if the subducting plates and overriding plates are separated by thin (or thick) layers of subducted sediments in most subduction zones, the behavior of the plate boundaries will be controlled by the mechanical properties of the sediments under shearing deformation and the effect of rocks types on the critical temperature will not be so large. Without subducted sediments, the behavior of plate boundaries will be controlled by the frictional properties of contacting rocks at the boundaries. The overall mechanical properties of rocks are controlled predominantly by weaker members, so that the onset of plastic deformation at subducting plate boundaries will be controlled by crustal rocks, rather than by the overriding wedge mantle. The variation in temperature for the onset of easy plastic deformation for crustal rocks is on the order of 100 to 200° C [e.g., Ranalli and Murphy, 1987].

However, such a variation in the critical temperature for the cutoff of seismicity does not invalidate the basic framework of our rheological model. We have only tried to show in this paper that the thermal state of a subducting plate boundary and the presence of abundant fluids at the shallow portion of the boundary can account for the diversity of subduction-zone seismicity and the lack of large and great earthquakes in some subduction zones. Further refinement of the rheological model is not possible until more detailed work is accumulated on the seismicity and the thermal state along various subducting plate boundaries and on the rheological properties of faults and plate boundaries at depths.

Acknowledgements. We thank Renata Dmowska for editing this special volume, two anonymous reviewers for critical comments, and Ali Lochhead for improving our English.

References

Andrews, D. J. and N. Sleep, Numerical modelling of tectonic flow behind island arcs, *Geophys. J. R. Astron. Soc.*, *38*, 237-251, 1974.

Batchelor, G. K., *An Introduction to Fluid Dynamics*, Cambridge Univ. Press, 224pp, 1967.

Bodri, L. and B. Bodri, Numerical investigation of tectonic flow in island arc areas, *Tectonophysics*, *50*, 163-175, 1978.

Bonilla, M. G., R. K. Mark, and J. J. Lienkaemper, Statistical relations among earthquake magnitude, surface rupture length, and surface fault displacement, *Bull. Seismol. Soc. Am.*, *74*, 2379-2411, 1984.

Brace, W. F. and J. D. Byerlee, Stick-slip as a mechanism for earthquakes, *Science*, *153*, 990-992, 1966.

Bridgman, P. W., Some implications for geophysics of high-pressure phenomena, *Bull. Geol. Soc. Am.*, *62*, 533-536, 1951.

Byerlee, J. D., Friction of rocks, *Pure Appl. Geophys.*, *116*, 615-626, 1978.

Chase, C. G., Extension behind island arcs and motions relative to hot spots, *J. Geophys. Res.*, *83*, 5385-5384, 1978.

Dewey, J. L., Episodicity, sequence and style at convergent plate boundaries, in *The Continental Crust and its Mineral Resources*, D. Strangway, ed., *Geol. Assoc. Canada*, *Spec. Pap.*, 20, 553-574, 1980.

Dieterich, J. H., Modeling of rock friction, 1, Experimental results and constitutive equations, *J. Geophys. Res.*, *84*, 2161-2168, 1979.

Forsyth, D. and Uyeda, S., On the relative importance of the driving forces of plate motion. *Geophys. J. R. Astr. Soc.*, *43*, 163-200, 1975.

Fukao, Y., Tsunami earthquakes and subduction processes near deep-sea trenches, *J. Geophys. Res.*, *84*; 2303-2314, 1979.

Fyfe, W. S., N. J. Price, and A. B. Thompson, *Fluids in the Earth's Crust*, *Elsevier, Amsterdam*, 383pp, 1978.

Griggs, D. T. and J. W. Handin, Observations

on fracture and a hypothesis of earthquakes, *Geol. Soc. Am. Memoir 79*, 347-364,1960.

Hasebe, K., N. Fujii, and S. Uyeda, Thermal processes under island arcs, *Tectonophysics 10*, 335-355, 1971.

Henry, S. G. and H. N. Pollack, Terrestrial heat flow above the Andean subduction zone in Bolivian and Peru, *J. Geophys. Res.*, *93*, 15,153-15,162, 1988.

Hiraga H. and T. Shimamoto, The textures of sheared halite and their implications for the seismogenic slip of deep faults, *Tectonophysics*, 144, 69-86, 1987.

Hitata, N., T. Kanazawa, K. Suehiro, and H. Shimamura, A seismicity gap beneath the inner wall of the Japan Trench as derived by ocean bottom seismograph measurement, *Tectonophysics*, *112*, 193-209, 1985.

Hobbs, B. E., A. Ord, and C. Teyssier, Earthquakes in the ductile regime? *Pure Appl. Geophys.*, *144*, 309-336, 1986.

Honda, S., Thermal structure beneath Tohoku, Northeast Japan - a case study for understanding the detailed thermal structure of the subduction zone, *Tectonophysics*, *112*, 69-102, 1985.

Honda, S. and S. Uyeda, Thermal process in subduction zones, A review and preliminary approach on the origin of arc volcanism, in *Arc Volcanism, Physics and Tectonics*, D. Shimozuru and I. Yokoyama, eds., *TERRAPUB*, *Tokyo*, 117-140, 1983.

Honda, S., H. Kawakatsu, and T. Seno, Centroid depth of the October 1981 off Chile outer-rise earthquake (M_s = 7.2) determined by a comparison of several waveform inversion methods, *Bull. Seism. Soc. Am.*, *80*, 69-87, 1990.

Hsui, A. T., X-M. Tang, and M. N. Toksöz, On the dip of subducting plates, *Tectonophysics 179*, 163-175, 1990.

Isacks, B., J. Oliver, and L. Sykes, Seismology and new global tectonics, *J. Geophys. Res.*, *73*, 5855-5899, 1968.

Kanamori, H., The energy release in great earthquakes, *J. Geophys. Res.*, *82*, 2981-2987, 1977a.

Kanamori, H., Seismic and aseismic slip along subduction zones and their tectonic implications, in *Island Arcs, Deep Sea Trenches and Back-Arc Basins*, M. Talwani and W. C. Pitman III, eds., *Maurice Ewing Ser.*, *Am. Geophys. Un.*, *1*, 162-174, 1977b.

Kanamori, H., Quantification of earthquakes, *Nature*, *271*, 411-414, 1978.

Karig, D., Evolution of arc systems in the Western Pacific, *Ann. Rev. Earth Planet. Sci.*, *2*, 51-76, 1974.

Kawakatsu, H. and T. Seno, Triple seismic zone and the regional variation of seismicity along the northern Honshu arc, *J. Geophys. Res.*, *88*, 4215-4230, 1983.

Kelleher, J. and W. Mccann, Buoyant zones, great earthquakes, and unstable boundaries of subduction, *J. Geophys. Res.*, *81*, 4885-4896, 1976.

Lay, T. and H. Kanamori, An asperity model of great earthquake sequences, in *Earthquake Prediction - An International Review*, D. W. Simpson and Richards, P. G., eds., *Maurice Ewing Ser.*, *Am. Geophys. Un.*, *4*, 579-592, 1981.

McKenzie, D. P., Speculations on the consequences and causes of plate motions, *Geophys. J. R. Astron. Soc.*, *18*, 1-32, 1969.

Minster, J. B. and T. H. Jordan, Present-day plate motion, *J. Geophys. Res.*, *83*, 5331-5354, 1978.

Mogi, K., Study of elastic shocks caused by the fracture of heterogeneous material and its relation to earthquake phenomena, *Bull. Earthquake Res. Inst., Univ. Tokyo*, *40*, 125-173, 1962.

Molnar, P. and T. Atwater, Interarc spreading and Cordilleran tectonics as alternates related to the age of subducted oceanic lithosphere, *Earth Planet. Sci. Lett.*, *41*, 330-340, 1978.

Molnar, P. and P. England, Temperature, heat flux and frictional stress near major thrust faults, *J. Geophys. Res.*, *95*, 4833-4856, 1990.

Nakamura, K., K. Shimazaki, and N. Yonekura, Subduction, bending and eduction, Present and Quaternary tectonics of the northern border of the Philippine Sea plate, *Bull. Soc. géol. France*, *26*, 221-243, 1984.

Newcomb, K. R. and W. R. McCann, Seismic history and seismotectonics of the Sunda arc, *J. Geophys. Res.*, *92*, 421-439, 1987.

O'Connell, R. J., C. W. Garble, and B. H. Hager, Toroidal-poloidal partitioning of lithospheric plate motions, in *Glacial Isostasy, Sea-Level and Mantle Rheology*, R. Sabadini et al., eds., 535-551, *Kluwer Academic Publishers*, 1991.

Peterson, E. T. and T. Seno, Factors affecting seismic moment release rates in subduction zones, *J. Geophys. Res.*, *89*, 10,233-10,247, 1984.

Ranalli, G. and Murphy, D. C., Rheological stratification of the lithosphere. *Tectonophysics*, *132*, 281-295, 1987.

Ricard, Y., C. Doglioni, and R. Sabadini, Differential rotation between lithosphere and mantle: a consequence of lateral mantle viscosity variations, *J. Geophys. Res.*, *96*, 8407-8415, 1991.

Ruff, L., Do trench sediments affect subduction zone earthquakes? *Pure Appl. Geophys.*, *129*, 263-282, 1989.

Ruff, L. and H. Kanamori, Seismicity and the subduction process, *Phys. Earth Planet. Inter.*, *23*, 240-252, 1980.

Rutter, E. H., Pressure solution in nature, theory and experiment, *J. Geol. Soc. London*, *140*, 725-740, 1983.

Rutter, E. H. and D. H. Mainprice, The effect of water on stress relaxation of faulted and unfaulted sandstone, *Pure Appl. Geophys.*, *116*, 634-654, 1979.

Sacks, I. S., Interrelationships between volcanism, seismicity, and anelasticity in the western South America, *Tectonophysics*, *37*, 131-138, 1977.

Sclater, J. G., U. G. Ritter, and F. S. Dixon, Heat flow in the southwestern Pacific, *J. Geophys. Res.*, *77*, 5697-5704, 1972.

Seno, T. and T. Eguchi, Seismotectonics of the western Pacific region, in *Geodynamics of the Western Pacific-Indonesian Region*, T. W. C. Hilde and S. Uyeda, eds., *Geodynamics Series*, *Am. Geophys. Un.*, *11*, 5-40, 1983.

Seno, T. and G. C. Kroeger, A reexamination of earthquakes previously thought to have occurred within the slab between the trench axis and double seismic zone, northern Honshu, *J. Phys. Earth*, *31*, 195-216, 1983.

Shimamoto, T., The origin of large or great thrust-type earthquakes along subducting plate boundaries, *Tectonophysics*, *119*, 37-65, 1985.

Shimamoto, T., Transition between frictional slip and ductile flow for halite shear zones at room temperature, *Science*, *231*, 711-714, 1986.

Shimamoto, T., Mechanical behaviours of simulated halite shear zones, implications for the seismicity along subducting plate-boundaries, in *Rheology of Solids and of the Earth*, S. Karato, and M. Toriumi, eds., *Oxford University Press, Oxford*, 351-373, 1989.

Shimamoto, T., The rheological basis of comparative subductology (*in Japanese*), *The Earth Monthly*, *"Uyeda Volume"*, *Suppl. 3*, *Kaiyo Shuppan, Tokyo*, 112-117, 1991.

Shimazaki, K., Pre-seismic crustal deformation caused by an underthrusting oceanic plate in eastern Hokkaido, Japan, *Phys. Earth Planet. Inter.*, *8*, 148-157, 1974.

Spiers, C. J., P. M. T. M. Schutjens, R. H. Brzesowsky, C. J. Peach, J. L. Liezenberg, and H. J. Zwart, Experimental determination of constitutive parameters governing creep of rocksalt by pressure solution, in *Deformation Mechanisms, Rheology and Tectonics*, R. J. Knipe and E. H. Rutter, eds., *Spec. Publs Geol. Soc. Lond.*, *54*, 215-227, 1990.

Stevenson, D. J. and J. S. Turner, Angle of subduction, *Nature*, *270*, 334-336, 1977.

Strehlau, J., A discussion of the depth extent of rupture in large continental earthquakes, *Geophys. Monogr. 37*, *Maurice Ewing Series*, *Am. Geophys. Un.*, *6*, 131-145, 1986.

Sykes, L. R. and R. Quittmeyer, Repeat times of great earthquakes along simple plate boundaries, *Maurice Ewing Series*, *Am. Geophys. Un.*, *4*, 217-247, 1981.

Thatcher, W. and J. B. Rundle, A model for the earthquake cycle in underthrust zones, *J. Geophys. Res.*, *84*, 5540-5556, 1979.

Tichelaar, B. W. and L. J. Ruff, Depth of seismic coupling along subduction zones, *J. Geophys. Res.*, *in review*.

Toksöz, N., J. W. Minear, and B. Julian, Temperature field and geophysical effects of a downgoing slab, *J. Geophys. Res.*, *76*, 1113-1138, 1971.

Toksöz, N. and A. Hsui, Numerical studies on back arc convection and the formation of marginal basins, *Tectonophysics*, *50*, 177-196, 1978.

Utsu, T., Seismological evidence for anomalous structure of island arcs with special reference to the Japanese region, *Rev. Geophys. Space Phys.*, *9*, 839-890, 1971.

Uyeda, S., Subduction zones, facts, ideas and speculations, *Oceanus*, *22*, 52-62, 1979.

Uyeda, S., Subduction zones, an introduction to comparative subductology, *Tectonophysics*, *81*, 133-159, 1982.

Uyeda, S., Subduction zones, their diversity, mechanism and human impacts, *Geojournal*, *8*, 381-406, 1984.

Uyeda, S. and K. Horai, Terrestrial heat flow in Japan, *J. Geophys. Res.*, *69*, 2121-2141, 1964.

Uyeda, S. and H. Kanamori, Back-arc opening and the mode of subduction, *J. Geophys. Res.*, *84*, 1049-1061, 1979.

Uyeda, S. and C. Nishiwaki, Stress field, metallogenesis and mode of subduction, in *The Continental Crust and Its Mineral Deposits*, D. W. Strangway ed., *Geological Association of Canada Special Paper 20*, 323-339, 1980.

Uyeda, S. and V. Vacquier, Geothermal and geomagnetic data in and around the Island Arc of Japan, in *The Crust and Upper Mantle of the Pacific Area*, L. Knopoff, C. L. Drake, and P. J. Hart, eds., *Geophys. Monogr.*, *Am. Geophys. Un.*, *12*, 349-366, 1968.

Uyeda, S. and T. Watanabe, Terrestrial heat flow in western South America, *Tectonophysics*, *83*, 63-70, 1982.

Wang, C. Y., Sediment subduction and frictional sliding in a subduction zone. *Geology*, *8*, 530-533, 1980.

Watanabe, T., M. Langseth, and R. N. Anderson, Heat flow in back arc basins in the western Pacific, in *Island Arcs, Deep Sea Trenches and Back-Arc Basins*, M. Talwani and W. C. Pitman III, eds., *Maurice Ewing Ser.*, *Am. Geophys. Un.*, *1*, 137-161, 1977.

Yamano, M. and S. Uyeda, Heat flow in the western Pacific, in *Handbook of Seafloor Heat Flow*, J. A. Wright and K. E. Louden, eds., *CRC Press*, 277-303, 1989.

Yoshii, T., A detailed cross-section of the deep seismic zone beneath northeastern Honshu, Japan, *Tectonophysics*, *55*, 349-360, 1979.

Zhang, J. and S. Schwartz, Depth distribution of moment release in underthrusting earthquakes at subduction zones, *J. Geophys. Res.*, *97*, 537-544, 1992.

Zhang, J. and T. Lay, The April 5, 1990 Mariana Islands earthquake and subduction zone stresses, *Phys. Earth Planet. Inter.*, *72*, 99-121, 1992.

T. Shimamoto and T. Seno, Earthquake Research Institute, University of Tokyo, 1-1-1 Yayoi, Bunkyo-ku, Tokyo 113, Japan.

S. Uyeda, Department of Geophysics, Texas A&M University, College Station, Texas 77863-3114.

Seismic Structure and Heterogeneity in the Upper Mantle

B.L.N. KENNETT

Research School of Earth Sciences, Australian National University, Canberra, ACT, Australia

The earliest models of the seismic velocity structure of the upper mantle were smooth. But, since the introduction of strong gradients near 400 km depth by Jeffreys to explain the '20° discontinuity' in observed travel times, there has been a steady accumulation of detail in mantle structure.

For a particular region, a smoothed and averaged representation of the seismic structure in the upper mantle can be derived from long-period body wave and higher mode surface wave observations. The vertical resolving power of such techniques is limited by the relatively long wavelengths. In contrast short-period observations offer potential resolution, but are susceptible to the influence of lateral heterogeneity.

Fortunately the major features of the upper mantle can be discerned but important questions for structural processes such as the detailed nature of the transitions near 410 and 660 km are generally inaccessible.

There is a natural tendency to overweight those observations on which particularly clear features are seen (as compared with the statistical anonymity of less spectacular data) which can lead to unwarranted generalizations of specific results. To reconcile different views of mantle structure requires us to address the purpose for which the mantle structures are to be used. For example, fine detail in a velocity model which is insignificant for travel time studies can have a profound effect on amplitudes and short-period seismic waveforms.

The variability in the patterns of body wave observations, especially at short periods, provides strong evidence for 1-2 per cent heterogeneity on scales around 200 km in the upper mantle. Such features are superimposed on larger scale and larger amplitude lateral variations which can be mapped using surface wave studies.

Much of the pattern of lateral variability in the upper mantle is likely to be due to thermal processes both directly by the influence of temperature and indirectly by compositional effects induced by flow patterns.

INTRODUCTION

The upper part of the mantle in the outer 700 km of the Earth contains some of the most important lateral heterogeneities in structure, such as those associated with subduction zones and major continental rifts. However, even in those regions away form the direct influence of active tectonics, there is a complex hierarchy of structural features in the mantle. The emphasis in this paper will be on features occurring in the depth range from 200 - 700 km.

Relating Geophysical Structures and Processes: The Jeffreys Volume
Geophysical Monograph 76, IUGG Volume 16
Published in 1993 by the International Union of Geodesy and Geophysics and the American Geophysical Union.

Early models of the seismic structure in the outer parts of the earth did not include significant structure until the work of Mohorovičić (1909) on the Kulpatal earthquake of 1909 revealed the existence of the transition which was subsequently inferred to represent the boundary between the earth's crust and the uppermost mantle. In a sequence of papers Jeffreys (1926, 1931, 1933) attempted to obtain a consistent crustal structure in the European region. With the more detailed information which is currently available we have to recognize that the degree of horizontal variability is such that any specific stratified structure will provide only a limited representation of the wave propagation processes in the crust in a particular area.

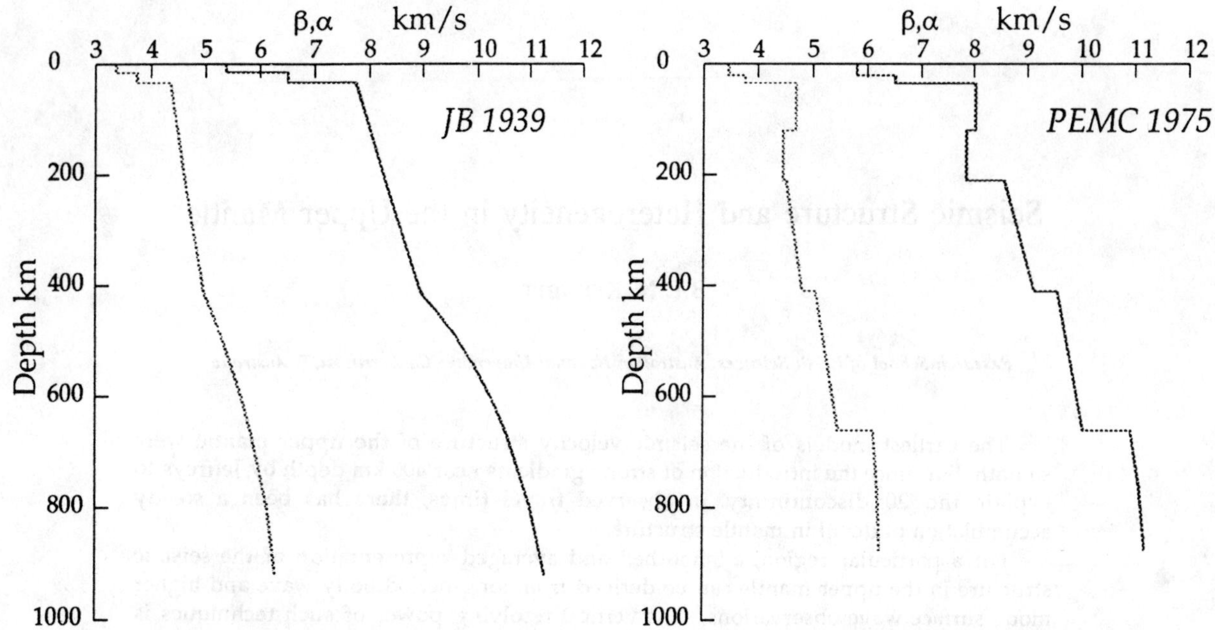

Fig 1. A comparison of radial models of upper mantle structure: a) Jeffreys (1939), b)Dziewonski, Hales and Lapwood (1975), showing the changes in style associated with higher resolution information and the recognition of the major discontinuities in the mantle.

Beneath the crust the early models of seismic velocity were entirely smooth (see e.g.. Walker 1915) but, with the recognition of the change of slope in seismic travel times near 20° epicentral distance, it was necessary to introduce additional structure. The '20° discontinuity' in the velocity model of Jeffreys 1939) imposes a distinct change of velocity gradient and curvature at a depth of 405 km (see fig 1). This feature has withstood the test of time, and in more recent studies has been supplemented by other discontinuities in the seismic wavespeed distribution that have a more subtle influence on the seismic travel times. Evidence for at least two such rapid changes in seismic wavespeed near 410 and 660 km came from direct measurements of the horizontal slowness of seismic waves at seismic arrays (Johnson 1967). Such results were confirmed by later studies and by 1975 a model of the upper mantle incorporated in a global model by Dziewonski et al (1975) shows a very different character from that of Jeffreys (1939).

It was about this time that the significance of horizontal variations in the structure of the earth began to be appreciated. This has led to a diversity of models in which the detailed character of the seismic wave field has been explained by complex one-dimensional models (e.g. Hales et al 1980) or alternatively by rather simpler radial structure with superimposed 2 or 3 dimensional structure on fairly small scales (e.g.. Kennett and Bowman 1990). The uppermost mantle, within the subcrustal lithosphere has proved to have a very complex structure (see e.g. Hirn et al, 1973; Yegorkin and Chernyshov 1983)

Recently it has become possible to construct 2 and 3-dimensional models of the seismic wavespeed distribution directly from observations of long-period S and surface waves (e.g.. Nolet 1990) but such methods can only determine the large spatial wave length components of the structure. In fig 2a we show the two-dimensional model WEPL3 determined by Nolet (1990) for the structure beneath the original NARS array in Western Europe. Figure 2b illustrates the model SW3p discussed by Kennett and Nolet (1990) which is based on WEPL3 but includes the class of smaller scale heterogeneity suggested by short-period body wave observations. For surface waves of 20 second period there is no effective difference between the response of the two structures illustrated in figure 2, but the differences increase as the period is reduced (Kennett and Nolet 1990).

SOURCES OF INFORMATION

There are a number of sources of information which can contribute to our knowledge about the seismic structure in the upper mantle, particularly the P and S wavespeeds, density can generally only be inferred indirectly e.g. by the impedance contrast at major discontinuities

Because the upper mantle down to 700 km represents only about 11 per cent of the radial extent of the earth it has only a modest influence on the shapes of the radial eigenfunctions of the lower frequency free oscillations of the Earth (more than 120 s period). Studies of the properties of such free oscillations do not place strong constraints on upper mantle structure, but recent high-

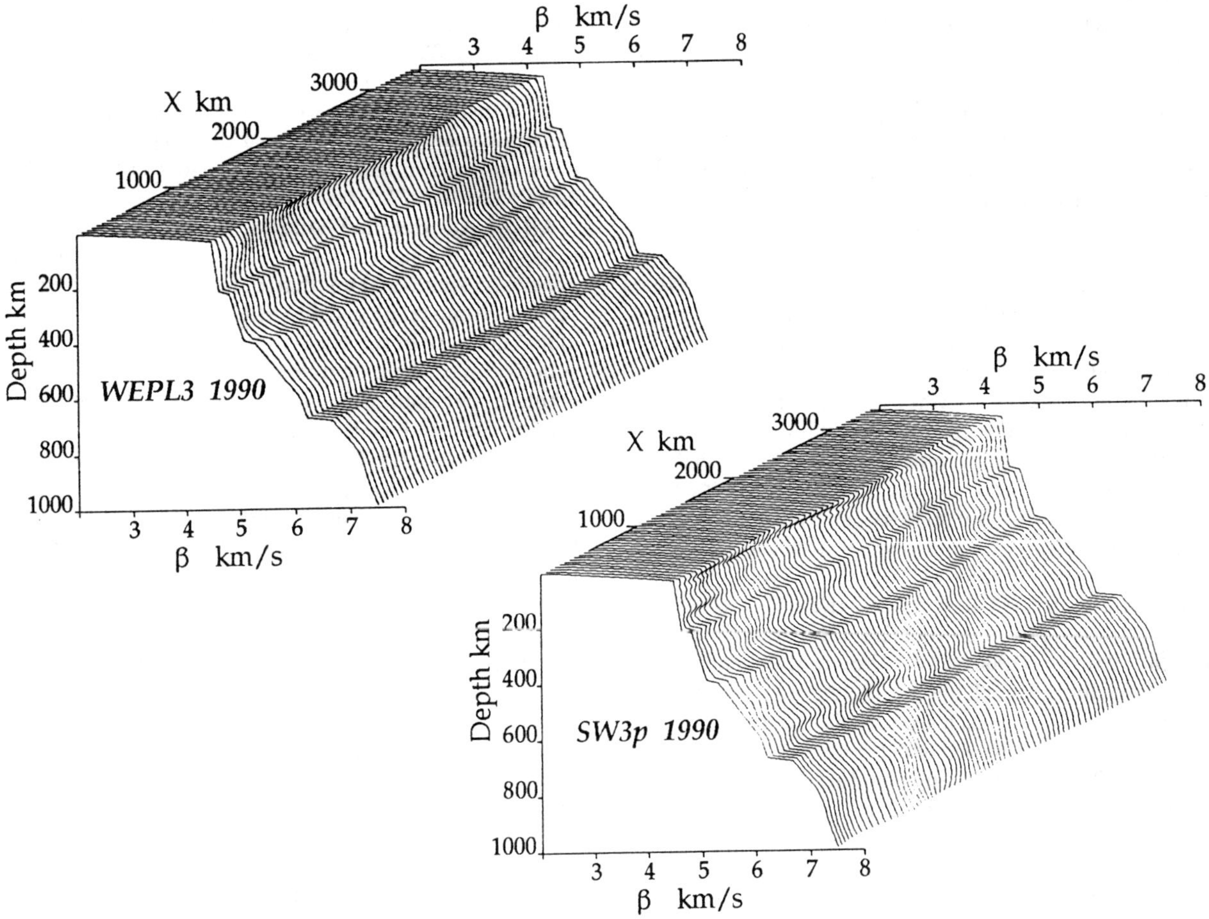

Fig. 2. Two dimensional models of upper mantle structure: a) model WEPL3 derived by Nolet (1990) for Western Europe by waveform inversion of S observations at the NARS array; b) model SW3p discussed by Kennett and Nolet (1990) which incorporates smaller scale heterogeneity required to satisfy body wave observation superimposed on the large scale structure of WEPL3.

precision determinations of free oscillation frequencies are best matched by models with upper mantle discontinuities (G. Masters - personal communication 1991). A number of studies suggest the presence of an anomaly of spherical harmonic degree 2 in S velocity and attenuation near the 660 km discontinuity, and recently Fukao et al (1992) have proposed that this velocity variation is associated with the fate of the subduction zones around the Pacific rim. Tomographic images of P wave velocity structure particularly in the northwestern Pacific region near Japan (van der Hilst et al, 1991; Fukao et al, 1992) suggest that material with high wavespeeds (which is likely to be cold) accumulates near 660 km before ultimately entering the lower mantle

At periods less than 120 s, observations of fundamental and higher mode surface waves are quite sensitive to the radial and horizontal structure in the upper part of the mantle. However, depth resolution is limited by the exponential decay of the amplitude of the surface wave modes below a depth which corresponds to the turning point for a geometric ray with the same slowness. The PREM model of Dziewonski and Anderson (1981) was constructed to fit a large suite of free oscillation periods, surface wave dispersion of Love and Rayleigh waves and also certain classes of travel time information. In order to reconcile the Love and Rayleigh wave dispersion information, transverse isotropy with a radial symmetry axis was introduced into the upper most part of the mantle (down to 210 km). Such a specialized class of anisotropy is likely to represent the averaged effect of more complex anisotropy within the upper mantle. Montagner and Anderson (1989) have investigated a range of models with transverse isotropy extending to different depths with the upper mantle and have concluded that only relatively weak anisotropy would be required below 210 km.

A number of techniques depend on the explicit recognition of body wave phases on long-period records. Much of such long-period studies have been based on the properties of refracted and reflected waves travelling at shallow angles to the horizontal in the upper mantle. Such studies have been carried out for P waves (see e.g. Burdick and Helmberger 1978) and for horizontally polarized SH waves (e.g. Helmberger and Engen 1974). Body wave studies have avoided analysis of vertically polarized SV waves in order to avoid the complications introduced by inter-conversion between P and SV waves. Where the source and receiver geometries are not so favorable, the volume of available data can be increased by combining direct mantle observations of P or S with the comparable surface multiples PP or SS (Grand and Helmberger 1984; LeFevre and Helmberger 1991) Because long period waves are used the wavelengths are of the order of 100 km for P waves and 60 km for S waves at a depth of about 400 km, and so it is not possible to determine the detailed nature of the velocity transitions in the mantle. But, the long wavelengths tend to reduce the influence of horizontal variations in mantle structure. Useful additional constraints on the depths and properties of upper mantle discontinuities are provided by observations of inter-conversions between P and S waves in transmission at epicentral distances greater than 60°. Faber and Müller (1980) have shown the benefits of detailed modelling of such observations.

Two new techniques which are sensitive to upper mantle structure make use of waves which travel much closer to the vertical. The first method introduced by Revenaugh and Jordan (1989) and subsequently elaborated by Revenaugh (1989) is based on the use of multiple ScS waves with SH polarization together with the secondary multiples associated with the presence of discontinuities in the mantle. Under the assumption of horizontal uniformity along the path from source to receiver, a matched filter procedure can be applied to estimate the impedance profile through the mantle. The major discontinuities at 410 and 660 km are clearly indicated with little variation in depth between profiles from different regions. Other weaker discontinuities are suggested particularly close to 520 km (Revenaugh and Jordan 1991). The alternative approach due to Shearer (1990) is to stack all available long-period seismograms from shallow sources recorded at the Global Digital Seismograph network in narrow bins according to epicentral distance and then display the stacks in various formats designed to emphasize particular phases e.g.. PP, SS. The presence of the mantle discontinuities gives rise to a sequence of arrivals which follow P and S and also to precursors to PP and SS. The effects of the major discontinuities can be clearly recognized and there is a clear indication of a weak feature at about 520 km (Shearer 1991).

Short period observations have the greatest potential resolving power, but are also the most susceptible to lateral heterogeneities. In this frequency band most studies have been of refracted and reflected waves using record sections of single events at a sparse array of stations (see e.g.. Hales et al 1980) or composite sections for many events recorded at smaller (but denser) arrays (see e.g.. Bowman and Kennett 1990). Pavlenkova and Yegorkin (1983) have been able to use controlled sources (Peaceful Nuclear Explosions) to obtain relatively high data density on very long profiles in Siberia. Some constraints on the properties of the major discontinuities have also been inferred from precursors to P′P′ (PKPPKP) (see Richards 1972, Davis et al 1989).

All of the detailed studies we have mentioned are directed at the structure of the mantle on scales which are generally smaller than that obtainable from large-scale seismic tomography using digital recordings from the global network of seismic stations. The S wave structure to spherical harmonic degree 6 has been determined for the whole upper mantle down to the 600 km discontinuity by Woodhouse and Dziewonski (1984) to give a horizontal resolution around 2000 km. Recent high resolution tomography by Zhang and Tanimoto (1991) using cellular models has achieved horizontal resolution of the order of 660 km for S wave velocities in the outer 300 km of the Earth.

GEOGRAPHICAL COVERAGE

When we discuss the structure of the upper mantle we have to take into account the relatively limited geographic coverage of the globe for which detailed studies are available. A number of velocity models are available for Northern America and Western Europe covering a broad range of frequencies. Elsewhere in Eurasia most studies penetrating to considerable depths are based on long-period seismic observations. There have been a number of studies in Northern Australia but these are all for short-period waves. Most of the oceans and much of South America and Africa are hardly sampled except in the global tomography work mentioned above. We therefore have to express some caution in interpreting the results of existing studies to represent the global pattern of behaviour within the upper mantle. The results of Shearer (1990,1991) from the global seismic network help to constrain the properties of the discontinuities in the upper mantle, but are not very sensitive to velocity gradients.

As an illustration of the information which can be generated by the combination of a range of different techniques we will consider two different regions: Western Europe and Northern Australia. In Europe the principal technique has been the analysis of multi-mode surface wave trains supplemented by some body wave observations. For northern Australia the proximity of the active earthquake belt in Indonesia and New Guinea has led to direct observations of the refracted wave field using mostly short-period instruments.

VELOCITY STRUCTURE FROM SURFACE WAVES - WESTERN EUROPE

As an example of the way in which knowledge of upper mantle structure has evolved using surface wave observations we consider the region of Western Europe. A

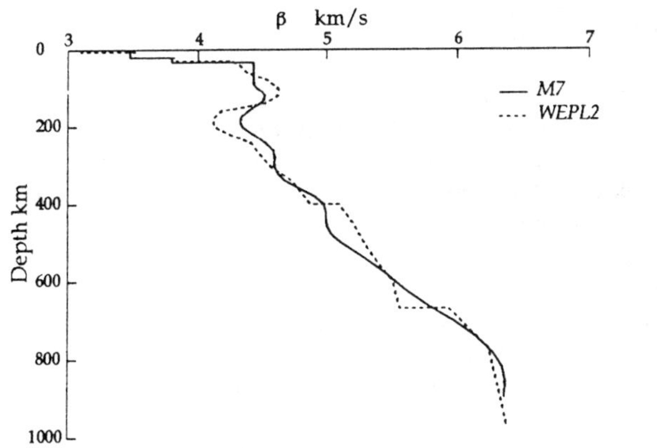

Fig. 3. Radial velocity models for the upper mantle under Western Europe. Model M7 (Nolet 1976) was derived from just the dispersion of higher mode surface wave. The later model WEPL2 (Dost 1986) incorporates body wave information to constrain the upper mantle discontinuities.

relatively dense network of permanent seismic stations enabled Nolet (1975, 1977) to determine the phase and group velocity of the fundamental and six higher Rayleigh modes as a function of frequency in the band 0.01-0.05 Hz using events in Japan. These data were then inverted to generate the shear wave and density model M7 (Nolet 1977) illustrated in fig 3. The success of the surface wave inversion techniques using a network of permanent stations, led to the establishment of a specially designed linear array of long-period stations (NARS) stretching from Malaga in southern Spain to Gothenburg in Sweden which has yielded a rich data set of long period observations. Using an analysis of the surface wave dispersion data from the NARS array with the addition of travel time constraints, Dost (1986) determined the velocity model WEPL2 for the region under the NARS array (fig 3). When compared with the earlier M7 model using just surface wave observations we note that it has been possible to resolve the location of upper mantle discontinuities. The vertical resolution attainable with even multi-mode surface wave dispersion cannot distinguish between a gradient zone and sharp discontinuity.

Nolet (1990) has shown how the observed S waveforms of two events along the great circle through the NARS array can be used to determine the 2-dimensional S velocity structure along the profile beneath the array (fig 4). The procedure which is termed "partitioned waveform inversion" comprises two stages: firstly, given a knowledge of the source parameters for the events, the best 1-dimensional model is determined by wave form inversion for the path between the source and each station. Secondly, the path averages are treated as providing linear constraints on the two-dimensional structure along

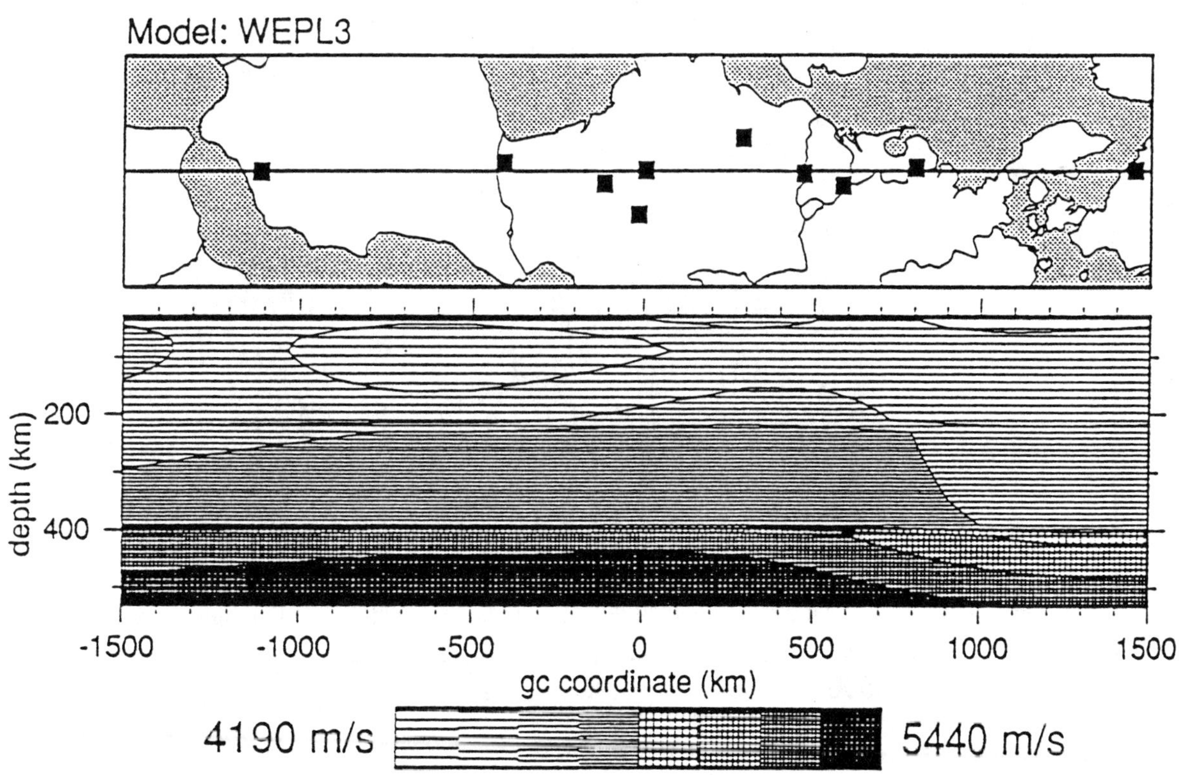

Fig 4. The two-dimensional velocity structure (WEPL3) under the NARS array determined using the technique of "partitioned waveform inversion" (Nolet 1990).

the line of the profile. A linear-inversion therefore can be used to extract the WEPL3 structure shown in fig 4 in which two-dimensional structure can be resolved to a depth of approximately 550 km. The major discontinuities in the model are maintained at fixed depth, although an alternative description of the structure on this profile could well include slight topography on the discontinuities without violating the constraints provided by the S-waveforms. This two stage technique for determining earth structure from surface observations has been extended to the three-dimensional structure beneath the Western European region by Zielhuis and Nolet (1991).

VELOCITY STRUCTURE FROM BODY WAVE OBSERVATIONS - NORTHERN AUSTRALIA

The major earthquake belt running through Indonesia and Papua New Guinea lies at the appropriate distance range to investigate upper mantle structure using recording stations in northern Australia. Since cultural noise is low in this sparsely populated region, it has proved possible to build up composite record sections for propagation over ranges from 1200-2800 km for many events with magnitudes between 4 and 4.5 with a deployment of portable instruments over a period of 3 months (see e.g. Bowman and Kennett 1990).

In order to cope with variations in source waveform and the influence of lateral heterogeneity, it is useful to stack the seismograms in a narrow distance range to enhance coherent signal. For this purpose a very stable measure of the seismic wavefield is provided by a stack of the complex envelopes of the waveforms over 10 km distance bins. Such composite record sections can show significant variation in character when constructed for sources in different regions recorded at the same seismometer array. For example, we compare in figure 5 composite record sections for events in the Flores Arc and in New Guinea

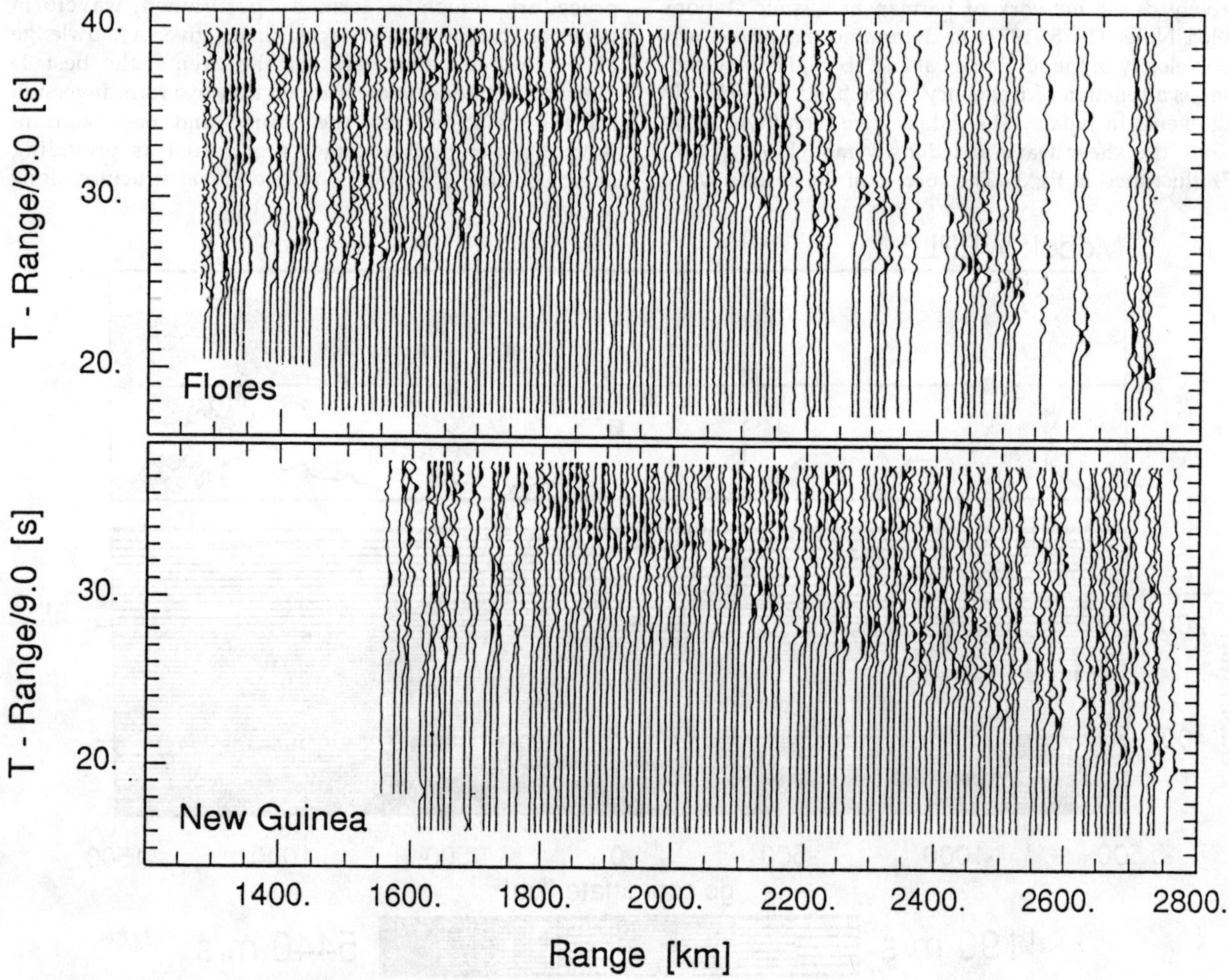

Fig. 5. Composite record sections generated by stacking the complex envelopes of seismograms from many events recorded at a large aperture array of portable stations in northern Australia: a) events in the Flores Arc, Indonesia to the northwest, b) events in New Guinea to the northeast.

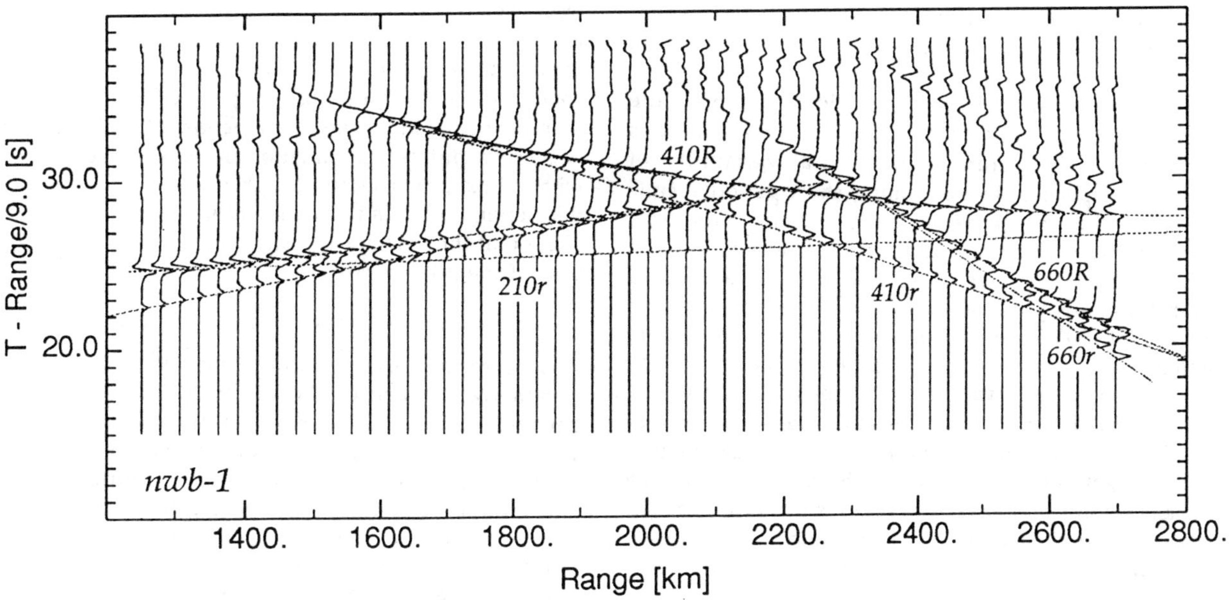

Fig. 6. Theoretical seismogram section for model NWB-1 (Bowman and Kennett 1990) for paths from the Flores arc to stations in northern Australia for a source at 33 km including free surface reflections (displayed with time corrections to surface focus). The calculations were performed using the WKBJ technique (Chapman 1978). For comparison with figure 5 the complex envelope of each trace is displayed and the travel time branches are denoted by the discontinuity with which they are associated (r - refraction below, R - retrograde reflection).

recorded across a portable arrays in the Northern Territory of Australia (Bowman and Kennett 1990, Dey 1989). The events used have magnitude between 4 and 5 and have shallow epicenters (less than 40 km deep); the times and ranges have been corrected to surface focus before the stacked sections were prepared. In each section we can clearly discern later phases corresponding to travel time branches from the major transitions in the upper mantle. However, the distribution of energy along the branches is significantly different (a key to the different branches is provided in figure 6). For sources in the Flores Arc, the refracted branch from below the 410 km transition (410r) is particularly prominent; such paths traverse the zone to the northwest of the recording array.. Whereas for sources in New Guinea the retrograde reflected branch (410R) is more significant; such paths cover the sector to the NNE of the array of portable recorders. The timing of the P phases is also different between the record sections for the two different source regions. This indicates a change in velocity structure over a horizontal distance of the order of 1000 km.

A very powerful tool for the interpretation of such record sections is the comparison of the envelopes of theoretical seismograms with the stacked data. The theoretical seismogram section for the model proposed by Bowman and Kennett (1990) calculated using the WKBJ technique is displayed in figure 6, together with labels for the different travel time branches. These theoretical seismograms give a good representation of the amplitude behaviour seen in the Flores section (fig 5a). The model itself is displayed in figure 7 together with an indication of

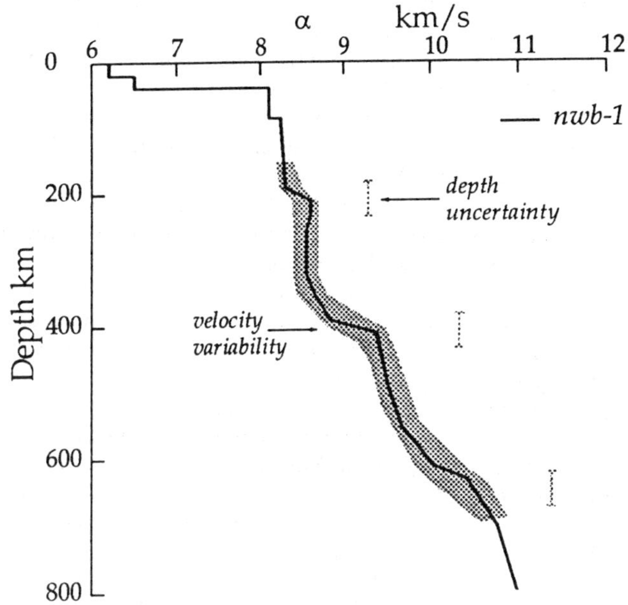

Fig. 7. The radial velocity model NWB-1 for the upper mantle under northwestern Australia derived for sources in the Flores Arc (Bowman and Kennett 1990). The shading shows the possible variation in the region sampled, inferred from waveform variability. Vertical bars show the uncertainty of the depth determination for the major discontinuities.

the degree of variation from the basic velocity model which could be tolerated without a major degradation of the amplitude fit.

The use of an array of stations allows not only the determination of radial structure but also provides constraints on the nature of the lateral variability of seismic wavespeeds. By combining short-period results from a number of portable array experiments with different array dimensions and station spacing, Kennett and Bowman (1990) have suggested the presence of P wave heterogeneity of the order of 1 per cent on scale lengths of 200-300 km superimposed on the main radial variation in velocity. The presence of such heterogeneity leads to amplitude and waveform variability for individual events recorded across arrays of aperture of 100 km or more. By combining observations from many events it is possible to get stable estimates of the major part of the wavefield (as in figure 5), the deviations from this pattern indicate the presence of medium scale variations in seismic wavespeed in addition to the larger scale variability illustrated in figure 5. This smaller scale heterogeneity has a significant influence on the amplitude distribution observed across the portable arrays, and may well be associated with localized anisotropy in the mantle. Similar length scales have also been indicated by detailed tomographic studies. The SW3p model in figure 2b was constructed by superimposing such a class of heterogeneity on the surface wave model WEPL3 (Kennett and Nolet 1990). The influence of medium scale heterogeneity with scale lengths of the order of 100 km is particularly important in zones where the radial gradients are small, e.g. above 200 km, since horizontal velocity gradients can then dominate the behaviour of the seismic wavefield (see e.g. Kennett 1991)

The long-period response of the two models SW3p and WEPL3 is very close for frequencies less than 0.05 Hz, even though their short-period response is quite different. The smaller scale features lead to significant focussing and defocussing of seismic energy which contributes to the complexity of the short-period wavefield.

For a region with a good distribution of seismic sources such as the belt through Indonesia and New Guinea it is possible to build up record sections for mantle propagation using only a single station, if the recording period is long enough. A preliminary study of this type is now underway using the broad-band seismometers at the Warramunga array in the Northern Territory of Australia (Goody 1991). The nearest seismicity is at about 1300 km and a reasonable coverage can be achieved to at least 3500 km. With three-component recording it has proved possible to separate the P, SV and SH components of the wavefield. The broad-band observations of P are generally consistent with the NWB-1 model for events in the Flores region but the retrograde reflections are marked by an increase in frequency content which suggests quite sharp transitions at the upper mantle discontinuities. The most striking feature arises for S wave propagation. High frequency S waves are clearly recorded out to 2000 km as the onset of the S waveform for both SV and SH waves; at greater distances the onset of S has no high frequency content even though a high frequency SS arrivals can be seen near 2800 km. The dramatic loss of high frequencies suggests significant attenuation in the zone below 210 km with guided propagation of high frequency S waves above this depth. For this predominantly shield region it seems likely that the velocity transition near 210 km marks the base of the lithosphere.

SUMMARY INFORMATION ON THE UPPER MANTLE

In a recent review Nolet and Wortel (1989) have summarized much of the available information on the seismic wavespeed distribution in the upper mantle. Although there is a general convergence as to the nature of the wavespeed variation with depth, there are considerable differences in the estimates of the seismic parameter contrasts at the major transitions in the upper mantle. This variability maps into estimates of the wavespeed gradients in the mantle which are far more varied than the actual wavespeeds. The uncertainties in the gradient estimates are quite large but there are suggestions of geographic variability.

Above 250 km, pronounced regional heterogeneity precludes the assessment of a reasonable average gradient. In a number of areas there are zones of very strong positive velocity gradients for P waves in the uppermost mantle which cannot easily be reconciled with present isotropic composition models (see e.g. Yegorkin and Chernyshov 1983), which suggests the presence of organized anisotropy.

At greater depth there is more consistency in the gradient estimates from different regions and a subjective summary of current constraints on seismic velocity gradients below 250 km is presented in figure 8, for both P and S waves. The domains of likely velocity gradient are shaded and the gradients adopted for the model *iasp91* (Kennett and Engdahl 1991 - see fig 10) are also indicated. The upper mantle portion of *iasp91* is constructed to provide a fit to averaged travel times for continental paths at the 30 degrees epicentral distance. Between discontinuities *iasp91* is represented by linear dependence on radius and so maps to constant gradient values for each zone.

Figure 8 also indicates likely bounds on the wavespeed contrasts across the major upper mantle transitions. The existence of relatively sharp transitions near 410 km and 660 km is well established even though there are significant variations in the estimates of the contrasts across these zones. Such variability may well reflect differences in structure in the regions sampled by different groups. There is also a natural tendency for modelling to be influenced by particularly clear observations of later phases which could lead to some inflation in estimates of the contrasts across the transitions. Secondary phases associated with the

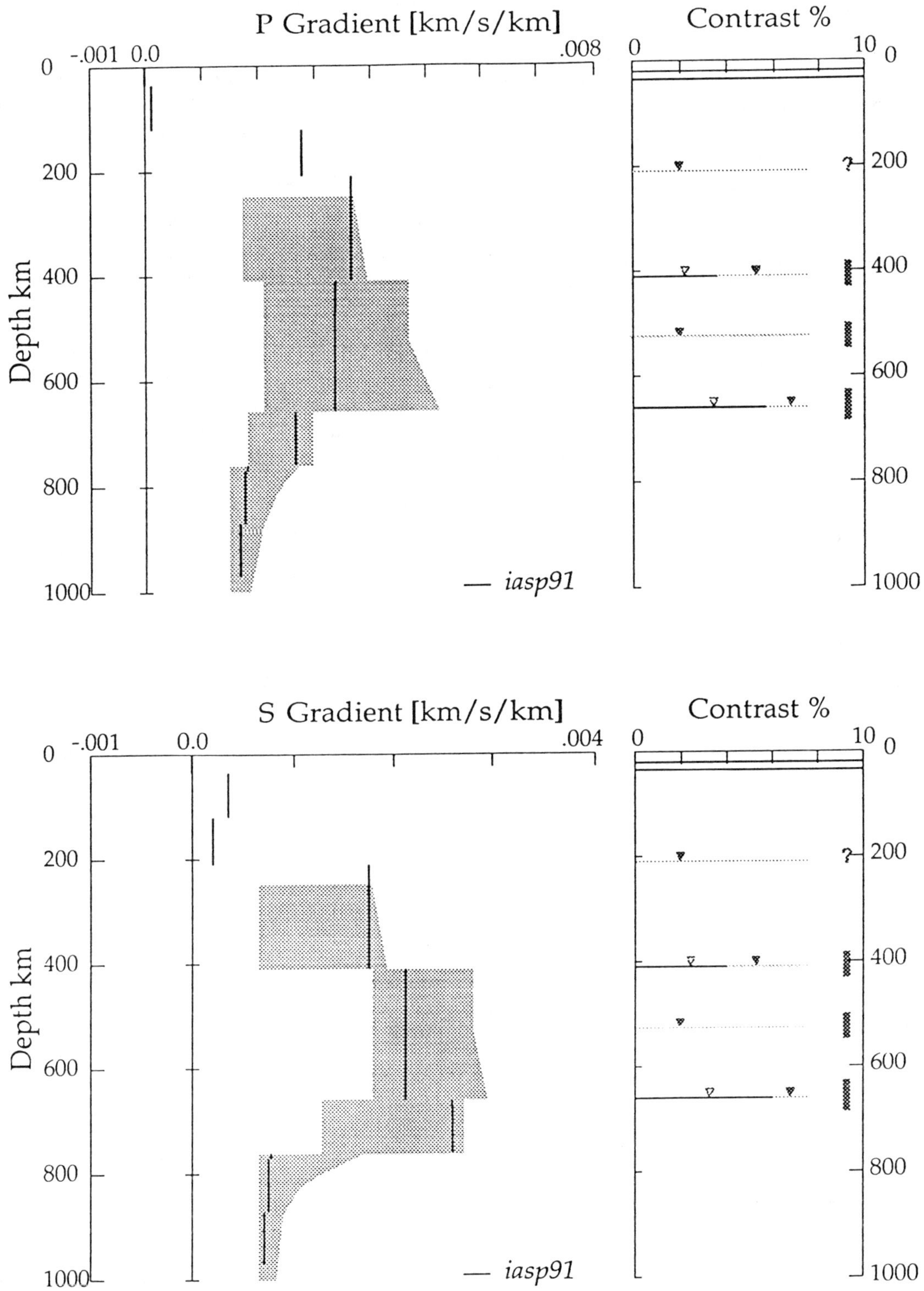

Fig. 8. Summary of constraints on upper-mantle velocity gradients below 250 km and the contrasts at upper mantle discontinuities. The shaded regions represent the most likely values for P and S gradients in the mantle. For reference, the gradients and discontinuity contrasts adopted for the model *iasp91* (Kennett and Engdahl 1991 - see also fig 10) are indicated. This model with linear gradients in radius was constructed to fit averaged travel times on continental paths to 30 degrees epicentral distance.

presence of the 410 and 660 km discontinuities are very clear in the global seismogram stacks presented by Shearer (1990,1991). These transitions are also the most prominent features in the SH impedance profiles presented by Revenaugh (1989). The likely range of seismic parameters for the transition zone between 410 and 660 km are summarized in Table 1.

The situation for the minor transitions in seismic velocity is less clear. A distinct change in the character of the wavespeed distribution at a depth close to 210 km has been observed in a number of locations e.g.. eastern North America (Lehmann 1962), northwestern Australia (Hales et al 1980, Bowman and Kennett 1990), beneath France (Steinmetz et al 1974) and in central Asia (Pavlenkova and Yegorkin 1983). However, there is no consistency in the nature of the proposed features near this depth. Anderson (1979) has proposed that a 210 km discontinuity is a global feature but the stacking results of Shearer (1990,1991) do not support this hypothesis since the expected secondary phases are absent. Some of the paths studied by Revenaugh (1989) show evidence for an impedance contrast near 210 km depth but the feature is geographically quite variable (see Revenaugh and Jordan 1991).

The other portion of the wavespeed distribution whose character is not yet resolved is the possible velocity transition near 520 km. From distinct, but small, secondary phases in the global stacks Shearer (1990) has inferred a contrast of about 2 per cent across this transition. The work of Revenaugh (1989) also favours the presence of an impedance contrast at this depth. However in a study of a very large number of events recorded in the range from 2000-3000 km on short-period instruments in northern Australia, Cummins et al (1991) have found little evidence to support the existence of a velocity transition near 520 km from refracted arrivals. This study exploited the distribution of events in depth along the earthquake zone through Indonesia and New Guinea to seek the optimum conditions for arrivals from a 520 km 'discontinuity' without finding any direct observations. One way of reconciling the different classes of information is to propose that the transition is spread over a depth interval of 50 km. Such a transition would still be sharp enough to generate arrivals for the wavelengths associated with the long-period observations of Shearer (1990) and Revenaugh (1989). However, as pointed out by Cummins et al (1992), an extended transition would have a very weak short-period signature.

TABLE 1. Seismic Parameters for the Upper Mantle Transition Zone

SEISMIC PARAMETER	P WAVES	S WAVES
Velocity contrast at 410 km transition (%)	2.5 - 5.8	2.7 - 5.7
Velocity contrast at at 660km transition(%)	3.6 - 7.3	3.0 - 7.5
Mean velocity gradient between 410 and 660 km	2.0-5.0 $\times 10^{-3}$	1.8-2.9 $\times 10^{-3}$

RELATION OF SEISMOLOGICAL AND LABORATORY INFORMATION

With improvements in high pressure techniques it has recently proved possible to synthesize candidate minerals under upper mantle conditions, recover the samples and then subject them to an examination of their physical properties. In such a way ultrasonic measurements of the wave speeds of the beta phase of $(Mg, Fe)_2\ SiO_4$ and the γ phase (spinel) have been determined under upper mantle pressures (Rigden et al 1991). These crystal structures are prime candidates for the mineral assemblage in the transition zone between the 410 and 660 km discontinuities.

The estimates of velocity gradients in this region derived from the pressure derivatives of the ultrasonic wavespeeds of the spinels are consistent with the seismic estimates for P waves summarized in fig 8 and Table 1. However, for S waves the velocity gradient estimates are significantly lower than any seismological model (Rigden et al 1991). Some measure of compositional gradation may be required between 410 and 660 km to reconcile the seismological and laboratory information. Further experiments are in progress on the temperature derivatives for the ultrasonic measurements at high frequency which could well alter the final interpretation.

PROCESSES AND STRUCTURE

We have discussed above both the radial velocity structure of the mantle and the two- and three-dimensional perturbations from such structure. We can illustrate these two different aspects of upper mantle structure by displaying the modified heterogeneity structure SW3p of figure 2b in a different form. Rather than a projective picture of a 2-dimensional velocity profile we show in figure 9 the radially stratified base model to the left and to the right the 2-dimensional variations away from the base model expressed as percentage perturbations. These two aspects of model SW3p (the radial and lateral structure) reflect the complementary roles of the different types of processes which help to shape the character of the seismic wavespeed distribution in the upper mantle.

Radial structure

Most of the work on upper mantle structure has been directed towards the elucidation of radial structure. In general, the radial gradients in velocity are substantially larger than horizontal gradients (Kennett 1991) and this favours the extraction of radial structure. The actual velocity profile represents a balance between the effect of

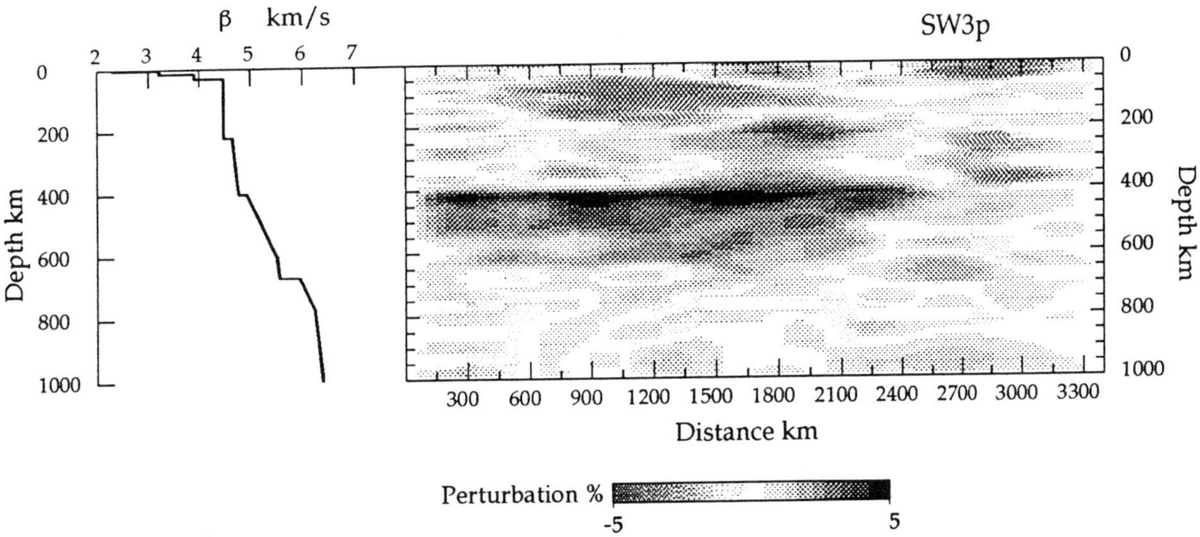

Fig. 9. Alternative representation of the heterogeneity model SW3p displayed in figure 2b. To the left is shown the radial velocity model and to the right, the perturbations from this model in per cent so that the range of scales of lateral heterogeneity can be appreciated.

pressure due to increasing lithostatic loading with depth and the influence of the increasing temperature with increasing depth (except near such features as continental rifts). For a single material below about 300 km depth the net effect of these influences is usually to give a steady increase in velocity with depth. But, this progression is interrupted by rapid increases in velocity and density as the crystal structures in the mineral assemblage in the upper mantle undergo phase transitions to accommodate the increasing load.

The principal candidate phase transformations (see e.g. Poirier 1991) are from olivine to the beta phase at 410 km (an endothermic change) and from the spinel structure to the perovskite structure at 660 km (an exothermic change). The influence of temperature can shift the depth at which the major proposed transitions can occur, but, away from the cold descending material in subduction zones, the anticipated depth variation would only be of the order of 20-30 km.

In the shallower parts of the upper mantle, particularly the astenosphere, the influence of temperature is of greater significance. In some areas incipient or actual partial melting may occur with lowered seismic velocities especially for S and also to increased attenuation of the seismic waves.

Lateral Variability

The deviations of seismic velocity structure from the purely radial are of particular significance because they can give a direct indication of processes operating in the mantle on a wide variety of scales.

Global tomographic images (see e.g. Woodhouse and Dziewonski, 1984) display velocity variations of several per cent on scales of 2000 km or more. Similar size deviations from a radial base model are seen in higher resolution images for specific regions, e.g. the WEPL3 structure of Nolet (1990) in Western Europe, at scales of 700 km or more. A number of lines of evidence support the existence of further heterogeneity at scales of 300 km or less (e.g. Kennett and Bowman 1990, Snieder 1988) although the estimates of the magnitude of the deviations differ somewhat. It is likely that such smaller scale features have 1-2 per cent variability superimposed on the 5 per cent or so variation associated with the larger scale structures. Heterogeneity certainly exists on even smaller scales and this smaller scale 'speckle' is a major contributor to the complexity of body wave codas. However, it is difficult to make much more than a stochastic description of such a heterogeneity field.

The high and low velocity features in the large scale structures determined by global tomography and the largest scale features in specific upper mantle models are likely to be thermal in origin. The influence of large-scale convective flow could also help to induce some compositional graduation which would also tend to modify the observed velocities.

Smaller scale structures can be produced by the substructure of a convectional flow regime. For example, Gurnis and Davies (1986) have demonstrated how a initial tracer can be distorted by flow to permeate a region. However, in areas which have been stable for times which are long compared with the thermal diffusion times for the corresponding scale length, it is likely that small scale structure represents genuine compositional variation (e.g. anisotropic effects inherited from former flow regimes).

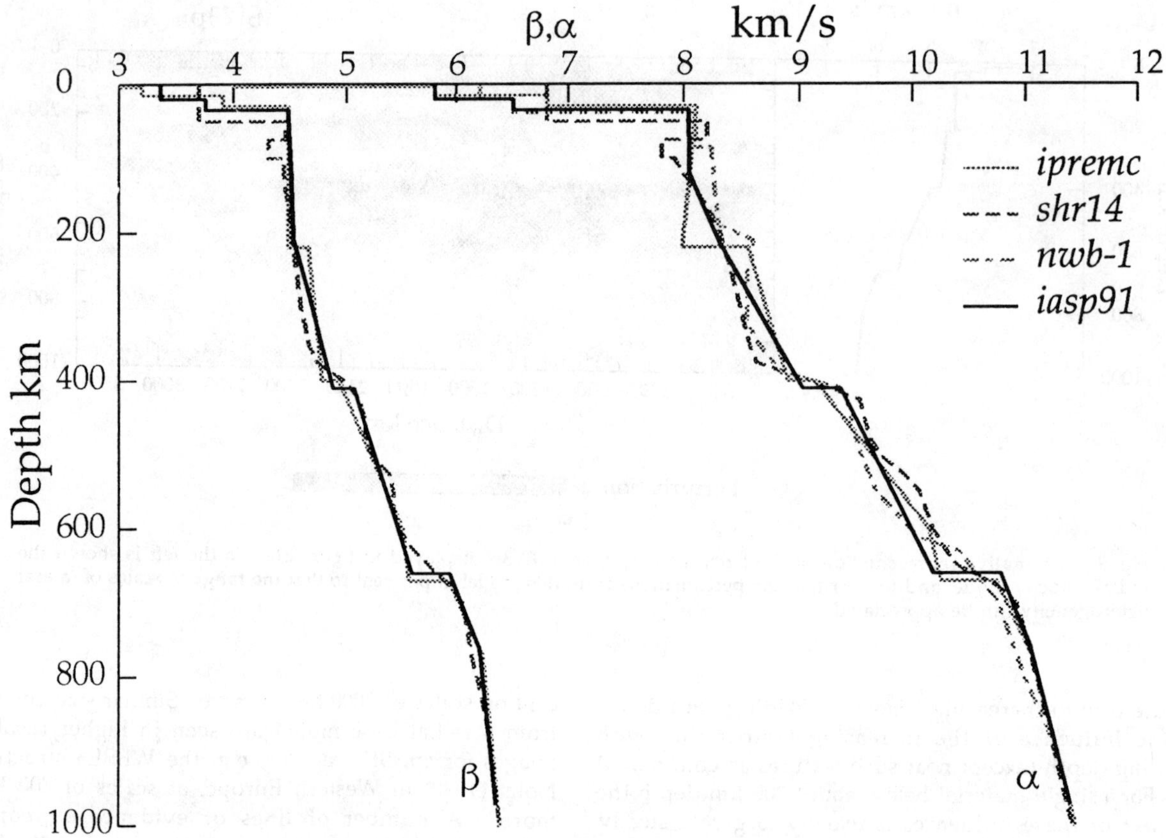

Fig. 10. Comparison of radial velocity models for the upper mantle using different techniques *prem* (Dziewonski and Anderson 1981) from free oscillations and travel times, *shr14* (Helmberger and Engen 1974) from analysis of long-period body waves, *nwb-1* (Bowman and Kennett 1990) from short-period body waves, *iasp91* (Kennett and Engdahl 1991) from travel time observations.

DISCUSSION: ONE MANTLE OR MANY?

We have seen how a wide range of techniques can bring information to bear on the seismic structure of the upper mantle. Each of the techniques has limitations in terms of potential resolution or sensitivity to lateral heterogeneity. In many ways the best view of our knowledge of upper mantle structure is as an ensemble of models derived from different classes of information.

The tangled skein of radial velocity profiles in figure 10 derived from different classes of information give a measure of the consistency of the general trends in seismic velocity and the diversity arising from geographic variability. Superimposed on such radial structure is lateral variability on a wide range of scales. The largest scale features can be described deterministically, but as the scale diminishes, it becomes more difficult to resolve the details of the horizontal structure within the upper mantle. For scales around 200 km a semi-stochastic description is currently the best that can be achieved, and a fully stochastic description is the most useful way of describing the finest scale structure.

With carefully designed experiments it may well be possible to resolve the finer details of the velocity structure and heterogeneity in particular regions. In general we will have to accept the fundamental limitations imposed by having only surface observations of an uneven geographic distribution of sources.

REFERENCES

Anderson, D.L., The deep structure of continents, *J Geophys Res, 84*, 7555-7560, 1979.

Bowman J.R. & B.L.N. Kennett, An investigation of the upper mantle beneath NW Australia using a hybrid seismograph array, *Geophys J Int, 101*, 411-424, 1990.

Burdick, L.J. & D.V. Helmberger, The upper mantle P velocity structure of the western United States, *J. Geophys Res, 83*, 1699-1712, 1978.

Cummins, P.R., B.L.N. Kennett, J.R. Bowman, & M.G. Bostock, The 520 km discontinuity? *Bull seism Soc Am, 82*, 323-326 1992.

Chapman, C.H., A new method for computing synthetic seismograms, *Geophys J R Astr Soc, 54*, 431-518, 1978.

Davis, J.P., R. Kind & I.S. Sacks, Precursors to P'P' re-examined using broad band data, *Geophys J Int, 99,* 595-604, 1989.

Dey S.C., Lateral variations in the upper mantle velocity structure under northern Australia, *Ph.D. thesis, The Australian National University,* 1989.

Dost B., Preliminary results from higher-mode surface-wave measurements in western Europe using the NARS array, *Tectonophysics, 128,* 289-301, 1986.

Dziewonski A.M. & D.L. Anderson, Preliminary Reference Earth Model, *Phys Earth Planet Inter., 25,* 297-356, 1981.

Dziewonski A.M., A.L. Hales & E.R. Lapwood, Parametrically simple Earth models consistent with geophysical data, *Phys Earth Planet Inter., 10,* 12-48, 1975.

Faber S. & G. Müller, Sp phases from the transition zone between the upper and lower mantle, *Bull seism Soc Am, 70,* 487-508, 1980.

Fukao, Y., M. Obayashi, H. Inoue and M. Nenbai, Subducting slabs stagnant inthe mantle transition zone, *J. Geophys Res., 97,* 4792-4809, 1992.

Goody, A., Broad-band studies of the upper mantle beneath northern Australia, *B.Sc. Hons. degree thesis, The Australian National University,* 1991.

Grand S. & D.V. Helmberger, Upper mantle shear structure of North America, *Geophys J R Astr Soc, 76,* 399-438, 1984.

Gurnis M. & G.F. Davies, Mixing numerical models of mantle convection incorporating plate kinematics, *J Geophys Res, 91,* 6375-6395, 1986.

Hales A.L., K.J. Muirhead & J.W. Rynn, A compressional velocity distribution for the upper mantle, *Tectonophysics, 63,* 309-348, 1980.

Helmberger D.V. & G.R. Engen, Upper mantle shear structure, *J Geophys Res, 79,* 4017-4028, 1974.

Hirn A., L. Steinmetz , R. Kind & K. Fuchs, Long range profiles in Western Europe II. Fine structure of the lithosphere in France (southern Bretagne), *Z. Geophys., 39,* 363-384, 1973.

Jeffreys H., On near earthquakes, *Mon Not R Astr Soc, Geophys Suppl, 1,* 22-31, 1926.

Jeffreys H., The times of P and S at short epicentral distances, *Mon Not R Astr Soc, Geophys Suppl, 2,* 399-407, 1931.

Jeffreys H., A rediscussion of some near earthquakes, *Mon Not R Astr Soc, Geophys Supple, 3,* 131-156, 1933.

Jeffreys H., The times of P, S and SKS and the velocities of P and S, *Mon Not R Astr Soc, Geophys Suppl, 4,* 498-536, 1939.

Johnson L.R., Array measurements of P velocities in the upper mantle, *J Geophys Res, 72,* 6309-6325, 1967.

Kennett B.L.N., Seismic velocity gradients in the upper mantle, *Geophys Res Lett, 18,* 1115-1118, 1991.

Kennett B.L.N. & J.R. Bowman, The velocity structure and heterogeneity of the mantle, *Phys Earth Planet Int, 59,* 134-144, 1990.

Kennett B.L.N. & E.R. Engdahl, Travel times for global earthquake location and phase identification, *Geophys J Int, 105,* 429-465, 1991.

Kennett B.L.N. & G. Nolet, The interaction of the S wavefield with upper mantle heterogeneity, *Geophys J Int, 101,* 751-762, 1990.

LeFevre, L.V. and Helmberger D.V., Upper mantle P velocity structure of the Canadian shield, *J. Geophys. Res., 94,* 17749-17765, 1989.

Lehmann I., Recent studies of body waves in the mantle of the Earth, *Geophys J R Astr Soc, 3,* 288-298, 1962.

Mohorovičić A., Das Beben von 8 Okt 1909. *Jahrb. meterol. Obs Zagreb (Agram), 9,* Tiel IV, 1909.

Montagner J-P and Anderson D.L., Constrained reference earth model, *Phys Earth Planet Int, 58,* 205-227, 1989.

Nolet G. & M.J.R. Wortel, Mantle, Upper Structure, in *Encyclopedia of Solid Earth Geophysics,* editd by D.E. James, 775-788, Van Nostrand Reinhold, New York, 1989.

Nolet G., Higher Rayleigh modes in Western Europe, *Geophys Res Lett, 2,* 60-62, 1975.

Nolet G., The upper mantle under western Europe inferred from the dispersion of Rayleigh modes, *J Geophys, 43,* 265-285, 1977.

Nolet G., Partitioned waveform inversion and 2D structure under the network of autonomously recording seismograph stations, *J Geophys Res, 95,* 8499-8512, 1990.

Pavlenkova N.I. & A.V. Yegorkin, Upper mantle heterogeneity in the northern part of Eurasia, *Phys Earth Planet Inter, 33,* 180-195, 1983.

Poirier J-P., *Introduction to the Physics of the Earth's Interior,* Cambridge University Press, Cambridge, 1991.

Revenaugh J.S., The nature of mantle layering from first-order reverberations, *Ph.D. thesis, Massachusetts Institute of Technology,* 1989.

Revenaugh J.S. & T.H. Jordan, A study of mantle layering beneath the western Pacific, *J. Geophys Res, 94,* 5787-5813, 1989.

Revenaugh J.S. and Jordan T.H., Mantle layering from ScS reverberations, 3, the Upper Mantle, *J Geophys Res, 96,* 19781-19811, 1991.

Richards P.G., Seismic waves reflected from velocity gradient anomalies with the Earth's upper mantle, *Z Geophys, 28,* 517-527, 1972.

Rigden S.M., G.D.C. Gwanmesia, J.D. FitzGerald, I. Jackson & R.C. Liebermann, Spinel elasticity and seismic structure of the transition zone of the mantle, *Nature, 354,,* 143-145, 1991.

Shearer P., Seismic imaging of upper-mantle structure with new evidence for a 520 km discontinuity, *Nature, 344,* 121-126, 1990.

Shearer, P.M., Constraints on upper mantle discontinuities from observations of long-period reflected and converted phases, *J. Geophys. Res, 96,* 18147-18182, 1991.

Snieder R.K., Large-scale waveform inversions of surface waves for lateral heterogeneity 2. Application to surface waves in Europe and the Mediterranean, *J. Geophys Res, 93*, 12067-12080, 1988.

Steinmetz L, A. Hirn & G. Perrier, Réflexions sismique à la base de l'asténosphère, *Ann Géophys, 30*, 173-180, 1974.

van der Hilst, R. Engdhal, R., Spakman, W. and Nolet, G., Tomographic imaging of subducted lithosphere below northwest Pacific island arcs, *Nature, 353*, 37-43, 1991

Walker M., *Modern Seismology*, Longman, London, 1915.

Woodhouse J.H. & A.M. Dziewonski, Mapping the upper mantle: three dimensional modelling of Earth structure by inversion of seismic waveforms. *J. Geophys Res, 89*, 5953-5986, 1984.

Yegorkin, A. V. & N.M. Chernyshov, Peculiarities of mantle waves from long-range profiles, *J. Geophys., 54*, 30-34, 1983.

Zielhuis, A. & Nolet G., A 3D velocity model for Europe obtained with partitioned waveform tomography, *EOS. supplement, 1991 AGU Fall Meeting*, 349

Zhang, Y.S. and Tanimoto T., Global Love wave phase velocity variations and its significance for plate tectonics, *Phys, Earth Planet. Int 66*, 160-202, 1991.

B.L.N. Kennett, Research School of Earth Sciences, Australian National University, P.O. Box 4, Canberra, ACT, Australia.

Seismic Tomography and Geodynamics

Adam M. Dziewonski, Alessandro M. Forte, Wei-jia Su and Robert L. Woodward

Department of Earth and Planetary Sciences
Harvard University, Cambridge, MA 02138

This paper presents recent developments in the investigation of the spectrum of lateral heterogeneity in the mantle, derivation of 3-D shear velocity models through a joint inversion of waveform and differential travel time data, and examples of the use of these models in explaining several geodynamic observables. A study of the spectrum of lateral heterogeneity is carried out using a set of SS data with a good global coverage of the mid-path surface reflection point. The conclusion is that the power in the spherical harmonic spectrum of the SS residuals begins to decrease rapidly beyond $\ell = 6$. A corner in the power spectrum is also observed between $\ell = 5$ and $\ell = 7$ for Love wave phase velocities, P-wave velocity anomalies in the lower mantle, ocean–continent function, and free air gravity anomalies. This implies a dominant horizontal scale of the anomalies with a half-wavelength exceeding 3,000–3,500 km. This result justifies the use of the low-order basis functions in modeling of the global scale heterogeneity. A set of waveforms consisting of mantle waves and long-period body waves selected from, roughly, 15,000 seismograms is used in conjunction with approximately 5,400 $SS - S$ and 2,600 $ScS - S$ differential travel times to derive two 3-D shear velocity models: $SH8/U4L8$ and $SH8/WM13$. Horizontal variations are modeled by spherical harmonics up to $\ell = 8$ and Chebyshev polynomials are used as the radial basis functions. Model $SH8/U4L8$ assumes a separate expansion for the upper and lower mantle, while variations are continuous across the 670 km discontinuity in model $SH8/WM13$. The two models show similar changes in the lateral heterogeneity across this discontinuity. This implies that the current data set is sufficient to resolve the structure in this region. The 670 km discontinuity is a barrier to material flux in some regions, but heterogeneities appear to be continuous across it in other areas, particularly those near spreading centers. The rms level of lateral heterogeneity is 2% near the surface, decreases to 0.35% in the middle mantle and begins to increase below 1900 km to reach 0.7% at the CMB. Below 2000 km depth the heterogeneity is dominated by the Y_2^2 harmonic. A viscosity function $\eta(r)$, including two low-viscosity zones: one just above 670 km discontinuity and the other above the CMB, and the proportionality factor $\alpha(r) = \partial ln\rho/\partial ln v_S$ are established empirically to model the dynamic response of the mantle to internal loads described by a 3-D model: $SH8/U4L8$. Some 68% of the variance of the observed geoid is explained, with the single largest discrepancy being in the Y_2^0 harmonic; when it is adjusted to match the observed value, the variance reduction increases to 82%. The predictions of the dynamic topography of the CMB leads to models with a relief of ±4.5 km and a pattern similar to that reported in some seismic studies. Modeling of the plate velocities leads to a variance reduction of 73% of the observed horizontal divergence field and 31% of the radial vorticity field. The observed and predicted surface dynamic topography show good agreement after the observed topography is adjusted for isostasy.

Introduction

Even though mapping of the earth's interior in three-dimensions (3-D) represents a significant challenge in its own right, the first major global tomographic study was clearly motivated by the need to address geodynamic problems. *Dziewonski et al.* [1977] assumed that perturbations in velocity and density are proportional and used their very coarse grid of mantle velocity anomalies to investigate the correlation between the predicted and observed gravity field. The satisfactory, statistically significant, answer was believed to accomplish two things: to indicate that the gravest terms of the geoid field originate in the lowermost mantle and to yield some credibility to the main thesis of the paper that lateral heterogeneities can indeed be mapped using very large, albeit noisy, data sets.

The interest in mapping the large-scale heterogeneity was revived by an incontrovertible proof of its existence. *Masters et al.* [1982] measured several thousand shifts of the eigenfrequencies of fundamental spheroidal modes in a period range from 200 to 500 seconds. When these shifts were plotted at the poles of the great circle source-receiver paths, they

Relating Geophysical Structures and Processes: The Jeffreys Volume
Geophysical Monograph 76, IUGG Volume 16

showed a dominating degree 2 pattern. It was clearly visible for different ranges of angular orders and from the measured frequency dependence the authors placed its source in the transition zone. This approach is sensitive only to the even orders of the heterogeneity. Nevertheless, it established that in the range of observations, the effect of the degree 2 anomaly was several times larger than the next even orders: 4 and 6. At the same time, *Nakanishi and Anderson* [1982] obtained a global expansion up to degree 6 of group velocity of mantle Rayleigh waves and they also confirmed the degree 2 dominance in the transition zone.

Clearly, the accumulating data from the early digital networks: GDSN [Global Digital Seismograph Network operated by the USGS, see *Peterson et al.*, 1976] and IDA [International Deployment of Accelerometers operated by the UC San Diego, see *Agnew et al.*,1976] and the millions of phase parameters collected by the ISC (International Seismological Centre) were a firm basis for a new major effort in global seismology.

Dziewonski [1982, 1984] used the ISC Bulletins to obtain degree 6 expansion of P-velocities in the lower mantle. His model *L*02.56 was defined using basis functions (spherical harmonics and Legendre polynomials in radius) with a total of 245 real coefficients. We shall call this a 'spectral' approach. Unlike in the studies of *Dziewonski* [1975] and *Dziewonski et al.* [1977], the earthquakes were relocated. An iterative procedure with several iterations was carried out to minimize the tradeoff between the lateral heterogeneity and the source origin time and location. Later, *Morelli and Dziewonski* [1991] solved a simultaneous inverse problem for both the lateral heterogeneity and the location of 'summary earthquakes'.

The large number of reporting stations (~3,000) and sources (~30,000 with more than 30 reporting stations) suggest — without accounting for the uneven distribution of the sources and the receivers — a possibility of retrieving the 3-D structure with high resolution. *Clayton and Comer* [1983] also used the P-wave arrival times from the ISC Bulletins. They divided the earth into a three-dimensional array of cells, with constant velocity perturbation in each shell. They had 29 spherical shells, each roughly 100 km thick, each divided into $5° \times 5°$ cells for a total of 2,592 elements in each shell and a grand total of 75,168 blocks. We shall call this a 'block' approach. In later experiments *Clayton and Comer* [1984] used a constant area discretization, corresponding to $5° \times 5°$ at the equator, reducing the nominal number of unknowns to 48,024. Even so, the ratio of the unknown parameters in the study by *Dziewonski* [1984] and that by *Clayton and Comer* — using virtually the same data base — is, roughly, 1 to 200.

The fact that such dramatically different approaches could be contemplated had to do with the fact that the power spectrum of lateral heterogeneity in the deep mantle was unknown. It also had to do with the proliferation of 'mathematical trickery', a term used by *Lancsoz* – the originator of the 'generalized inverse' [1961] – to describe methods allowing one to obtain solutions to numerically unstable inverse problems. Often, these involve the introduction of an 'objective function', which leads to a solution that is a compromise between an optimal fit to the data and one based upon minimizing some norm imposed on the resulting model.

Actually, the approach adopted by *Clayton and Comer* [1984] used an iterative matrix inversion technique in which incremental contributions to the solution are obtained by considering only the inverse of the diagonal of the inner product matrix. This process is, in the limit, equivalent to evaluation of the exact inverse and it converges only if the matrix is well conditioned. Otherwise, the solution 'blows up', even though contributions from individual iterations are finite. It is, of course, possible to stop this process after some number of iterations, but the results can be noisy. The 3-D P-velocity model of *Inoue et al.* [1990], also derived from the ISC data base, is an example of a stable result obtained for a large number of unknown parameters (32,768) with application of the appropriate smoothing conditions. The question of the spectrum of the lateral heterogeneity will be discussed in Section 2.

Studies involving the analysis of waveform data from digital stations commonly adopted the spectral approach. Although the number of digital stations has been increasing rapidly, even today, there are fewer than 100 globally distributed digital stations with the appropriate long-period response. The number of sources large enough to generate mantle waves with amplitudes needed by the analysis also numbers in the hundreds. An early study, following those of *Masters et al.* [1982] and *Nakanishi and Anderson* [1982], is by *Woodhouse and Dziewonski* [1984] which not only presented a model of the upper mantle (M84C), which has become a *de facto* reference model, but also set up the formalism for waveform inversion of complete, low-pass filtered seismograms. This will be briefly discussed in Section 3 of this paper. Other papers using either mantle wave analysis or waveform inversion include: *Nataf et al.* [1984, 1986], *Woodhouse and Dziewonski* [1986, 1989], *Tanimoto* [1987, 1990], *Romanowicz et al.* [1987], *Roult et al.* [1990], *Romanowicz* [1990], *Montagner and Tanimoto* [1990, 1991], *Su and Dziewonski* [1991]. A recent review of seismic tomography of the earth's mantle has been published by *Romanowicz* [1991].

An important recent development was the introduction of travel time data derived from the analysis of long-period body wave data recorded by the GDSN stations. Their response is poor for periods shorter than 15 s and reading of arrival times of individual phases is very inaccurate. There are two ways to circumvent this drawback. One is to determine the differential travel time between two phases by measuring the time lag of the maximum in their cross-correlation function: $SS - S$ [*Woodward and Masters*, 1991a; with the SS subjected to the Hilbert transform] or $ScS - S$ [*Woodward and Masters*, 1991b]. The other method is to estimate the 'absolute' travel time residuals by cross-correlating the observed pulse, SS, for example, with the synthetic computed for a reference earth model [*Su and Dziewonski*, 1991]. In the latter approach the answers are more sensitive to the precision with which the source parameters are known, but the global coverage of observations may be better.

Introduction of these data into the process of derivation of earth models is significant for several reasons. First, with a cut-off period of 32 s our SS ray paths average a volume which is at least 1000 times larger than at the 1 Hz frequency, characteristic of the P-waves reported in the ISC Bulletins. This filters out most of the 'noise' associated with small-scale structures. This is why a single, high quality measurement of a long-period travel time may be worth hundreds of short period data. For example, explanation of a spectrum from a single source–station pair may, for a low order normal mode, require integration over the entire volume of the earth. Long-period, large wavelength data are very effective volume integrators and their availability is essential for studying the earth on a planetary scale.

Second, the patterns in the data itself, without the need for a 3-D inversion, may be able to provide us with information on certain properties of the lateral heterogeneity. For instance, the $ScS - S$ data of *Woodward and Masters* [1991b] clearly show the presence of large-scale, large amplitude velocity anomaly in the lowermost mantle, rather similar to that of model $L02.56$ of Dziewonski in a depth range from 2000 km to the CMB. Similarly, the pattern of $SS - S$ residuals correlates well with the velocity anomalies in a depth range 100-300 km in model $M84C$ of *Woodhouse and Dziewonski* [1984].

Third, the theory used to interpret the travel time data and waveform data are different. The algorithm for the interpretation of waveform data developed by *Woodhouse and Dziewonski* [1984] invokes the 'path average approximation'. This assumes that the observed waveforms can be explained by perturbing the average structure along the minor arc path and, for mantle waves, also along the complete great circle. In this approximation the three-dimensional structure is collapsed into one dimension: average radial structure along the great circle path. A much different theory is used for the travel times. The rays are traced through a spherically symmetric reference Earth, but the travel time anomaly due to aspherical velocity perturbations is integrated in two dimensions along this path [*Dziewonski*, 1984]. At the same time, the geometrical ray theory is used to interpret data with the wavelengths of the order of 200–300 km. By using these two data sets simultaneously we test their compatibility and the accuracy of the approximations involved. Earth models obtained recently by simultaneous inversion of travel time and waveform data are presented in Section 3, preceded by an outline of the methods and description of data used in their derivation.

Studies of splitting of normal mode eigenfrequencies are considered separately, because of the different theory involved. Such studies were primarily responsible for the discovery of the anomalous properties of the core [*Masters and Gilbert*, 1981; *Ritzwoller et al.*, 1986; *Woodhouse et al.*,1986; *Widmer*, 1991] and are beyond the scope of this paper.

Modes whose energy is principally confined to the mantle are also anomalously split, but the effect, unlike that for the core modes, is not predominantly axi-symmetric. Also, these modes have lower Q and the individual spectral lines overlap significantly. Furthermore, the estimation of the frequencies of individual spectral lines provides incomplete information on the heterogeneity. The effect is fully described by the splitting matrix [*Dahlen*, 1968, 1976], while the singlet frequencies are only its eigenvalues, which are invariant with respect to rotations.

To deal with this problem, *Woodhouse and Giardini* [1985] proposed a procedure leading to the retrieval of the splitting matrix and the derivation of the splitting function, which has a geographical representation. The resulting map predicts values akin to the 'local eigenfrequencies' [*Jordan*, 1978a] and these represent linear constraints on structure. Applications of this approach can be found in *Giardini et al.* [1987, 1988], *Ritzwoller et al.* [1988], *Li et al.* [1991].

To the first order, the splitting depends only on the even harmonic degree part of the lateral heterogeneity, and is therefore of limited use. However, it provides useful auxiliary information, such as on the possible frequency dependence of the effects caused by the lateral heterogeneity. More recently, *Lognonne and Romanowicz* [1990] proposed a new method of studying the coupled normal modes, which – in principle – depend also on the odd-order terms of the expansion.

Normal modes used in the sources quoted above span a range of frequencies from 0.3 mHz to 5 mHz; the body wave waveforms have peak energy at about 20 mHz; the waveforms used to derive differential travel times – 50 mHz; the P-wave arrival times from the ISC Bulletins are typically 1 Hz data. It has been demonstrated [*Dziewonski and Woodhouse*, 1987, Figure 9; *Woodhouse and Dziewonski*, 1989, Plate 2 b, e, h and c, f, i] that very similar patterns of heterogeneity are obtained, for even harmonic degrees, from each of these data types.

This also confirms that spatial aliasing is not an important problem: one of the advantages of the normal mode data is that they represent excellent averages over the entire volume of the mantle. On the other hand, mantle waves have wavelengths on the order of 1,000 km, while 1 Hz P-wave paths have a ray tube that is only some tens of kilometers in diameter.

Section 4 explores application of seismic tomographic models to modeling several important geodynamic processes. For this purpose, we must establish the proportionality factor, $\alpha(r) = d\ln\rho(r)/d\ln v(r)$ and the viscosity profile: $\eta(r)$. The different geodynamic variables are sensitive to density contrasts at different depths. The dynamic core–mantle boundary (CMB) topography is most sensitive to density contrasts in the lowermost mantle. The gravest harmonics of the geoid are fairly uniformly sensitive to density perturbations at all depths. Plate velocities are highly sensitive to the upper and middle mantle structure, including the transition zone. The surface dynamic topography is most sensitive to the shallow features. The fact that we can obtain a fairly good fit to all these variables with a single tomographic model and the same radial functions $\eta(r)$ and $\alpha(r)$ increases the likelihood that our approach is principally valid. Of course, one does not expect a perfect fit; there must be some effect due to lateral variations in viscosity as well as in the coefficient $\alpha(r)$. Although we proceed with a working hypothesis that the thermal effects are dominating, the presence of chemical heterogeneity cannot be ruled out.

SPECTRUM OF HETEROGENEITY

The comparison, in the Introduction, of the approaches to representation of lateral heterogeneity in P-velocity by *Dziewonski* [1982, 1984] and *Clayton and Comer* [1983, 1984] illustrates why there has been uncertainty with regard to the spectrum of heterogeneity. On the one hand, the spectrum is *a priori* truncated at a low harmonic degree On the other, there are so many unknown parameters that, without appropriate regularization, a finding that the spectrum is white might be indicative of the instability of the inverse problem rather than the nature of the heterogeneity.

Because of the uneven distribution of sources and receivers, the low order models may be biased, in addition to being incomplete, if there is significant power in the rejected part of the spectrum. Two recent papers suggest that this may be the case. *Gudmundsson et al.* [1990] presented results of a statistical analysis of the P-wave travel time residuals and concluded that in the upper mantle the spectrum of lateral heterogeneity is white, down to wavelengths as short as 100-200 km. In the second paper, *Snieder et al.* [1991] suggested that the origin of what appear to be large-scale features in both upper and lower mantle tomographic models may be an artifact of low-pass filtration of relatively narrow structures such as mid-oceanic ridges or subduction zones. In reality, they say, the highly anomalous properties may be limited to narrow zones, whose width is an order of magnitude less than indicated by the tomographic models. If either of the above statements is correct, then much of the tomographic work may be wrong or, at best, misleading. In turn, this could invalidate conclusions such as those of *Olson et al.* [1990], which convey an impression that the basic question of geodynamics — the driving mechanism of plate tectonics — has been conceptually answered.

The Global Spectrum of SS Residuals

One way to answer this question is to take a step back, in order to complete a missing stage in the direct progression of seismology from one to three dimensions. *Su and Dziewonski* [1991] investigate a 2-D surface projection of 3-D lateral heterogeneity. Seismic tomography may be compared to the inference of 3-D shapes of objects from the shadows they cast on a 2-D screen (measurements on the earth's surface). If the space is not densely packed, a simple inspection of the 'shadows' would indicate whether there are relatively few large objects or many small ones. If the former is true, then the low order expansion is justified, and a global 3-D inversion can be attempted with a good chance of success. If the shadows indicate predominance of small-scale objects or a uniform mixture of sizes (white spectrum), the problem is much more difficult.

Su and Dziewonski [1991] use long-period body wave data, and make precise measurements of the travel-time residuals of the SS phase. The advantage of using body wave data is that an adequate signal-to-noise ratio can be achieved for earthquakes whose magnitude is many times smaller than that produce mantle waves of an amplitude suitable for the analysis. The ability to use smaller events leads to better geographical distribution of sources in regions without large earthquakes. For the details of their measurement technique, data selection and analysis, see *Su and Dziewonski* [1992].

Figure 1a shows the distribution of SS phase mid-path reflection points at the Earth's surface. There are 3,313 mea-

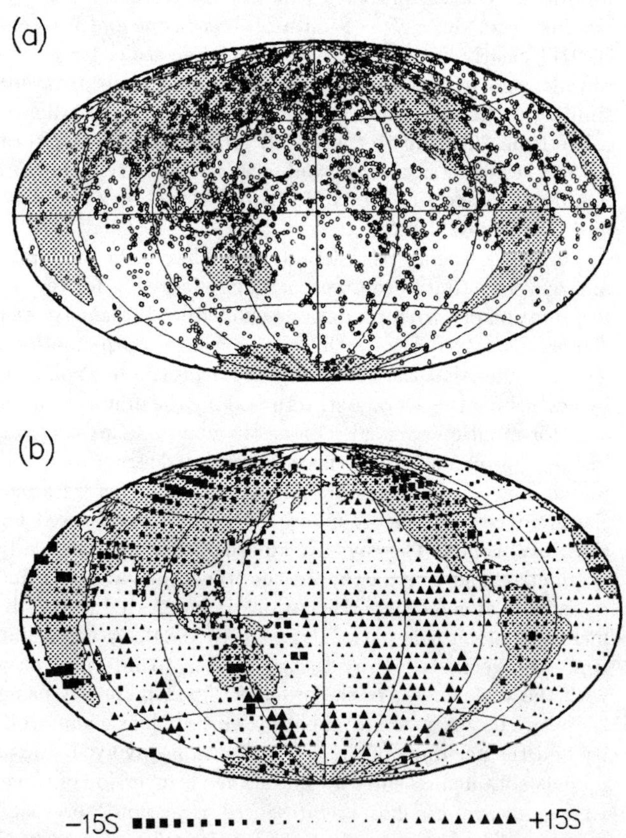

−15S ▪▪▪▪▪▪▪▪ · · · · · · · · · · · ▴▴▴▴▴▴▴▴ +15S

Fig. 1. (a) The distribution of SS phase mid-path reflection points at the Earth's surface. Each circle represents one of the 3,313 measurements. (b) Travel-time residuals for SS phase projected on an equally spaced grid ($5° \times 5°$ at the equator). The residuals are smoothed by a moving cap of radius $5°$. The values shown are corrected for topography, bathymetry and crustal thickness.

surements of the SH component of the SS phase recorded between 50° and 150°. The coverage is denser in the Northern Hemisphere, but even in the Southern Hemisphere there are no large areas without data. The residuals range from -15 s to +15 s. The standard deviation of SS residuals is about 6 seconds for SS residuals. Since the typical measurement error is of the order of 1.0 s, there is considerable signal to be explained.

To obtain a better view of the measurements, we average the SS residuals in 5° radius caps to smooth the data. The mean of the measurements which fall into the cap is calculated and the resulting value is plotted at the center of a cap. The cap is then moved in a 5° increment in latitude and an increment in longitude which corresponds to 5° of arc; this provides a half cap radius overlap of the adjacent caps. So the number of smoothed values at higher latitudes decreases proportionally to the area. There is no minimum number of measurements for a cap: as long as there is at least one measurement in the cap, the value is taken as the smoothed value.

Figure 1b shows the smoothed residuals plotted at the center of caps. The data coverage is very good: more than 90% of possible cap locations are filled with measurements. The SS reflection points cover the earth well, especially in areas like the Pacific Ocean which is not satisfactorily sampled by the direct P or S phases. The SS residuals reveal a large-scale pattern of heterogeneity. The pattern is dominated by features clearly related to the known tectonic features on the surface.

Su and Dziewonski [1991, 1992] use spherical splines [*Shure et al.*, 1982] to obtain a spherical harmonic expansion, for degree 0 through 36, of the data from the 1452 grid points shown in Figure 1b. The associated Legendre functions are normalized:

$$p_l^m = (2 - \delta_{m,0})^{1/2}(2l+1)^{1/2}\left[\frac{(l-m)!}{(l+m)!}\right]^{1/2} P_{lm} \qquad (1)$$

so that the rms value over the surface of the sphere is equal to one for each sine or cosine harmonic. The spherical splines are used here, even though with the coverage as complete as that shown in Figure 1b a direct least squares fit would be stable. In choosing the accuracy of the fit to the data we consider two cases. First, the 'exact' solution; the data should be fit as accurately as possible ($\chi^2 \to 0$). Second, the 'damped' solution: the data should be fit with an rms error

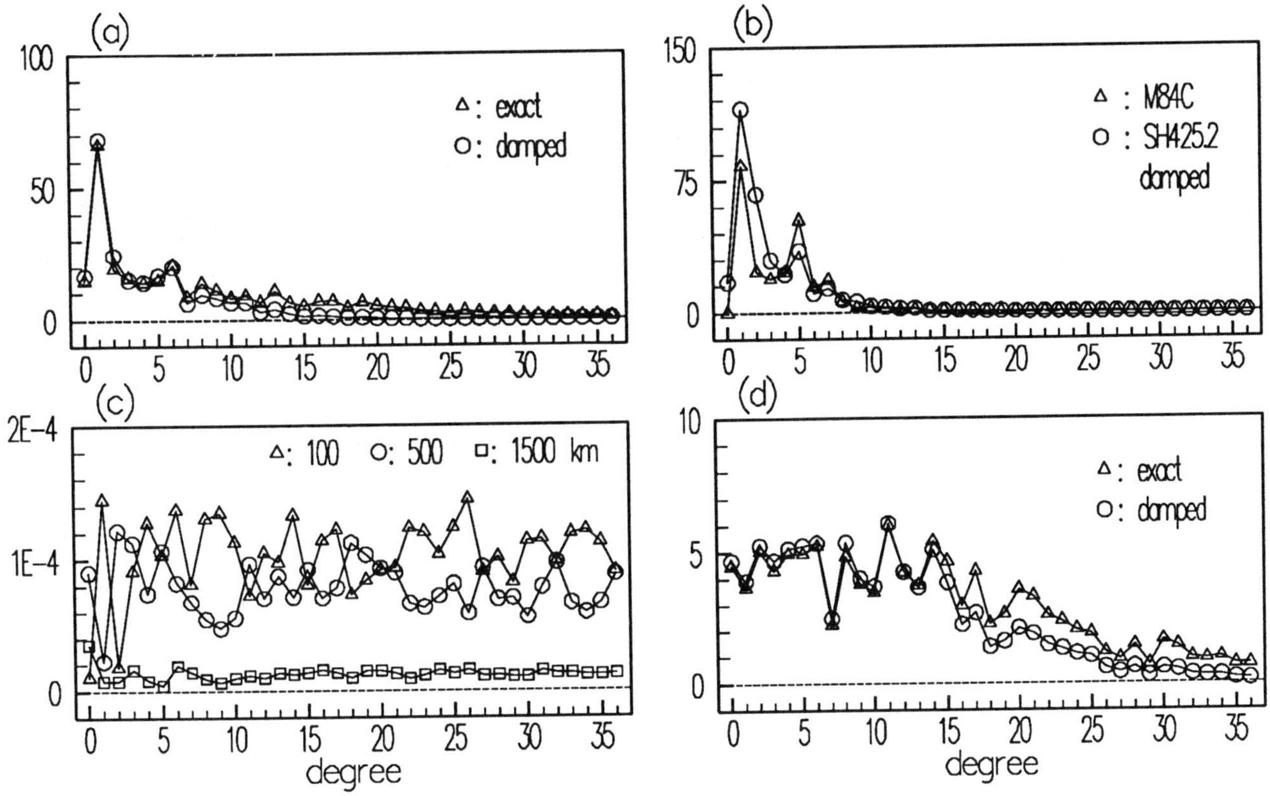

Fig. 2. Power spectrum, as a function of harmonic degree, defined by equation (2), for the spherical harmonic expansions of: (a) two different levels ('damped' and 'exact') of fit to the observed SS travel-time residuals; (b) 'damped' fit to SS travel-time residuals predicted by the upper mantle model $M84C$ and the whole-mantle S wave model $SH425.2$ (Su and Dziewonski, 1991); (c) velocity perturbations ($\delta v/v$) at three depths for the random model of degree 36, the units are dimensionless; (d) 'exact' and 'damped' fits to the SS travel-time residuals for the random model.

of 1 second ($\chi^2 = 1452$), which seems quite conservative considering the potential sources of error.

Figure 2a shows the sums:

$$D_l = \sum_{m=0}^{l} A_{lm}^2 + B_{lm}^2 \qquad (2)$$

for both solutions, where $A_{\ell m}$ and $B_{\ell m}$ are the spherical harmonic expansion coefficients.

There is very little power in the 'damped' solution beyond degree 10 or so. The power decreases less rapidly for the 'exact' fit, but even in this case it approaches zero for degrees greater than 20. The degree 1 term has the largest power, which probably results from the uneven distribution of continents in the Northern and Southern Hemisphere. The source of degree one heterogeneity may thus be located in the top few hundreds of kilometers of the upper mantle. The degree 2 term has the next largest power, although all degrees for 2 to 6 are of comparable size. The next peak is at degree 6, its source has been argued to be related to the hotspot distributions, special ocean-continent topography patterns, and current distribution of continents, oceans and ridges [*Tanimoto*, 1991].

Synthetic Tests

It is instructive to compute the SS residuals for existing earth models and compare their spectra with those in Figure 2a. We use two earth models: the first one is the upper mantle model $M84C$ of Woodhouse and Dziewonski (1984); the other is a whole mantle shear velocity model $SH425.2$ of *Su and Dziewonski* [1991].The power spectra for both models are shown in Figure 2b. The power spectra are quite similar to those for the data, except that both models $M84C$ and $SH425.2$ predict a strong local maximum at degree 5, while degree 5 and 6 in the spectrum of the observed residuals have comparable powers. The predicted power spectra for both models $M84C$ and $SH425.2$ are dominated by low angular orders. This is not surprising, because one would expect such spectra for the models truncated at degree 8. What is striking, however, is the similarity of the overall structure of the predicted power spectra to that obtained from the data.

However, integration of anomalies along a ray path is a smoothing process and one could still suppose that the actual structure has a substantial power at high orders. To answer this question we create a velocity model in which the anomalies are random at each depth and the rms level of perturbations changes with depth in the same way as the power in model $SH425.2$. The random variations have been scaled such that at each depth the expected total power for each degree should remain constant; spherical harmonics up to degree 36 are included. This is shown in Figure 2c for three different depths.

Using this hypothetical model, we calculate the SS residuals in exactly the same way as described above and Figure 2d shows the spectral power for the 'damped' and 'exact' spherical harmonic fits. Calculation of the SS residuals for the random structure is equivalent to an empirical evaluation of the response of a complex filter. If the filter were 'all-pass', the spectra in Figure 2d would be roughly flat for all degrees from 0 to 36. Instead, the response is 'all-pass' up to $\ell = 15$, or so, and then it begins to decrease gradually. This decrease reflects the smoothing effect of integration along the ray path and averaging of data over the spherical caps. It is clear that these spectra are distinctly different from those in Figure 2a, where the abrupt decrease in power occurs at $\ell = 6$. They have much more energy at high harmonics, so that the smoothing through integration

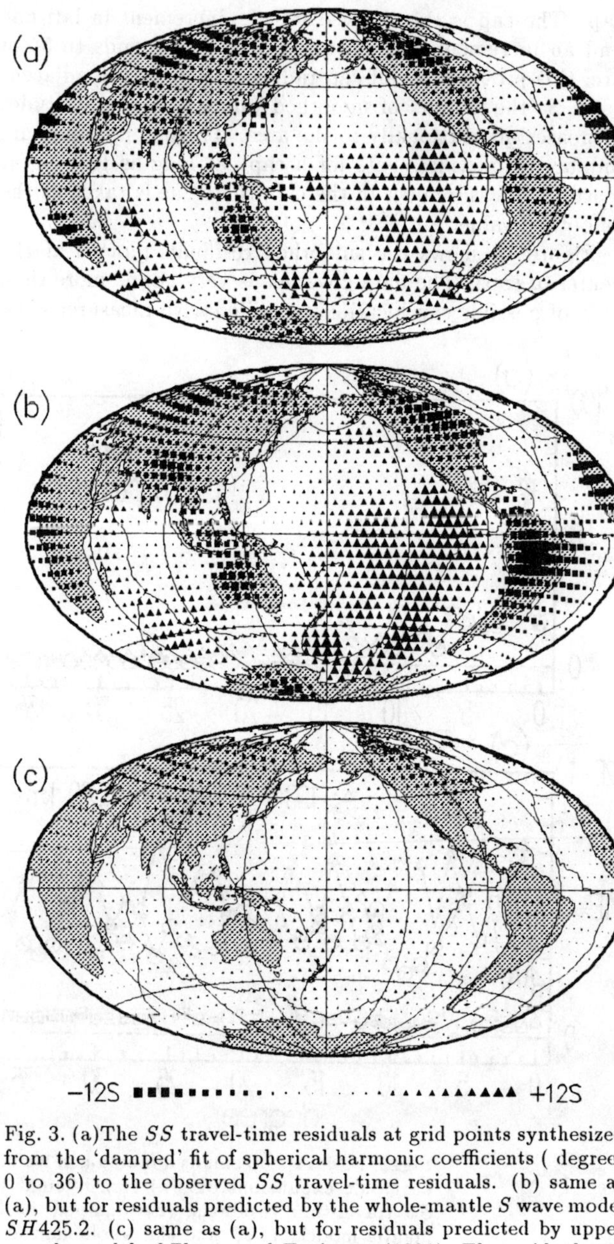

Fig. 3. (a)The SS travel-time residuals at grid points synthesized from the 'damped' fit of spherical harmonic coefficients (degrees 0 to 36) to the observed SS travel-time residuals. (b) same as (a), but for residuals predicted by the whole-mantle S wave model $SH425.2$. (c) same as (a), but for residuals predicted by upper mantle model of Zhang and Tanimoto (1991). The residuals are about 3 times smaller than the residuals in (a) and (b).

is not sufficient to mask the presence of high wavenumber structure.

Figure 3a shows a map derived by the synthesis of the spherical harmonic coefficient for degrees 1–36 obtained in a 'damped' fit to the observed residuals [see Table 1 in *Su and Dziewonski*, 1992]. Figure 3b is the same as Figure 3a but for the residuals predicted by model $SH425.2$. The overall patterns are quite similar, with model $SH425.2$ predicting somewhat larger residuals. The principal difference between the observed and predicted residuals is limited to the East Pacific Rise and South America. The overall correlation coefficient for degrees from 1 to 36 is 0.74, but it is as high as 0.86 for degrees from 1 to 6. The correlation coefficient is above 0.79 for each degree from 1 through 6; it drops suddenly to 0.29 for degree 7 and to 0.20 for degree 8, indicating that the power beyond degree 6 might be too low to be resolved on the global scale.

It should be pointed out that the SS travel-time residuals were not explicitly used in deriving model $SH425.2$. Even though about 60% of the events used were the same, the method of *Woodhouse and Dziewonski* [1984, 1986] uses entire waveforms, which consist of mantle waves and many other body wave phases in addition to SS [see Figures 3 and 5 in *Dziewonski and Woodhouse*, 1987]. The satisfactory agreement between the observed and predicted SS travel-time residuals indicates that the 'path average approximation' [*Woodhouse and Dziewonski*, 1984] works rather well not only for surface waves but also for body waves.

Recently, *Zhang* [1991] and *Zhang and Tanimoto* [1992] derived a high resolution model of shear velocity in the upper mantle based on an extensive study of propagation of mantle waves. We thought it would be instructive to compute SS residuals for this model, which has a nominal resolution similar to our data. Figure 3c are the synthetic residuals predicted for a degree 36 upper mantle shear velocity model of *Zhang* [1991]. The predominant pattern of the sign of the residuals is the same, with the continents being faster than average and oceans slower. But with the scale in Figures 3a-c being the same, it is clear that the residuals for *Zhang*'s model are roughly 3 times smaller than those observed or those predicted by model $SH425.2$. The shear velocity anomalies in the model of *Zhang*, filtered to degree 8, are similar to those of $SH425.2$ at 50 km depth. However, they decrease very rapidly below 100 km. *Zhang and Tanimoto* [1992] base their geodynamic interpretation of this model on the depth dependence of the velocity anomalies associated with mid-oceanic ridges and hotspots. *Su et al.* [1992] point out several discrepancies between the observed data and those predicted by the model of *Zhang and Tanimoto* [1992].

Other Observations and Geodynamic Functions

The importance of the upper mantle structure below the lithospheric depth can also be shown in a different way. Even visual examination of Figure 3a-c indicates that there is a general decrease of amplitude of the negative residuals with the increasing distance from the mid-ocean ridge. To examine the correlation of the residuals with seafloor age we divide the oceans into $5° \times 5°$ cells with each cell assigned an age value. The averages of residuals which fall into each age group are calculated. Figure 4a shows the mean residuals against the square root of ocean age. The regression tests show that a linear relationship is apparent for age groups up to 150 my, with a linear correlation coefficient of −0.82. Such linear correlation has been found in other studies [e. g. *Woodward and Masters*, 1991a; *Sheehan and Solomon*, 1991; *Zhang and Tanimoto*, 1991].

The slope of the line is $-0.41 \pm 0.047 s/My^{1/2}$. This value is slightly smaller than $-0.51 \pm 0.01 s/My^{1/2}$ for the slope of the $SS - S$ residuals obtained by *Woodward* [1989] and $-0.68 \pm 0.08 s/My^{1/2}$ by *Sheehan and Solomon* [1991] for the North Atlantic. We calculate the residuals predicted by $SH425.2$ [*Su and Dziewonski*, 1991] for the same source-receiver pairs and treat the synthetic travel times for each age group in the same way as we treat the observed residuals.

Figure 4b shows the best fit straight line and mean residuals for each age group. It has a slope of −0.39 ±

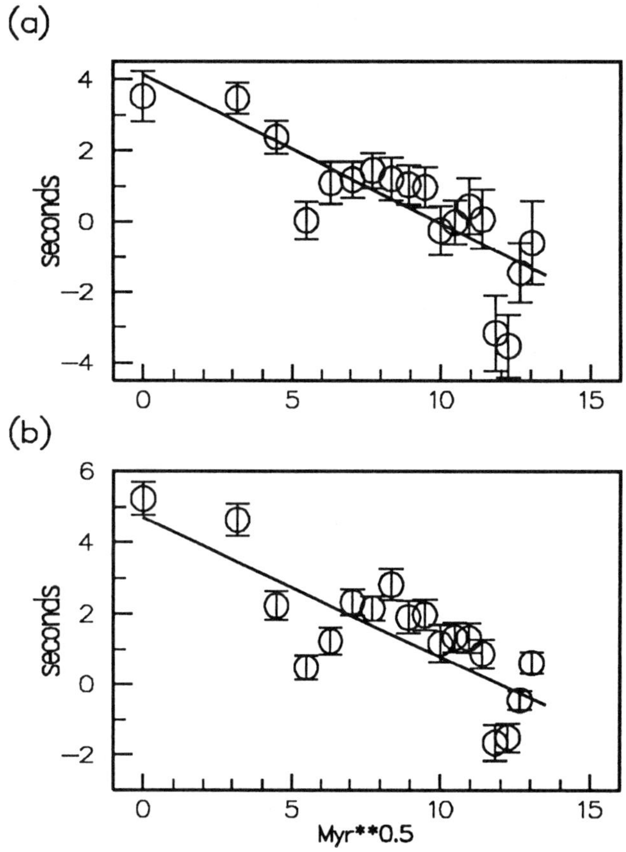

Fig. 4. (a) Relationship of SS residuals to the square root of seafloor age at the mid-path reflection point. Each symbol represents the average of all the residuals in each age bin. The increment of age bin is 10 million years. (b) the same as (a), but for the residuals predicted by model $SH425.2$.

$0.029s/My^{1/2}$, which is very close to the value for observed residuals (−0.41). It is interesting to see that the misfit of the synthetic residuals to the straight line is, point by point, very similar to that for the observations; for example there are large departures at 3, 5.5, 11.8 and 12.2 $My^{\frac{1}{2}}$ both in Figure 4a and 4b. This indicates that the observed SS residuals can be modeled much better than the signal predicted by the cooling of the lithosphere.

With the good correlation between the residuals and age of the ocean floor, one would expect a large variance reduction by age-residual correction. From Figure 4a and 4b one can see that the residuals predicted by age–residual correlation are much less than the observed residuals (Figure 2b). This indicates that the residuals are not only sensitive to the lithospheric structure but also the upper mantle structure beneath the lithosphere. The age-residual relation can only produce ~10% variance reduction [*Woodward*, 1989]. This low variance reduction by the age–residual relation also indicates that cooling of the lithosphere alone can not explain the SS travel-time residuals. The thermal signature of the spreading centers represents only a fraction of the thermal anomalies introduced by convection into the upper mantle.

Some researchers [e. g. *Tanimoto*, 1991; *Zhang and Tanimoto*, 1991; *Zhang*, 1991] report the ℓ^{-1} dependence of rms spectral amplitudes with harmonic degree. In Figure 5a, we show the power of spherical harmonics for SS residuals as a function of harmonic degree (ℓ), using a log-log scale. Note that we plot the total power in a spherical harmonic degree ℓ — see eq. (3) — while *Tanimoto* or *Zhang* plot an average amplitude of all spherical harmonic coefficients with degree ℓ; both representations yield the same result when the dependence is proportional to ℓ^{-1}. The variation of slope with harmonic degree is apparent. For degrees 2-8, the power shows less than ℓ^{-1} decrease (slope is −0.64); but for degrees 9-20, the slope is much steeper (slope: −3.69), indicating a decrease much more rapid then ℓ^{-1}. The slope is even steeper for degrees 21-36, but this could be the effect of smoothing due to the averaging of the residuals over 5°

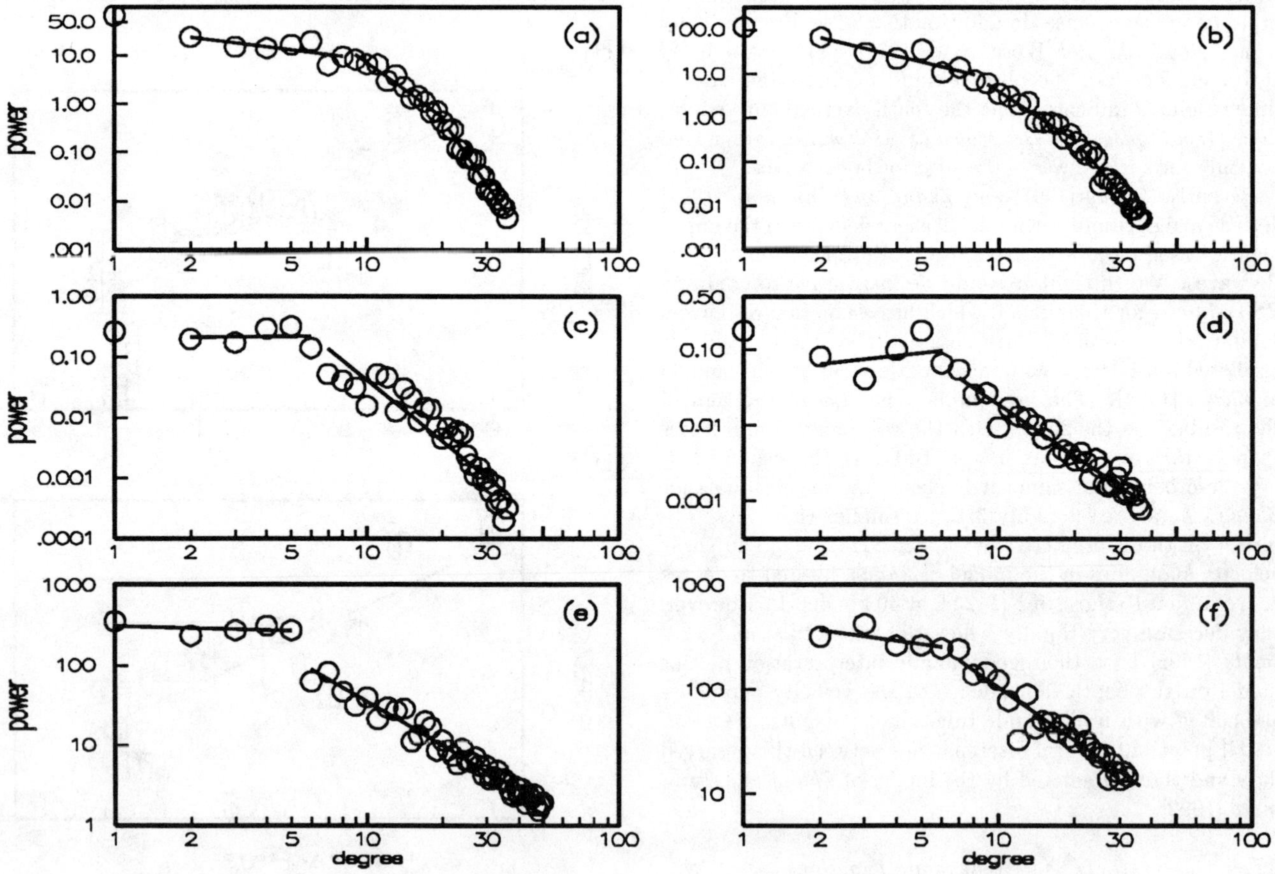

Fig. 5. The power of spherical harmonics as a function of harmonic degree (ℓ) on a log-log scale for (a) the 'damped' solution for observed SS residuals; (b) the 'damped' solution for $SH425.2$ predicted residuals; (c) P-velocity model (Inoue *et al.*), layer 16 (depth 2566km – CMB); (d) phase velocity of model G1 (Zhang and Tanimoto, 1991); (e) ocean function and (f) free-air gravity anomaly.

caps and integration of the anomalies along the ray paths. Figure 5b is the same as Figure 5a, but for the residuals predicted by model $SH425.2$. It has a similar decrease pattern even though the model is truncated at degree 8.

Figure 5c is the log-log plot of the spectrum of P-velocity anomalies in Layer 16 (2566–2890 km) of *Inoue et al.* [1990]. The plot shows a distinct change of slope for $\ell > 6$. There is some increase in the corner degree for the mid-mantle layers: 800 – 2253 km [see Figure 12 in *Su and Dziewonski*, 1992]. Figure 5d is the power spectrum of the local Love wave phase velocity at 100 s adapted from Figure 13 of *Zhang and Tanimoto* [1991]. For degrees 2-8, it shows a decrease less than ℓ^{-1} (slope: -0.38), while for degrees 9-20 and 21-36 the spectrum decreases much more rapidly: the slopes are -2.27 and -2.52, respectively.

The examples given so far are more or less directly related to the lateral heterogeneity of the mantle. In Figure 5e we show the spectrum of the continent–ocean function (continental shelves are counted as continents). Its value, translated in terms of perturbation of crustal thickness, is -11.02 km for the oceans and $+17.53$ km for the continents. The same function used in *Woodhouse and Dziewonski* [1984] and in the current study to obtain corrections for perturbation in crustal thickness. There is a distinct drop in power between $\ell = 5$ and $\ell = 6$ and then relatively rapid decrease (slope: -1.75). We feel that the continent-ocean function better represents the principal pattern of the heterogeneity than topography used by *Tanimoto* [1991]. Even though there is a distinct difference in the topographic function between the oceans and continents, with continents (high) being fast and oceans (low) — slow, the variations in topography within each of the two basic types have the opposite sense: the elevated regions, such as spreading centers or mountains, have lower velocities.

The gravity field is an important geodynamic function, particularly in that it can be related to lateral heterogeneity in seismic velocities [*Dziewonski et al.*, 1977; *Hager et al.*, 1985]. The geoid field is very smooth, but differentiation with respect to radius (and correcting for the distance to the geoid) yields the free air gravity field, and it is this spectrum, shown in Figure 5f, that has a distinct corner beyond $\ell > 6$.

The examples shown involved data sensitive to the earth structure in different ranges of depths, from the surface (Figure 5e), through average properties of the upper mantle (Figure 5a and d), aggregate properties of the entire mantle (Figure 5f) to a layer in the lowermost mantle layer (Figure 5c). It appears therefore that dimensions greater than about 3,000 km (at the surface) are characteristic of the large amplitude lateral variations in the seismic velocity, density and, by inference, the temperature field throughout the mantle. It is perhaps not a coincidence that this scale of 3,000 km is also the thickness of the mantle.

A similar picture of large-scale mantle convection comes from recent numerical experiments. *Solheim and Peltier* [1990] show that in the region of steady or weakly time-dependent thermal convection in spherical shells, the spectrum of mantle heterogeneity is dominated by long-wavelengths. *Gurnis and Zhong* [1991] performed numerical experiments of a convecting mantle with different plate sizes on top of a cylindrical Earth. They showed that even with small plates the system has a strong tendency to produce significant long-wavelength heterogeneity. Significant heterogeneity at the gravest harmonics is a fundamental observational constraint on the dynamics of the mantle convection process.

The conclusion important to the next section of this paper is that the low-order expansion approach to modeling of the heterogeneity is justified by the results of our spectral analysis of SS travel time residuals and other functions. Because of the rapid decrease in the spectral power beyond degree 6, or so, the danger of aliasing of the models through the abrupt truncation of the expansion appears insignificant.

This does not mean that smaller-scale features do not exist or are not important. Figures 6a-c are intended to illustrate this point. Figure 6a is the synthesis of the free air gravity anomaly for harmonics from 2 to 8. This is the same order of expansion as used in our seismic models and the rate of the decrease in the spectral power for $\ell > 6$ is between those for the 'damped' and the 'exact' solutions for the SS residuals. Figure 6b is the synthesis of the free air anomaly for harmonics up to $\ell = 16$ and in Figure 6c includes harmonics up to ℓ=32. The approximate ratio of the number of the coefficients used to obtain these maps is 1:4:16. While the amount of detail visible in these maps is increasing, the underlying importance of the low order harmonics is clear: even though for $2 \leq \ell \leq 8$ they involve less than 7% of the coefficients used in the synthesis of Figure 6c, they account for 65% of the signal energy in it. The overall areal extent of the regions of positive and negative anomalies is primarily defined by the low order field. At the same time, the smaller-scale features provide important information on the regional scale and much more clearly define the correlation of the gravity field with tectonic features.

Furthermore, the experiment of *Su and Dziewonski* [1992], in which the power of the spherical harmonic expansion of velocity anomalies in the Benioff zones does not vary with the harmonic degree, demonstrates that the rapid decrease of the spectral power shown in Figure 5 must level off. In the total rms of the heterogeneity, the contribution of the lithospheric velocity anomalies with the characteristic wavelength of 100 km may be comparable to that of the planetary scale anomalies ($\ell \leq 8$). The difference in power per degree — a factor of 50, say — is offset by the range of the harmonic degrees spanned by the short-wavelength anomalies: from 1 to 8 *vs.* from 10 to 400.

There is no doubt that it is important to extend the resolution of the tomographic analyses beyond the degrees considered in this study. This, however, is going to be difficult because of the number of parameters involved, their low average amplitude and the lesser sensitivity of the observables. At the same time, we can accomplish now the essential task of describing the primary planetary-scale effects, knowledge

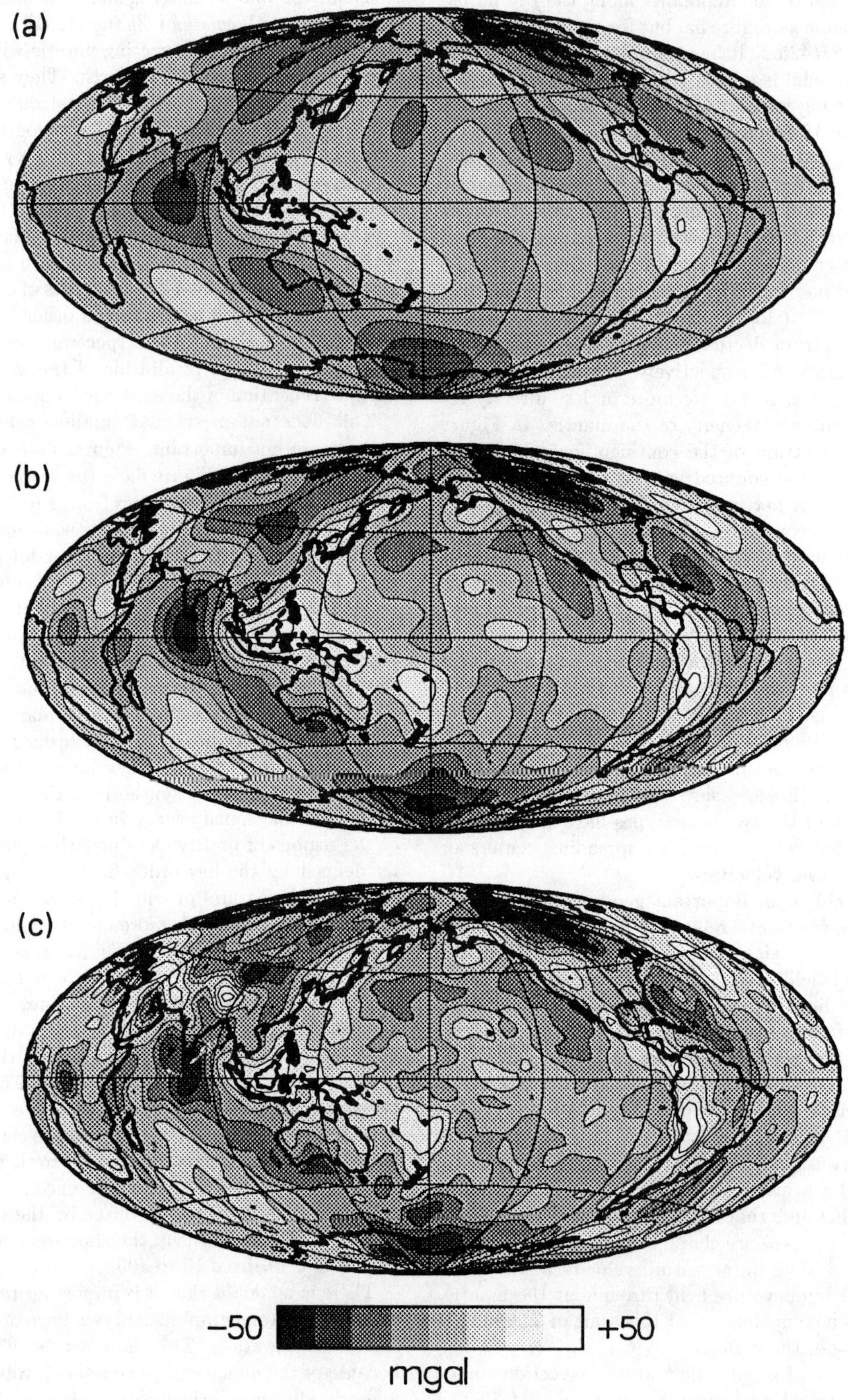

Fig. 6. Synthesized nonhydrostatic GEM-T2 free-air gravity anomaly for (a) degree ℓ=2–8; (b) degree ℓ=2–16; (c) degree ℓ=2–32.

of which is essential to the understanding of the inner workings of the earth.

3-D EARTH MODELS

This section will consider the problem of definition of the model, the theory required to obtain a model from observations, the data set used to obtain the model and the modeling results.

Model Representation

The evidence presented in the previous section provides justification of the approach to the modeling of lateral heterogeneity using low order basis function expansion of the 3-D structure [*Dziewonski*, 1984; *Woodhouse and Dziewonski*, 1984]:

$$\delta v(r,\vartheta,\varphi)/v_0(r) = \sum_{k=0}^{K}\sum_{\ell=0}^{L}\sum_{m=0}^{\ell} f_k(r)\, p_\ell^m(\vartheta)(\,{}_kA_\ell^m \cos m\varphi + {}_kB_\ell^m \sin m\varphi); \quad (3)$$

where δv is the perturbation in the seismic velocity, v_0 is the spherically symmetric reference velocity model, p_ℓ^m is the normalized associated Legendre polynomial defined in equation (1) and $f_k(r)$ is the radial basis function. It is possible to use this expansion to describe velocity perturbations in more than one spherical shell. For example, *Dziewonski and Woodward* [1992] derive separate sets of coefficients ${}_kA_\ell^m$ and ${}_kB_\ell^m$ for the upper and lower mantle without requiring the continuity between the two regions. On the other hand, *Woodward et al.* [1993] use a single set of coefficients to describe the entire mantle. Legendre polynomials were used in the earlier studies to represent the radial basis functions $f_k(r)$; we now find Chebyshev polynomials to have more desirable properties, particularly at the ends of the $(-1,1)$ interval:

$$f_k(r) = \bar{T}_k(x); \quad (4)$$

where $T_k(x)$ is defined in the interval (-1, 1) through the recurrence relationship: $T_{k+1} = 2xT_k - T_{k-1}$ with $T_0 = 1$ and $T_1 = x$ [*Abramowitz and Stegun*, 1965]. $\bar{T}$ is normalized,

$$\bar{T}_k = \left[\frac{(2k)^2-1}{(2k)^2-2}\right]^{1/2} T_k \quad (5)$$

so that

$$\int_{-1}^{1} [\bar{T}_k(x)]^2\, dx = 1. \quad (6)$$

The mapping of the variable r into x is:

$$x = (2r - r_a - r_b)/(r_b - r_a); \quad (7)$$

where r_a and r_b are the radii at the bottom and the top of the shell. For example, *Dziewonski and Woodward* [1992] use $r_a = r_{670}$ and $r_b = r_{Moho}$ for the upper mantle shell and $r_a = r_{CMB}$ and $r_b = r_{670}$ for the lower mantle.

Perturbation theory

As mentioned in the Introduction, we are deriving models using two different types of data: travel times and complete waveforms. The approximations used to establish the relationship between the model and the data is slightly different in each case.

Travel times. The formulae necessary to relate perturbations in travel times to the coefficients of a 3-D model described by equation (3) have been presented by *Dziewonski* [1984]. In the following development it is more concise to use complex notation; thus, the real coefficients ${}_kA_\ell^m$ and ${}_kB_\ell^m$ are replaced by:

$${}_kc_\ell^m = (-1)^m(2\pi)^{1/2}({}_kA_\ell^m - i\,{}_kB_\ell^m); \quad (8)$$

for $m > 0$. For $m = 0$: ${}_kc_\ell^0 = (4\pi)^{1/2}\ {}_kA_\ell^0$ and for $m < 0$: ${}_kc_\ell^m = (2\pi)^{1/2}({}_kA_\ell^{|m|} + i\,{}_kB_\ell^{|m|})$.

The perturbation in travel time is:

$$\delta t = \sum_{k=0}^{K}\sum_{\ell=0}^{L}\sum_{m=-\ell}^{\ell} {}_kc_\ell^m \int_{\mathbf{x}_s}^{\mathbf{x}_r} G(\xi)\, f_k[r(\xi)]\, Y_\ell^m[\vartheta(\xi),\varphi(\xi)]d\xi; \quad (9)$$

where $\mathbf{x}_s$ and $\mathbf{x}_r$ are the coordinates of the source and the receiver, respectively, ξ is the running arc distance along the ray path with the transformation $r = r(\xi)$ determined numerically, $G(\xi)$ is the differential kernel computed for the spherical reference earth model, $f_k(r)$ a radial basis function and Y_ℓ^m a fully normalized spherical harmonic. The integral along the ray path is evaluated numerically. The most efficient approach involves rotation of the coordinate system such that the ray path is in the equatorial plane; details can be found in *Dziewonski* [1984]. In this way, the observed travel time perturbation for the i-th station and the j-th source is expressed as:

$$\delta t_{ij} = \sum_{k,\ell,m} {}_kc_\ell^m\,(\,{}_k\alpha_\ell^m)_{ij}; \quad (10)$$

where $({}_k\alpha_\ell^m)_{ij}$ is the integral along the ray path in (9). This can be used to build the equations of condition as well as to predict path anomalies for a known 3-D structure, described by a set of coefficients ${}_kc_\ell^m$. In case that the observed travel time anomaly, δt, represents a residual for a differential travel time, say, $SS - S$, the kernel $G(\xi)$ is $G_{SS}(\xi) - G_S(\xi)$. The normal equations have the form:

$$\sum_{k'\ell'm'} A_{k\ell m;k'\ell'm'} \cdot {}_{k'}c_{\ell'}^{m'} = b_{k\ell m}$$

The approximation involved in this representation, in addition to the linearization invoking Fermat's principle, is the assumption that the geometrical ray path representation applies to waves with periods as long as 50 seconds. Partly for this reason, most of the applications use data from the distance range in which the rays bottom in the lower mantle. Because the structure there has relatively small gradients, it is believed that the approximation does not lead to substantial errors.

Waveforms. *Dziewonski and Steim* [1982] described a waveform inversion method appropriate for determination of perturbation of average elastic and anelastic structure for a particular great circle. This corresponds to determination of only the even-ℓ part of the 3-D structure. *Woodhouse and Dziewonski* [1984] give a detailed account of their method of derivation of 3-D earth model (both even- and odd-ℓ terms in the spherical harmonic expansion) from inversion of waveform data. This includes both mantle wave data (multi-orbit waveforms) as well as body wave data, representing – most commonly – the minor arc arrivals. Only the first kind of data was used to derive models M84A and M84C; more general application was used later by *Woodhouse and Dziewonski* [1986]. Their approach involves the 'path average approximation': assuming that the effect of 3-D structure can be approximated by the average perturbation over the minor arc or a complete great circle. *Romanowicz* [1987] gives a formal derivation, based on asymptotic properties of normal modes, of the formulae proposed by *Woodhouse and Dziewonski* [1984]. If we use basis functions $f_k(r)$ to describe variations of the model with radius, then the average relative perturbation in velocity along the minor arc between the s-th source and r-th receiver is:

$$\delta\tilde{v}^{sr}(r)/v_0(r) = \sum_{k=0}^{K} \tilde{c}_k^{(sr)} f_k(r); \tag{11}$$

and over a complete great circle:

$$\delta\hat{v}^{(sr)}(r)/v_0(r) = \sum_{k=0}^{K} \hat{c}_k^{(sr)} f_k(r). \tag{12}$$

When the synthetic seismograms and differential seismograms are calculated by superposition of normal modes [*Gilbert and Dziewonski*, 1975; *Dziewonski et al.*, 1981], we may write for the reference model – omitting the instrumental response:

$$u(t)^{sr} = \sum_n \exp(i\omega_n t)\cdot[\mathbf{a}_n^{sr}(\Delta)\cdot\mathbf{f}] \tag{13}$$

where ω_n is the eigenfrequency of the n-th normal mode and it has a small imaginary component representing the attenuation, $\mathbf{a}^{sr}$ is a length six vector of the excitation coefficients for the sr source–receiver path and $\mathbf{f}$ is a length six vector of the seismic moment tensor. The equivalent equation for an aspherical model with the minor arc coefficients $\tilde{c}_k$ and great circle coefficients $\hat{c}_k$ – we omit path identification (sr) for brevity – is:

$$u(t) = \sum_n \exp\left[i(\omega_n + \delta\omega_n^e + \sum_k C_{kn}\hat{c}_k)t\right] \times \left[\mathbf{a}_n[\Delta + \delta\Delta^e + \sum_k D_{kn}(\hat{c}_k - \tilde{c}_k)]\cdot\mathbf{f}\right]; \tag{14}$$

where C_{kn} represent the perturbation in the n-th eigenfrequency corresponding to a unit model perturbation $f_k(r)$, the terms in brackets following Δ (epicentral distance) have the sense of a fictitious shift in the source location for the n-th mode and

$$D_{kn} = \frac{a\Delta}{(\ell+1/2)U_n} C_{kn}; \tag{15}$$

where a is the earth's radius, U_n the group velocity of the n-th mode of angular degree ℓ. The terms with a superscript e represent the effect of ellipticity.

Differential seismograms for unit changes in the $\tilde{c}_k$ and $\hat{c}_k$ can be evaluated by differentiation of equation (14) and the inverse problem formulated by imposing the least squares condition:

$$\int_{t_1}^{t_2}\left[u_{obs} - \left(u_{th}^0 + \sum_k \frac{\partial u}{\partial\tilde{c}_k}\cdot\tilde{c}_k + \frac{\partial u}{\partial\hat{c}_k}\cdot\hat{c}_k\right)\right]^2 dt = \min, \tag{16}$$

where u_{obs} is the observed seismogram and u_{th}^0 – the synthetic seismogram for the reference spherically symmetric earth model; t_1 and t_2 define the time window for a particular waveform. Upon performing the differentiation and integration indicated in (16), we obtain a system of normal equations which consists of two groups of equations:

$$\sum_{k'} \tilde{a}_{kk'}\tilde{c}_{k'} + \bar{a}_{kk'}\hat{c}_{k'} = \sum_{k'}\tilde{b}_{k'};$$
$$\sum_{k'} \breve{a}_{kk'}\tilde{c}_{k'} + \hat{a}_{kk'}\hat{c}_{k'} = \sum_{k'}\hat{b}_{k'}; \tag{17}$$

where we define:

$$\tilde{a}_{kk'}^{sr} = \int_{t_1}^{t_2}\frac{\partial u^{sr}}{\partial\tilde{c}_k}\cdot\frac{\partial u^{sr}}{\partial\tilde{c}_{k'}}dt;\quad \bar{a}_{kk'}^{sr} = \int_{t_1}^{t_2}\frac{\partial u^{sr}}{\partial\tilde{c}_k}\cdot\frac{\partial u^{sr}}{\partial\hat{c}_{k'}}dt;$$
$$\breve{a}_{kk'}^{sr} = \int_{t_1}^{t_2}\frac{\partial u^{sr}}{\partial\tilde{c}_{k'}}\cdot\frac{\partial u^{sr}}{\partial\hat{c}_k}dt;\quad \hat{a}_{kk'}^{sr} = \int_{t_1}^{t_2}\frac{\partial u^{sr}}{\partial\hat{c}_k}\cdot\frac{\partial u^{sr}}{\partial\hat{c}_{k'}}dt; \tag{18}$$
$$\tilde{b}_{k'}^{sr} = \int_{t_1}^{t_2}u_{obs}^{sr}\cdot\frac{\partial u^{sr}}{\partial\tilde{c}_{k'}}dt;\quad \hat{b}_{k'}^{sr} = \int_{t_1}^{t_2}u_{obs}^{sr}\cdot\frac{\partial u^{sr}}{\partial\hat{c}_{k'}}dt;$$

The problem can be highly nonlinear and it is necessary to solve it through iterations, such that after finding $\tilde{c}_k^0$ and $\hat{c}_k^0$ in the initial iteration we substitute their values into equation (14) to obtain u_{th}^1 and then solve (16) for δc_k to obtain $\tilde{c}_k^1$ and $\hat{c}_k^1$ and so on, until convergence is achieved.

Most often more than one seismogram is available for a given source–station pair and equation (16) includes summation over all available components. If both mantle wave data and body wave data are used, care must be taken to establish the appropriate weights. Usually the number of parameters K is too large to obtain a stable solution for an individual path; the small eigenvalues are rejected in forming the generalized inverse. Woodhouse and *Dziewonski* [1984] obtained structures for 870 paths and used the great circle average part of these structures to predict the eigenfrequencies of ${}_0S_{30}$ and ${}_0T_{30}$: see their Figures 1a and 1b.

We see little conceptual difference between this procedure and that proposed by *Nolet* [1990] as 'path integral representation' (which we call 'path average approximation') and 'partitioned inversion', which we avoid, because we believe it does not lead to the optimal model. In 'partitioned inversion' we would express $\tilde{c}_k$ and $\hat{c}_k$ in terms of the spherical harmonic expansion:

$$\tilde{c}_k^{(sr)} = \sum_{\ell,m} {}_k c_\ell^m \, \tilde{y}_{sr\ell m}; \tag{19}$$

where

$$\tilde{y}_{sr\ell m} = \frac{1}{\Delta} \int_{\mathbf{X}_s}^{\mathbf{X}_r} Y_\ell^m \, ds \tag{20}$$

and

$$\hat{c}_k^{(sr)} = \sum_{\ell,m} {}_k c_\ell^m \, \hat{y}_{sr\ell m}; \tag{21}$$

where

$$\hat{y}_{sr\ell m} = \frac{1}{2\pi} \oint_{\mathbf{X}_s,\mathbf{X}_r} Y_\ell^m \, ds. \tag{22}$$

Using $\tilde{c}_k^{(sr)}$ and $\hat{c}_k^{(sr)}$ as data it is possible, in principle, to solve the inverse problem for the coefficients ${}_k c_\ell^m$:

$$\sum_{sr}\sum_{k} \left(\tilde{c}_k^{(sr)} - \sum_{\ell m} {}_k c_\ell^m \, \tilde{y}_{sr\ell m} \right)^2 + \left(\hat{c}_k^{(sr)} - \sum_{\ell m} {}_k c_\ell^m \, \hat{y}_{sr\ell m} \right)^2 = \min. \tag{23}$$

Yet, after *Woodhouse and Dziewonski* [1984] had examined the roughness of their Figures 1a and 1b, they decided that a more reliable global model could be derived by solving for ${}_k c_\ell^m$ directly. This seems impractical, because unlike in the path by path approach, where K (minor arc data only) or $2K$ (minor and great circle data) differential seismograms need to be constructed, one would need to build $K(L+1)^2$ or $2K(L+1)^2$ differential seismograms. For $L = 8$ this represents a nearly two order of magnitude increase in the computational effort. It turns out, however, that this increase can be avoided.

The solution to the problem is presented in equations from (27) to (31) in *Woodhouse and Dziewonski* [1984]. The global inverse problem requires solution of the normal equations:

$$A_{k\ell m;k'\ell'm'} \cdot \delta_{k'} c_{\ell'}^{m'} = b_{k\ell m}; \tag{24}$$

however, by substituting (19) and (21) into (17) and summing over all paths rs we find that:

$$A_{k\ell m;k'\ell'm'} = \sum_{rs} w_{rs} (\tilde{y}_{rs\ell m} \, \tilde{y}_{rs\ell'm'} \, \tilde{a}_{kk'}^{rs} + \tilde{y}_{rs\ell m} \, \hat{y}_{rs\ell'm'} \, \bar{a}_{kk'}^{rs} + \hat{y}_{rs\ell m} \, \tilde{y}_{rs\ell'm'} \, \breve{a}_{kk'}^{rs} + \hat{y}_{rs\ell m} \, \hat{y}_{rs\ell'm'} \, \hat{a}_{kk'}^{rs});$$

$$b_{k\ell m} = \sum_{rs} w_{rs} (\tilde{y}_{rs\ell m} \, \tilde{b}_k + \hat{y}_{rs\ell m} \, \hat{b}_k) \tag{25}$$

where w_{rs} are the weights chosen to equalize appropriately the amplitudes of waveforms from earthquakes spanning a wide range of magnitudes and the matrices a and vectors b are defined in (18). Thus, we can test different orders L of expansion in spherical harmonics without recalculating our synthetic seismograms or differential seismograms. Of course, we still solve the problem iteratively by substituting the initial solution for ${}_k c_\ell^m$ into equations (19) and (21) to obtain $\tilde{c}_k$ and $\hat{c}_k$ and then use equation (14) to calculate the synthetic seismogram for the next iteration. Usually two or three complete iterations are required for the process to converge.

The weakness of the average path approximation is that it is insensitive to the angular position of the anomalies between the source and receiver in the great circle plane. An anomaly can be moved laterally within such a pie slice, but as long as the average remains the same, the synthetic seismogram and differential seismograms will not change. This is satisfactory with regard to surface waves, but can lead to misfits of body wave phases. This is not a fatal flaw, since with a large data set covering different pie slices in the same great circle plane it is possible to resolve the position of the anomalies. *Li and Tanimoto* [1992] proposed a method that deals with this issue, but it requires isolation of a particular phase and, in a way, is an improvement of the travel time approach outlined above.

Inversion and the Resulting Models

The theory presented above allows us to combine in a single inversion the travel time and waveform data. The weights with which these two data sets are combined are chosen empirically so that the contribution of these two types of data is commensurate. *Woodhouse and Dziewonski* [1984, 1986] solved the inverse problem by retaining only the 'significant' eigenvalues [*Gilbert*, 1971]. This approach makes it difficult to influence the process such that the resulting model has 'desirable' characteristics, which we choose on the basis of prior knowledge.

In the current generation of models we choose to stabilize the inverse by requiring, in addition to the reduction of variance of the data misfit, that the resulting model should have, with some weighting factor, a squared gradient which, when integrated over the entire volume of the mantle, is as small as possible. After n iterations we know coefficients ${}_k c_\ell^m$ and seek the incremental perturbation $\delta_k c_\ell^m$:

$$\left[A_{k\ell m;k'\ell'm'} + \eta^2(k,\ell)\mathbf{I}\right] \delta_{k'} c_{\ell'}^{m'} + \eta^2(k,\ell) \, {}_k c_\ell^m = b_{k\ell m} \tag{26}$$

where $\mathbf{I}$ is the unit matrix and $\eta^2(k,\ell) = \gamma_0 + \gamma_\ell \ell(\ell+1) + \gamma_k k^2$. The constants γ_0, γ_ℓ and γ_k are chosen empirically.

All of the data considered here are taken from the long-period recordings of the Global Digital Seismograph Network [*Peterson et al.*, 1976], which consists of several sub-networks, and the International Deployment of Accelerometers [*Agnew et al.*, 1976]. The waveform data set used here was collected some time ago and its preliminary interpretation can be found in *Woodhouse and Dziewonski* [1986, 1989] and *Dziewonski and Woodhouse* [1987]. In addition, model SH425.2 of *Su and Dziewonski* [1991] was also obtained from these data. This data set consists of two fundamental parts.

First, mantle wave records, which are low-passed seismograms (cut-off frequency of 1/135 Hz) of up to 4 1/2 hours duration. These recordings are dominated by the fundamental mode Rayleigh and Love surface waves which make one or more orbits around the globe. Surface waves enable a good mapping of the upper mantle, but do not have sufficient sampling of the lower mantle to resolve features in this region. There are approximately 6,000 mantle wave recordings in this part of the data set.

The second portion of the waveform database consists of roughly 9,000 recordings of body waves. These are low passed (cut-off frequency of 1/45 Hz) recordings of the seismic phases which travel through the interior of the Earth. Aliasing of signal from 'small-scale' structure into the large-scale structure recovered in the global inversion is avoided, or significantly reduced, by using waves with wavelengths of the order of several hundred kilometers. This effectively low-pass filters the short-wavelength structure.

The second major class of data we have used are data sets of $SS - S$ and $ScS - S$ differential travel times observed by *Woodward and Masters* [1991a,b]. These differential times are robust in the sense that they are insensitive to the source origin time and have reduced sensitivity to source mislocation and small-scale structure in the source and receiver regions. It has been shown that the $SS - S$ times are particularly sensitive to upper mantle structure [*Woodward and Masters*, 1991a], while the $ScS - S$ times are dominantly sensitive to structure in the lowermost mantle [*Woodward and Masters*, 1991b]. Approximately 5,400 measurements of $SS - S$ and 2,600 of $ScS - S$ were used.

We present here two models obtained from the same data set but using different radial parameterizations; in both models spherical harmonics up to degree 8 are used to describe variations with the geographical coordinates. In the first case, like in the earlier studies, the upper and lower mantle are expanded in separate sets of radial basis functions. *Dziewonski and Woodward* [1992] used K=4 in the upper mantle and K=8 in the lower mantle. This model, called SH8/U4L8, is listed in Table 1.

The entire mantle is represented by a single set of coefficients in the other model, which utilizes Chebyshev polynomials up to order 13, which *Woodward et al.* [1993] call SH8/WM13; this model is listed in Table 2. Corrections for the thickness of the crust have been considered in the same way as in *Woodhouse and Dziewonski* [1984]. Three complete iterations have been carried out for both models.

Both models provide an equally good fit to the waveform data and the differential travel times. Figure 7a shows 5° cap averages derived from the original $SS-S$ measurements; we find noteworthy the smoothness of the data and the contrast between the residuals for the continental and oceanic areas. Figure 7b is the smoothed version of the $SS - S$ residuals predicted by the model SH8/WM13. The variance reduction for the individual data is 55% (χ^2 = 8877 for 5388 observations) and for the averaged and smoothed data it is 78%. To demonstrate that the differential travel time data provide a powerful discriminant, we show in Figure 7c the smoothed residuals of $SS - S$ predicted for model MDLSH of *Tanimoto* [1990]; here the variance reduction is negative (−43%; this means that after subtracting model predictions from the observed residuals, the variance of these differences is 43% larger from that of the data alone).

Figure 8a shows a smoothed version of the 2605 $ScS - S$ observations which we used in the inversion. Figure 8b shows the $ScS - S$ predictions of the model. The variance reduction for the individual data is 53% (χ^2 = 1694 for 2605 observations). Comparison of the smoothed predictions to the smoothed observations yields a variance reduction of 80%. The χ^2 values indicate that the $SS - S$ data are being fit to within roughly 1.3 standard deviations, while the $ScS - S$ data are being fit to better than 1 standard deviation. Figure 8c are the predictions by model MDLSH of *Tanimoto* [1990]; both the pattern and amplitude of the predicted residuals do not predict observations well: the variance reduction is only 17%.

Variance reductions for the waveform data are roughly 65% or more for the mantle wave data, while the variance reductions provided by PREM [*Dziewonski and Anderson* [1981] alone are on the order of 30%. The top of Figure 9a shows the comparison of an observed mantle wave record to the synthetic predicted by the spherically symmetric model PREM. The bottom of Figure 9a shows the same observed seismogram compared to the synthetic predicted by our model. Figures 9b and 9c show similar comparisons for body wave records. For the body wave data the variance reduction averages roughly 55%, where PREM alone produces about a 28% variance reduction.

Experiments with different combinations of subsets of data and, in particular, investigation of how much the final model changes if either the waveform data or differential travel times are removed, lead us to conclude that these two subsets of data are compatible, despite the different theories employed (see above). Thus, there is empirical evidence that the path average approximation does not lead to systematic errors in the interpretation of the structure from a large set of overlapping paths. This, of course, does not mean that individual waveforms would not be better matched if a more advanced theory were developed.

Figure 10a summarizes for the models $SH8/U4L8$ and $SH8/WM13$ the radial variations of the rms amplitudes of heterogeneities with different angular order number; the rightmost panel shows the rms variations for the entire model; it is compared there with the equivalent function for model $MDLSH$ of *Tanimoto* [1990]. The radial functions are defined as:

$$g_\ell(r) = \left\{ \sum_m \left[\left(\sum_k f_k(r)\ {}_kA_\ell^m \right) \right]^2 + \left[\left(\sum_k f_k(r)\ {}_kB_\ell^m \right) \right]^2 \right\}^{1/2} \tag{27}$$

with the total rms variation being $G(r) = \left[\sum g_\ell^2(r)\right]^{1/2}$.

Table 1. Coefficients of the three-dimensional upper and lower mantle model of shear velocity perturbations, SH8/U4L8.
Velocity perturbations may be synthesized according to eq. (3); for details see text.
For the upper mantle, the model is defined between $r_a = r_{670}$ and $r_b = r_{Moho}$, and for the lower mantle, $r_a = r_{CMB}$ and $r_b = r_{670}$; see eq. (7). Units are $10^3 \times \delta v/v$.

		CRUST		UPPER MANTLE									
				$k=0$		$k=1$		$k=2$		$k=3$		$k=4$	
ℓ	m	A	B	A	B	A	B	A	B	A	B	A	B
0	0	0.00		-0.57		1.80		-0.06		0.10		0.36	
1	0	3.92		3.26		2.64		0.31		0.00		-0.06	
	1	3.04	1.83	4.86	2.39	0.62	0.76	-0.50	0.32	-0.24	-0.02	-0.04	-0.12
2	0	3.16		1.14		2.47		-0.23		-0.36		0.00	
	1	1.35	1.70	-1.00	-0.59	0.62	-1.88	-0.16	-2.19	-0.16	-0.70	0.08	0.25
	2	-1.66	-1.00	2.12	-4.27	1.38	0.16	1.40	0.28	0.35	0.15	-0.04	0.05
3	0	-1.02		0.70		0.18		0.17		0.05		0.01	
	1	-1.08	0.18	1.35	0.61	0.36	-0.10	0.50	-0.42	0.07	-0.16	-0.12	0.02
	2	-2.00	2.79	-0.35	1.68	-0.81	1.73	0.47	0.04	0.19	-0.11	-0.01	0.00
	3	0.54	2.71	0.19	1.38	-1.59	-0.19	-0.48	-0.10	0.09	0.16	0.05	0.02
4	0	2.12		1.76		2.05		0.70		-0.05		-0.17	
	1	-1.03	-0.44	-0.52	0.16	0.34	0.38	0.53	0.34	0.10	0.10	-0.05	-0.03
	2	-2.13	0.53	-2.41	0.44	-1.26	0.41	0.04	-0.56	0.22	-0.36	0.08	0.01
	3	1.55	-0.07	-0.06	0.20	-1.38	0.14	-0.49	0.20	0.03	0.07	0.07	-0.03
	4	-0.61	3.36	-0.21	3.39	0.59	2.92	-0.04	0.37	0.00	-0.28	0.00	-0.20
5	0	-2.73		-2.14		-1.02		0.34		0.15		-0.01	
	1	-0.10	-0.42	1.71	0.28	1.13	-0.02	0.02	0.27	-0.08	0.00	0.01	-0.09
	2	-0.58	-0.86	-1.69	-0.05	-1.66	-0.30	-0.13	-0.13	0.12	-0.04	0.01	0.01
	3	0.35	0.72	-1.11	1.56	-0.17	1.52	0.11	0.14	0.09	-0.05	0.01	0.03
	4	3.21	-0.90	3.32	-2.15	2.26	-1.77	-0.05	-0.11	-0.21	0.14	-0.04	0.03
	5	0.07	1.11	1.24	0.97	0.66	0.26	0.03	0.24	0.02	0.01	-0.01	-0.04
6	0	1.24		0.65		0.95		0.25		0.02		-0.05	
	1	-0.08	-0.32	-0.78	-0.83	-0.50	-0.84	-0.05	0.11	0.06	0.19	0.05	0.04
	2	-0.38	-0.11	-1.33	0.42	-1.35	1.13	-0.55	0.47	0.01	-0.04	0.13	-0.08
	3	0.20	1.16	0.07	2.19	0.20	1.00	0.11	-0.34	0.09	-0.18	0.03	0.00
	4	0.72	-0.54	0.15	0.92	0.90	0.34	0.48	0.10	0.00	0.07	-0.12	-0.01
	5	-0.29	-0.79	1.40	0.29	1.62	0.84	0.43	0.11	-0.13	-0.07	-0.11	0.03
	6	0.23	0.43	0.25	0.72	-0.40	0.43	-0.27	-0.10	-0.11	-0.05	0.00	0.00
7	0	-1.15		-1.03		-0.44		0.08		0.06		0.00	
	1	-0.11	0.90	-0.32	1.17	-0.22	0.53	-0.04	-0.10	-0.03	-0.03	-0.01	-0.03
	2	0.62	0.36	0.04	1.16	0.05	0.76	-0.15	-0.01	-0.08	-0.09	0.02	-0.04
	3	-0.52	0.42	-0.30	0.78	-0.63	0.40	-0.06	-0.16	0.12	-0.02	0.02	0.05
	4	-1.09	0.11	0.36	-0.13	-0.40	0.63	-0.08	0.17	0.09	0.05	0.02	0.05
	5	0.43	-0.35	0.47	-0.98	0.36	-0.58	0.07	-0.27	-0.07	-0.02	-0.01	0.04
	6	-0.21	-1.11	-0.23	-1.09	-0.18	-0.68	-0.04	0.09	-0.01	0.13	0.03	0.02
	7	-0.31	-0.74	-0.68	0.02	-0.32	-0.03	0.06	-0.03	0.04	0.00	0.02	-0.01
8	0	-0.72		-1.11		-0.91		-0.03		0.08		0.05	
	1	-0.07	-0.30	-0.70	-0.40	-0.53	-0.51	-0.14	-0.11	-0.03	0.00	0.06	0.05
	2	0.27	0.11	-0.75	0.16	-0.73	0.11	-0.14	-0.04	0.06	0.00	0.07	0.00
	3	0.17	0.32	0.36	0.62	0.70	0.57	0.31	0.05	0.05	0.00	-0.05	0.04
	4	0.05	0.84	-0.24	1.03	0.40	0.24	0.43	-0.16	0.04	-0.01	-0.04	-0.04
	5	-0.41	-0.14	-0.76	-0.20	-0.41	-0.09	0.12	-0.02	-0.02	0.01	0.00	0.00
	6	0.57	0.57	-0.34	0.61	-0.09	-0.34	0.12	-0.27	-0.04	-0.07	-0.04	-0.01
	7	0.99	0.38	0.77	-0.37	-0.32	0.04	-0.66	-0.01	-0.02	0.01	0.14	0.03
	8	-0.22	-0.39	0.23	-0.12	-0.09	-0.34	-0.12	-0.22	0.03	0.04	0.02	0.07

		LOWER MANTLE																	
		$k=0$		$k=1$		$k=2$		$k=3$		$k=4$		$k=5$		$k=6$		$k=7$		$k=8$	
ℓ	m	A	B	A	B	A	B	A	B	A	B	A	B	A	B	A	B	A	B
0	0	-0.49		0.50		1.16		0.66		0.60		0.56		0.42		0.26		0.17	
1	0	0.39		-0.46		-0.08		-0.43		0.01		0.03		-0.03		0.04		0.08	
	1	-0.97	0.37	0.12	0.06	-0.22	0.68	-0.57	0.05	0.22	0.26	-0.09	0.08	-0.28	-0.01	-0.12	0.04	-0.07	-0.09
2	0	0.26		-1.18		0.08		0.09		-0.03		-0.22		-0.21		-0.04		0.02	
	1	0.22	-0.14	0.04	-0.20	0.28	-0.38	-0.03	-0.30	-0.01	-0.13	-0.04	-0.03	-0.03	0.04	0.09	0.02	0.04	-0.05
	2	-2.23	-1.63	2.00	0.59	0.19	0.15	0.05	0.11	0.16	0.02	0.27	0.03	0.23	0.06	-0.02	0.08	-0.15	0.02
3	0	-0.84		0.63		0.09		-0.04		0.11		0.01		-0.03		0.00		-0.02	
	1	-0.75	-0.56	0.90	-0.17	0.09	-0.13	0.27	-0.17	0.06	0.01	0.10	0.21	0.07	0.02	-0.03	-0.21	-0.02	-0.10
	2	0.36	1.45	0.04	-0.33	0.35	0.10	0.60	-0.37	0.10	-0.20	0.07	0.05	0.02	0.06	0.08	0.15	0.05	-0.11
	3	-0.56	0.08	0.07	0.51	0.15	0.22	-0.10	-0.13	-0.03	0.18	0.19	-0.12	0.06	-0.01	0.10	-0.08	0.14	-0.08
4	0	-0.34		0.42		-0.09		-0.19		-0.06		0.01		-0.12		-0.02		0.00	
	1	0.03	-0.07	-0.08	-0.37	-0.10	-0.06	0.31	0.34	0.06	0.01	0.26	-0.07	0.13	0.08	0.03	-0.04	-0.01	-0.02
	2	0.45	-0.52	-0.45	-0.45	0.24	-0.25	-0.21	-0.34	-0.23	0.28	0.03	0.18	0.06	-0.03	0.12	0.07	0.13	0.08
	3	0.18	0.72	-0.19	0.04	0.55	0.37	0.07	0.04	-0.07	0.14	0.19	0.06	0.05	-0.07	-0.03	-0.02	0.00	-0.08
	4	0.02	0.78	0.33	0.12	-0.56	0.52	-0.10	-0.01	-0.08	-0.49	-0.12	-0.06	0.22	-0.04	-0.01	0.12	-0.05	0.07
5	0	0.04		-0.04		0.05		0.12		0.16		0.09		-0.12		0.11		0.05	
	1	0.80	-0.11	-0.34	0.18	0.47	0.26	0.43	0.33	-0.42	0.04	0.02	-0.05	0.10	0.08	-0.02	0.04	-0.05	0.02
	2	-0.03	-0.30	0.11	-0.18	0.29	0.05	-0.06	-0.35	0.11	-0.11	-0.08	0.17	-0.05	0.21	-0.02	0.20	0.03	0.04
	3	-0.44	0.81	0.06	-0.02	-0.23	0.28	-0.08	-0.06	-0.11	-0.24	0.03	-0.06	0.08	-0.13	-0.08	0.15	-0.02	0.02
	4	0.83	0.82	0.12	-0.16	-0.02	-0.37	0.11	0.24	-0.23	-0.28	-0.05	0.00	0.15	0.06	-0.10	-0.14	-0.13	0.03
	5	-0.14	0.15	0.07	0.26	-0.15	-0.32	-0.14	0.10	-0.01	0.12	-0.03	0.05	-0.06	0.19	-0.09	-0.07	-0.01	-0.18
6	0	0.00		-0.12		-0.10		-0.02		0.01		0.10		0.07		0.04		-0.11	
	1	-0.08	-0.28	0.02	0.25	0.23	-0.04	0.20	-0.09	-0.09	0.10	-0.16	0.04	0.14	0.05	0.04	0.00	-0.07	-0.01
	2	-0.15	0.33	-0.21	-0.18	0.30	-0.12	0.04	0.03	-0.02	-0.14	-0.21	-0.07	-0.06	0.02	0.13	0.00	0.03	-0.02
	3	0.19	-0.17	-0.04	0.24	0.06	-0.29	0.04	-0.08	0.02	0.14	0.03	0.11	-0.03	0.05	-0.01	-0.15	-0.09	-0.03
	4	0.34	-0.17	0.09	-0.34	-0.08	0.11	-0.02	0.28	-0.07	-0.04	-0.07	-0.06	-0.05	0.05	-0.02	0.03	0.00	-0.01
	5	-0.22	-0.24	0.42	-0.33	0.22	-0.10	0.00	-0.18	0.01	-0.07	-0.02	0.02	0.00	0.19	-0.06	0.12	-0.06	-0.05
	6	0.16	-0.16	0.08	0.23	-0.02	0.06	0.08	-0.10	-0.01	-0.15	-0.09	-0.02	0.04	0.00	0.07	-0.03	-0.01	0.10
7	0	-0.18		-0.08		0.00		0.09		0.04		-0.08		-0.02		0.00		-0.04	
	1	-0.35	0.35	0.11	0.14	0.09	-0.11	0.00	-0.04	0.08	-0.05	-0.10	0.00	0.11	0.10	0.08	0.02	-0.04	-0.09
	2	-0.25	0.14	0.16	0.13	0.13	-0.09	0.23	0.13	0.08	-0.04	-0.19	-0.10	-0.09	-0.04	-0.07	-0.03	0.04	-0.02
	3	0.00	-0.10	0.18	0.14	0.39	-0.21	-0.02	0.08	-0.16	0.08	0.04	-0.11	0.03	0.02	0.03	-0.02	-0.06	0.07
	4	0.18	0.52	0.00	-0.28	0.07	-0.12	0.22	-0.09	0.10	-0.20	0.03	0.04	-0.13	-0.03	-0.19	0.12	-0.11	0.02
	5	-0.42	-0.82	0.23	-0.42	-0.06	-0.40	-0.04	-0.18	0.32	0.21	0.07	0.08	0.06	0.14	-0.02	0.07	-0.12	0.04
	6	-0.14	0.15	-0.17	-0.47	-0.16	-0.22	-0.13	0.10	0.02	-0.26	-0.03	0.02	-0.02	0.14	0.06	0.12	0.10	0.15
	7	-0.86	-0.17	0.22	0.01	0.00	-0.01	-0.05	0.06	0.20	0.11	-0.10	-0.08	-0.07	-0.08	-0.02	0.07	0.08	0.00
8	0	-0.04		0.16		-0.15		-0.32		0.03		0.09		0.10		0.08		0.02	
	1	0.14	0.43	0.30	0.10	-0.18	0.05	0.00	0.14	0.18	-0.03	-0.22	0.04	-0.16	0.06	0.01	-0.04	0.01	-0.12
	2	-0.06	0.24	-0.19	-0.01	-0.05	-0.18	-0.06	0.14	0.03	0.03	0.18	-0.08	-0.01	-0.04	0.01	-0.09	-0.01	0.00
	3	0.21	0.45	-0.06	0.29	-0.03	0.13	0.20	0.04	0.07	-0.01	-0.01	-0.05	-0.10	-0.12	0.00	-0.08	-0.04	-0.05
	4	0.04	0.01	-0.26	0.29	-0.27	0.13	0.13	-0.07	0.13	0.04	0.05	-0.02	0.04	0.02	-0.13	0.09	-0.05	-0.05
	5	-0.11	0.13	-0.10	-0.17	0.22	0.10	0.16	0.05	-0.01	-0.04	-0.01	-0.04	0.09	-0.01	0.04	-0.02	-0.11	-0.01
	6	-0.21	0.20	0.25	-0.01	-0.36	0.36	0.03	0.32	0.31	-0.16	-0.05	-0.01	-0.06	-0.07	-0.03	-0.14	0.09	0.01
	7	0.06	0.16	-0.17	-0.26	0.15	0.00	0.15	0.19	-0.09	0.03	0.02	-0.17	0.01	-0.09	0.02	0.06	0.00	-0.04
	8	0.13	0.32	-0.24	-0.21	-0.17	-0.02	-0.02	-0.08	0.28	-0.16	0.14	-0.05	-0.08	0.04	0.10	-0.01	-0.02	0.02

Table 2. Coefficients of the three-dimensional whole-mantle model SH8/WM13. See Table 1 for the crustal coefficients and conventions.

		$k=0$		$k=1$		$k=2$		$k=3$		$k=4$		$k=5$		$k=6$	
ℓ	m	A	B	A	B	A	B	A	B	A	B	A	B	A	B
0	0	-0.66		0.62		0.80		0.41		0.22		0.27		0.13	
1	0	1.91		1.27		1.57		1.04		0.72		0.12		-0.18	
	1	0.90	1.04	2.07	0.81	1.00	0.98	0.13	0.18	-0.14	0.18	-0.78	0.04	-0.53	0.08
2	0	0.97		0.28		1.40		0.82		0.24		-0.17		-0.31	
	1	-0.09	-0.56	-0.13	-0.65	0.17	-1.13	0.04	-1.11	0.26	-1.10	0.16	-1.09	-0.02	-0.83
	2	-1.23	-2.62	2.87	-0.31	0.43	-0.56	0.69	0.04	0.73	0.45	0.64	0.50	0.45	0.19
3	0	-0.35		0.79		0.22		0.08		0.14		0.07		0.07	
	1	-0.27	-0.15	1.22	0.23	0.25	0.14	0.30	-0.10	0.12	-0.27	0.21	-0.36	0.16	-0.37
	2	-0.02	2.01	-0.45	0.23	-0.09	0.83	-0.09	0.43	-0.25	0.49	0.21	0.23	0.37	-0.15
	3	-0.62	0.25	-0.08	0.72	-0.40	-0.03	-0.72	-0.25	-0.58	-0.05	-0.38	-0.17	-0.22	0.08
4	0	0.67		1.22		0.87		0.81		0.67		0.38		0.01	
	1	-0.01	0.13	-0.05	-0.01	0.19	0.40	0.58	0.50	0.26	0.01	0.31	0.05	0.13	0.14
	2	-0.45	-0.04	-1.35	0.25	-0.37	0.32	-0.51	0.18	0.05	0.09	0.28	-0.49	0.15	-0.49
	3	-0.06	0.56	-0.59	-0.17	-0.11	0.26	-0.82	-0.08	-0.57	0.15	-0.18	0.14	-0.10	0.13
	4	0.05	1.92	0.36	1.54	-0.29	1.62	0.44	0.74	0.31	0.45	0.12	0.52	-0.01	-0.04
5	0	-0.84		-0.97		-0.73		-0.46		-0.06		0.12		0.22	
	1	1.37	-0.02	0.34	0.17	1.11	0.18	0.41	0.10	-0.14	-0.09	0.32	0.12	0.05	0.21
	2	-0.89	-0.22	-0.94	-0.11	-0.84	-0.01	-0.92	-0.33	-0.30	-0.02	-0.15	-0.09	0.12	-0.21
	3	-0.75	1.39	-0.14	0.64	-0.36	0.92	0.03	0.45	0.07	0.43	0.10	0.39	0.03	0.04
	4	1.91	-0.15	1.47	-1.67	1.20	-1.34	0.97	-0.62	0.35	-0.60	0.22	0.01	-0.10	0.07
	5	0.45	0.45	0.75	0.59	0.42	0.04	0.35	0.44	0.24	0.18	0.02	-0.01	-0.05	0.06
6	0	0.37		0.41		0.46		0.47		0.29		0.14		0.03	
	1	-0.34	-0.67	-0.31	-0.13	-0.13	-0.43	-0.19	-0.26	-0.19	0.02	0.07	0.02	0.17	0.11
	2	-0.67	0.64	-0.85	0.25	-0.34	0.53	-0.69	0.60	-0.44	0.40	-0.31	0.31	-0.01	0.09
	3	0.21	0.67	-0.08	1.07	0.04	0.36	0.01	0.37	0.03	0.02	0.09	-0.28	0.07	-0.29
	4	0.46	0.28	0.22	0.27	0.28	0.64	0.45	0.33	0.40	-0.17	0.32	0.01	0.14	0.13
	5	0.41	0.18	1.04	0.14	0.67	0.46	0.46	0.32	0.39	0.23	0.24	0.03	0.06	-0.10
	6	0.11	0.08	-0.15	0.58	-0.28	0.21	-0.28	0.08	-0.34	0.04	-0.23	0.05	-0.07	-0.07
7	0	-0.60		-0.44		-0.35		-0.23		-0.16		-0.05		0.13	
	1	-0.51	0.72	0.03	0.58	-0.15	0.26	-0.14	0.30	-0.09	0.12	-0.13	0.01	0.08	-0.09
	2	-0.27	0.58	0.22	0.72	0.04	0.46	-0.02	0.50	-0.14	0.14	-0.12	0.07	0.05	0.01
	3	-0.36	0.24	-0.26	0.53	-0.24	0.13	-0.58	0.32	-0.28	0.00	0.09	-0.16	0.10	-0.04
	4	0.13	0.58	-0.08	-0.28	-0.10	0.18	-0.13	0.33	-0.29	0.31	-0.15	0.29	0.07	0.00
	5	-0.21	-0.98	0.56	-0.40	0.20	-0.35	0.26	-0.10	0.23	-0.13	-0.14	-0.46	-0.02	-0.27
	6	-0.17	-0.17	-0.05	-0.95	-0.06	-0.38	0.05	-0.09	0.05	-0.20	-0.10	0.13	-0.05	0.03
	7	-1.10	-0.09	0.11	-0.01	-0.34	-0.08	-0.18	-0.02	-0.06	-0.05	-0.16	-0.09	0.09	0.07
8	0	-0.55		-0.47		-0.70		-0.45		0.01		-0.08		0.00	
	1	-0.21	0.14	-0.26	-0.35	-0.63	-0.29	-0.18	-0.17	-0.04	-0.16	-0.17	0.04	0.09	0.03
	2	-0.30	0.27	-0.60	-0.02	-0.36	-0.08	-0.37	0.17	-0.21	-0.07	-0.14	-0.09	-0.15	0.03
	3	0.42	0.52	0.21	0.23	0.31	0.06	0.48	-0.02	0.24	0.04	0.18	0.09	0.16	0.10
	4	0.19	0.29	-0.12	0.60	0.11	0.21	0.55	-0.05	0.25	0.02	0.11	-0.05	0.12	-0.01
	5	-0.38	0.13	-0.41	-0.18	-0.06	0.14	-0.15	0.02	-0.12	-0.01	0.05	0.04	0.08	0.02
	6	-0.34	0.26	0.09	-0.06	-0.42	0.17	0.18	-0.28	0.14	-0.53	-0.15	-0.06	0.04	-0.03
	7	0.31	0.13	0.05	-0.33	0.18	0.15	-0.22	0.19	-0.50	-0.01	-0.29	-0.03	-0.18	0.09
	8	0.25	0.29	-0.17	-0.36	-0.03	-0.09	0.01	-0.16	-0.01	-0.15	-0.26	-0.05	-0.11	-0.08

		$k=7$		$k=8$		$k=9$		$k=10$		$k=11$		$k=12$		$k=13$	
ℓ	m	A	B	A	B	A	B	A	B	A	B	A	B	A	B
0	0	0.01		0.05		0.03		0.11		0.05		-0.01		-0.03	
1	0	-0.12		-0.04		-0.02		-0.01		-0.07		-0.09		-0.03	
	1	-0.07	0.06	-0.01	-0.06	0.02	-0.07	0.00	-0.13	0.05	-0.08	0.00	-0.03	-0.03	0.03
2	0	-0.31		-0.23		-0.03		0.13		0.11		0.04		0.00	
	1	-0.15	-0.46	-0.21	-0.06	-0.04	0.19	0.05	0.23	0.09	0.18	0.03	0.13	-0.03	0.10
	2	0.29	-0.05	0.20	-0.09	-0.02	-0.04	-0.18	0.06	-0.13	-0.02	0.02	-0.02	0.01	-0.02
3	0	0.12		0.03		0.01		-0.05		-0.03		0.03		-0.02	
	1	0.10	-0.11	0.02	0.12	-0.12	0.04	-0.14	-0.02	-0.09	0.05	0.00	0.05	0.01	0.03
	2	0.27	-0.23	0.08	-0.16	-0.03	0.07	-0.03	0.06	-0.03	0.00	0.01	0.01	0.03	0.01
	3	0.11	0.10	0.18	0.09	0.10	0.06	0.06	-0.06	-0.04	0.02	-0.07	0.00	0.00	-0.04
4	0	0.00		-0.11		-0.08		-0.05		-0.05		0.00		-0.04	
	1	0.06	0.02	0.04	0.05	-0.07	-0.01	-0.09	-0.05	-0.09	0.00	-0.01	0.00	0.02	-0.01
	2	0.10	-0.20	0.05	-0.11	0.05	0.00	0.04	0.06	-0.01	0.04	-0.08	0.02	-0.04	0.06
	3	0.10	0.12	0.14	-0.02	0.04	-0.04	-0.02	-0.07	0.00	0.00	0.01	-0.01	0.01	-0.03
	4	-0.25	-0.27	-0.03	-0.29	0.04	-0.13	0.02	0.02	0.00	-0.02	-0.03	-0.04	-0.01	0.00
5	0	0.27		-0.02		-0.06		-0.01		-0.05		0.00		-0.01	
	1	-0.14	0.03	-0.04	-0.05	0.00	-0.07	-0.02	-0.06	0.02	-0.02	0.05	0.00	0.02	0.04
	2	0.13	-0.08	0.01	0.04	-0.03	0.13	0.00	0.04	0.02	-0.06	0.00	-0.04	0.02	0.03
	3	0.00	-0.06	0.10	-0.18	0.05	-0.02	-0.03	0.06	-0.01	0.02	-0.01	0.00	-0.05	-0.04
	4	-0.31	0.05	-0.08	0.15	0.00	0.02	-0.03	0.01	0.02	0.01	0.02	0.01	-0.01	-0.02
	5	0.01	-0.01	0.03	0.07	0.01	0.03	-0.04	-0.12	-0.02	-0.07	0.00	0.05	-0.05	0.04
6	0	0.00		-0.02		0.02		-0.06		-0.07		0.02		0.01	
	1	-0.07	0.10	-0.01	0.09	0.05	0.04	-0.02	-0.03	0.01	-0.01	0.01	-0.01	0.03	-0.01
	2	0.02	-0.07	-0.07	-0.09	0.08	-0.06	0.10	-0.04	0.06	-0.04	0.01	0.00	-0.01	0.00
	3	0.07	-0.11	0.04	0.09	0.04	0.00	-0.02	-0.02	0.00	0.04	0.00	0.04	-0.04	0.00
	4	-0.04	0.08	-0.10	0.07	-0.09	0.03	-0.07	-0.02	-0.04	-0.05	-0.02	-0.05	-0.02	-0.02
	5	-0.07	-0.16	-0.11	-0.02	-0.10	0.12	-0.07	0.03	0.00	0.01	0.04	0.00	0.00	0.01
	6	-0.06	-0.07	-0.03	0.00	0.04	-0.01	0.06	0.07	0.03	0.01	0.01	-0.03	0.05	-0.01
7	0	0.03		-0.01		0.01		-0.01		0.00		0.03		0.02	
	1	-0.05	-0.14	-0.05	-0.01	0.03	0.05	0.01	-0.02	-0.02	-0.03	0.01	0.00	0.05	0.02
	2	-0.03	-0.08	-0.07	-0.08	-0.06	-0.03	0.04	0.00	0.05	0.02	0.02	0.03	-0.02	-0.03
	3	0.08	-0.02	0.06	0.06	0.04	0.02	-0.01	0.02	-0.05	0.03	-0.02	-0.03	0.01	-0.04
	4	0.19	-0.03	0.12	-0.06	-0.03	0.06	-0.06	0.06	0.02	-0.01	0.07	-0.04	0.01	-0.06
	5	0.01	-0.08	-0.01	0.09	-0.02	0.12	-0.04	0.07	-0.01	0.01	0.05	0.00	0.07	0.02
	6	-0.01	0.00	0.00	0.07	0.03	0.05	0.04	0.06	0.03	-0.05	-0.02	-0.06	-0.02	-0.02
	7	0.10	0.07	0.05	-0.05	-0.03	-0.01	0.00	0.00	-0.01	-0.02	0.00	0.01	-0.02	0.01
8	0	0.01		0.04		0.06		0.01		-0.01		-0.01		0.03	
	1	0.01	-0.05	-0.10	0.01	-0.02	0.01	0.08	-0.01	0.05	0.00	0.03	0.05	0.02	0.04
	2	0.09	0.01	0.08	0.01	0.06	-0.02	0.04	0.02	0.00	0.05	0.03	0.00	0.01	-0.01
	3	0.12	0.08	-0.04	-0.01	-0.07	-0.02	-0.07	0.01	-0.06	0.04	-0.03	0.02	-0.02	-0.02
	4	0.04	-0.02	0.05	-0.05	-0.07	0.02	-0.12	-0.02	0.00	-0.06	0.04	-0.02	-0.02	0.03
	5	-0.02	-0.01	-0.01	0.00	0.02	-0.01	-0.03	-0.02	-0.03	0.00	0.04	-0.03	0.06	-0.01
	6	0.05	0.00	-0.03	0.01	-0.10	-0.08	-0.02	-0.02	0.02	0.05	0.01	0.07	-0.01	0.02
	7	-0.01	0.00	0.08	-0.10	0.10	0.02	0.09	0.03	0.02	0.03	0.01	0.01	0.00	-0.01
	8	0.16	-0.08	0.04	0.04	0.05	0.08	0.00	0.02	-0.02	0.07	0.00	-0.01	0.01	-0.04

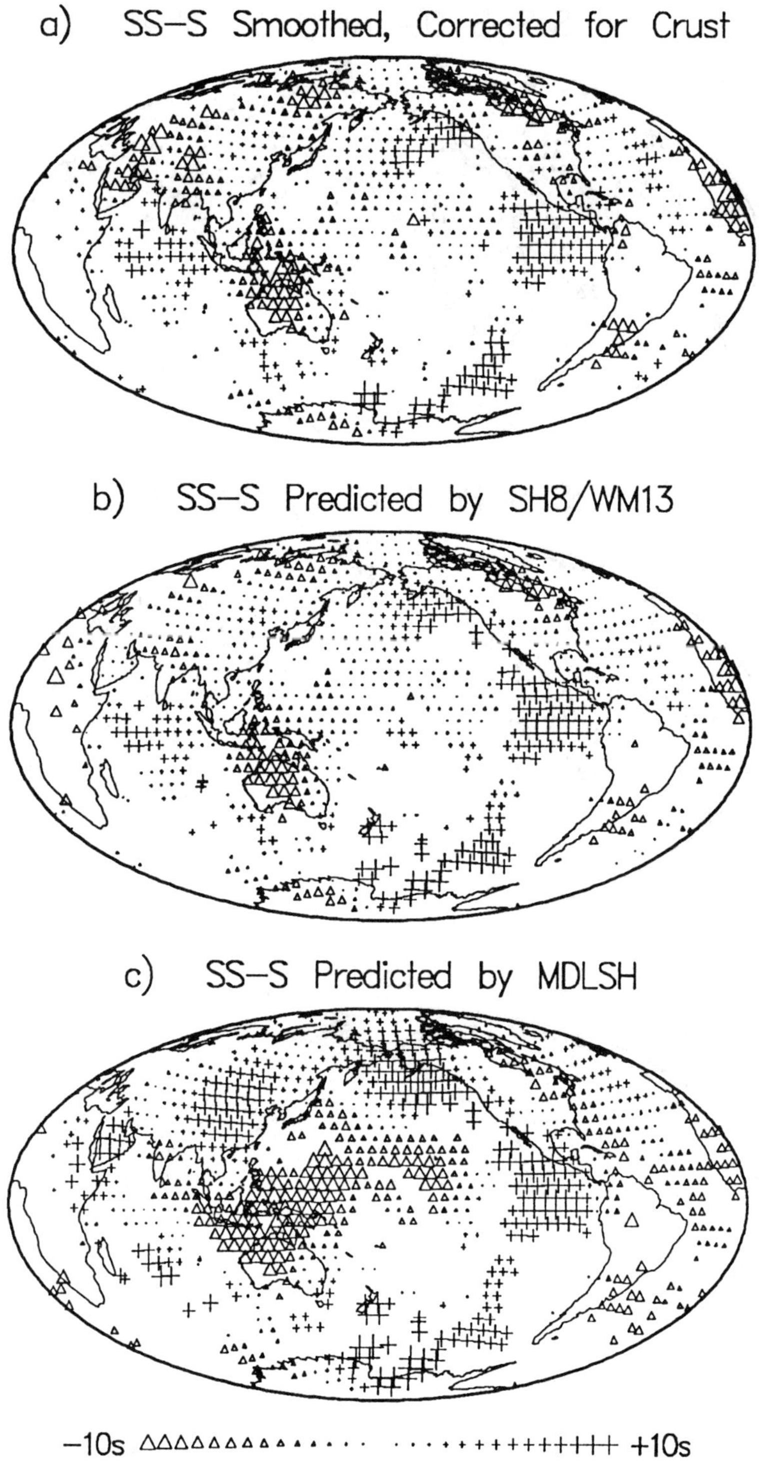

Fig. 7. (a) Observed $SS - S$ differential travel times averaged in 5° spherical caps compared with (b) the predictions of model SH8/WM13 derived in this study and (c) the predictions of model MDLSH of Tanimoto (1990). Predictions of the model SH8/WM13 remove, roughly, 80% of the variance from these cap averages, while the predictions of MDLSH actually increase the variance by 43%. Notice that a very long-wavelength pattern predominates both observations and model predictions.

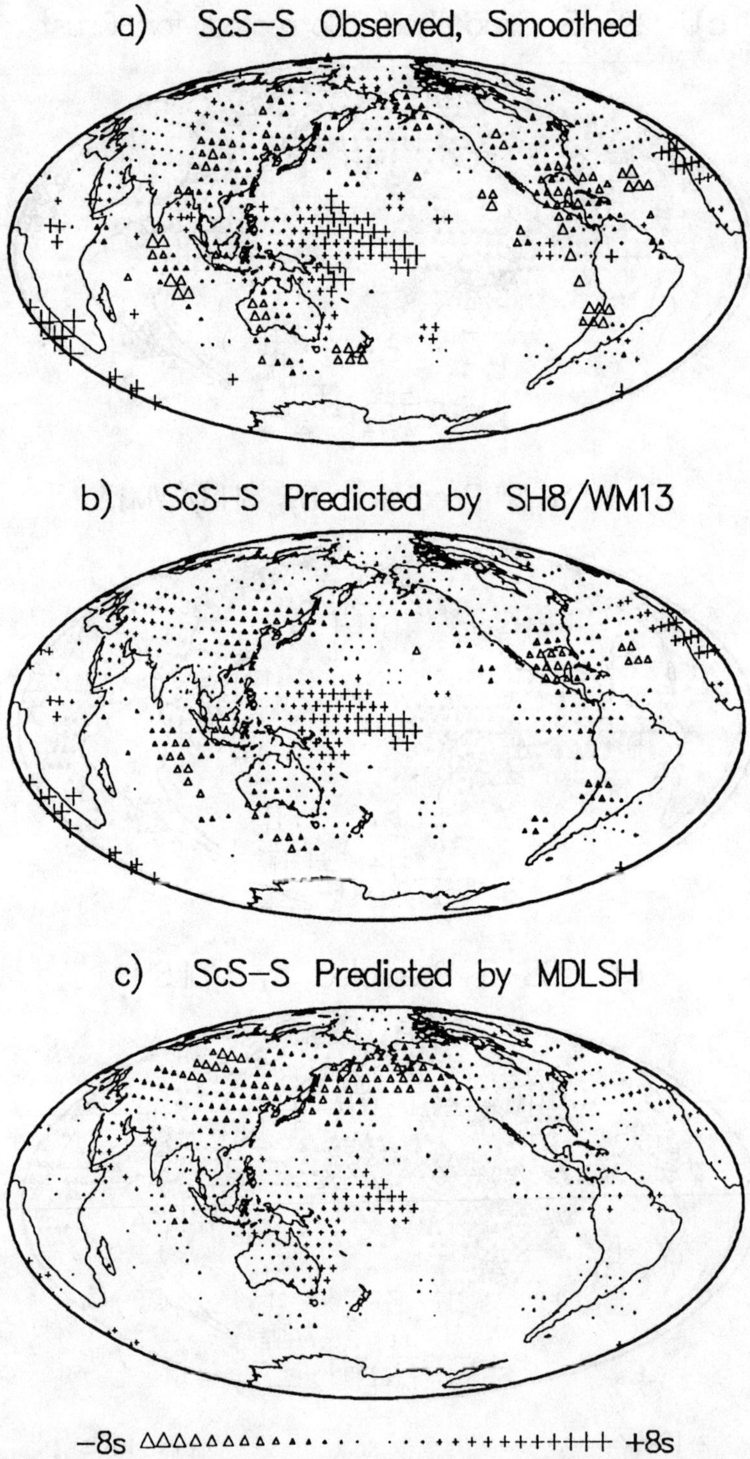

Fig. 8. Same as figure 7, but for $ScS-S$ observations. (a) Observed $ScS-S$ differential travel times averaged in 5° spherical caps compared with (b) the predictions of model SH8/WM13 derived in this study and (c) the predictions of model MDLSH of Tanimoto (1990). Predictions of the model SH8/WM13 remove, roughly, 80% of the variance from these cap averages, while the predictions of MDLSH actually decrease the variance by 17%. Notice that a very long-wavelength pattern predominates both observations and model predictions.

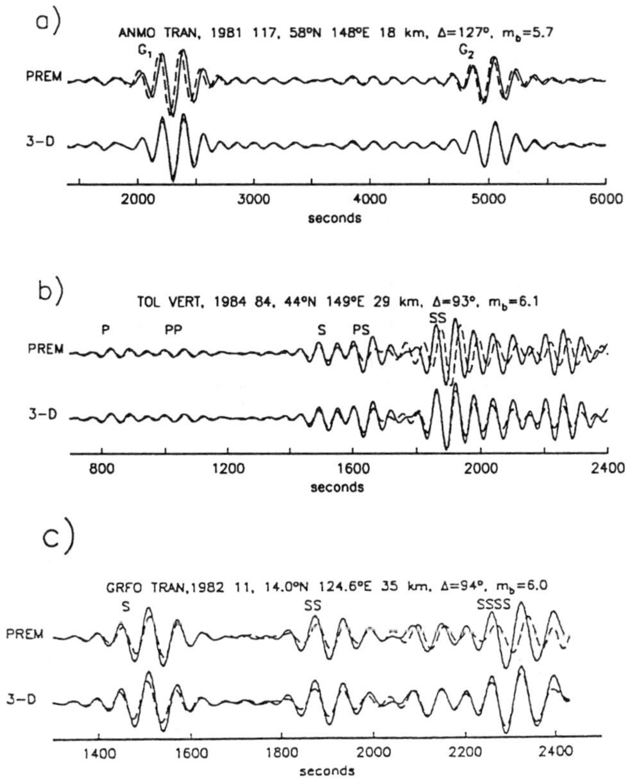

Fig. 9. Comparison of observed waveforms (solid line) with the synthetic seismograms (broken line) for a spherically symmetric model PREM (top) and a 3-D model (bottom). a) Mantle waves (Love waves) with periods in excess of 135 s; b) and c) body waves, identified by the appropriate code, with periods longer than 45 s.

Figure 10b shows radial variations of the cross-correlation coefficients between models SH8/U4L8 and SH8/WM13 for the individual harmonic degrees as well as for the entire model.

Both the models *SH8/U4L8* and *SH8/WM13* were obtained subject to roughness penalty function η^2 (26). However, model *SH8/U4L8* placed no constraint on how continuous or discontinuous structure across the 670 km discontinuity could be, whereas model *SH8/WM13* penalized rapid changes in structure at all depths in the mantle. Clearly, the parameterization of model *SH8/U4L8* contains significant extra degrees of freedom. The fact that the structure in both models is virtually the same demonstrates that the structure at 670 km depth is demanded by the data.

Discussion of the Models

Average Radial Properties. Examination of Figure 10a shows that the heterogeneity is largest near the surface, with an rms value of 2.0% at the Moho. The amplitude of heterogeneity decreases with depth throughout the upper mantle. In addition, the spectrum is changing, so that in the transition zone the spectrum is dominated by degrees 1 and 2. The latter is consistent with the observation of *Masters et al.* [1982]. By 900 km depth the the heterogeneity is reduced to a rather consistent level of 0.35%. In this portion of the mid-mantle, from roughly 900 km depth to 1900 km, the spectrum is relatively uniformly distributed among harmonic degrees from 1 to 8. At about 1900 km depth the amplitude begins to increase rapidly, mostly at harmonic degree 2, and to a lesser extent also 3, 4 and 5, and reaches an rms value of roughly 0.7% at the CMB. There are significant differences between our total rms curves and that of *Tanimoto* [1990] , particularly in the transition zone and the lowermost mantle; *Tanimoto* derived his model from waveform data alone.

Maps of lateral variations. Figures 11a–h are maps of shear velocity anomalies obtained by synthesizing, according to equation (3), the A and B coefficients of Table 1. The maps are 'split' along 60°E, because few significant 3-D features span this longitude. The heterogeneity near the surface (Figure 11a) is related to tectonics. The dominant fast features are the continental shields and the dominant slow features are the mid-ocean spreading centers; back arc basins also show as slow regions. The old (and fast) oceanic lithosphere in the west Pacific shows up as a relatively fast feature.

At a depth of 300 km in the upper mantle (Figure 11b) the signal is dominated by the difference between continents and oceans. At this depth there is no indication that the relatively slow velocities under the oceans vary with the age of the oceanic lithosphere. At the same time, the mantle under the continents is substantially faster; this yields support to *Jordan*'s hypothesis of the continental tectosphere [1975, 1978b, 1979].

A gradual change takes place in the transition zone. At 670 km, on the upper mantle side, most of Asia is slow, but there appears a strong velocity high centered on the Philippine plate and extending meridionally far north and south. At the same time, there are still high velocities associated with North and South America and Africa while most of the Pacific remains slow. There is a strong shift in the spectrum towards lower harmonics.

A change takes place across the 670 km discontinuity: Figure 11d is a map on the lower mantle side. There is a significant shift of power towards the higher harmonic degrees, although the total power is decreasing. A strong, nearly meridional linear feature develops from the north of Hudson Bay to the south of Tierra del Fuego. Another linear anomaly develops between the Arabian Peninsula and the Macquarie triple junction. The pattern of anomalies in the Pacific changes significantly with the development of a velocity high in its central part. At the same time, there are certain features that remain continuous: the low velocities under Siberia or high velocities under Africa, for example.

Our results are consistent with the recent regional tomographic inversions of the P-wave travel time anomalies [*van der Hilst et al.*, 1991; *Fukao et al.*, 1992]. The width and amplitude of the velocity high under the north-western

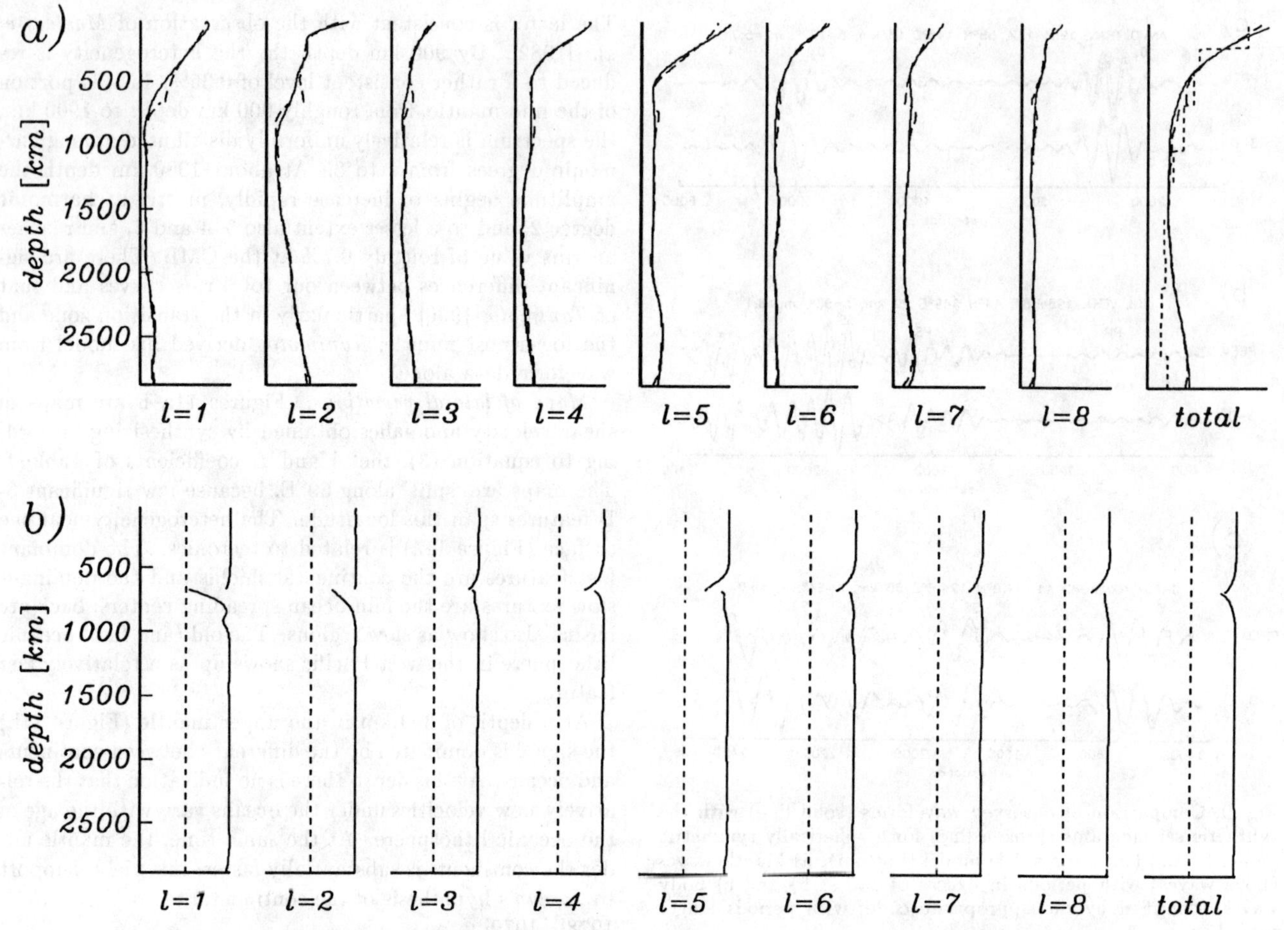

Fig. 10. (a) Variation with depth of the root-mean-square amplitudes of shear velocity perturbations for different harmonic degrees ℓ; for definition see eq. (5). Model SH8/SW13 is shown with a solid line; model SH8/U4L8 with a dashed line. The amplitude at each ℓ is plotted on a horizontal scale of 0.0% to 1.0%. The total amplitude is shown in the rightmost panel, and is plotted on a horizontal scale of 0.0% to 2.0%. The rms amplitudes of model MDLSH (Tanimoto, 1990) are shown with a dotted line (b) The cross-correlation between models SH8/WM13 and SH8/U4L8 plotted as a function of depth and harmonic degree. For each panel, the horizontal scale is -1.0 to 1.0, and the dashed line is zero. The total correlation, for harmonic degrees 1 to 8, is shown in the rightmost panel.

Pacific and easternmost Asia are too large to be a result of a low-pass filtration of a narrow velocity increase associated with a subduction zone, but are entirely consistent with the interpretation that the flow becomes horizontal in that region. Our results also confirm the result of *Fukao et al.* [1992] that the flow associated with the Indonesian subduction zone extends through the 670 km discontinuity. All this is consistent with the observation of *Ekström et al.* [1990] of deep earthquakes occurring at distances of up to 200 km away from subduction zones. One such earthquake occurred in 1989 in Argentina, a region for which there are no regional tomographic studies, but for which our global inversion shows a significant horizontal width of a positive velocity anomaly in the transition zone.

There is a significant degree of similarity between the map at 670 km, below the discontinuity, and at 1200 km (Figure 11e), although some features disappear (low velocity under Siberia and high under Africa) and new ones develop such as the Indian high, although this feature is continuous with the Indonesian high in Figure 11d. There is also a slight rearrangement of the pattern in the central Pacific.

A shift in the spectrum towards longer wavelengths is clearly visible at 2000 km (Figure 11f). At this depth the ring of high velocities around the Pacific is already well pronounced. This trend is reinforced at 2500 km depth (Figure 11g), where the amplitude of the anomalies has increased by roughly 50%. The amplitude of heterogeneity increases still further towards the core-mantle boundary (Figure 11h). This overall character of heterogeneity in the lowermost mantle has been observed in compressional veloc-

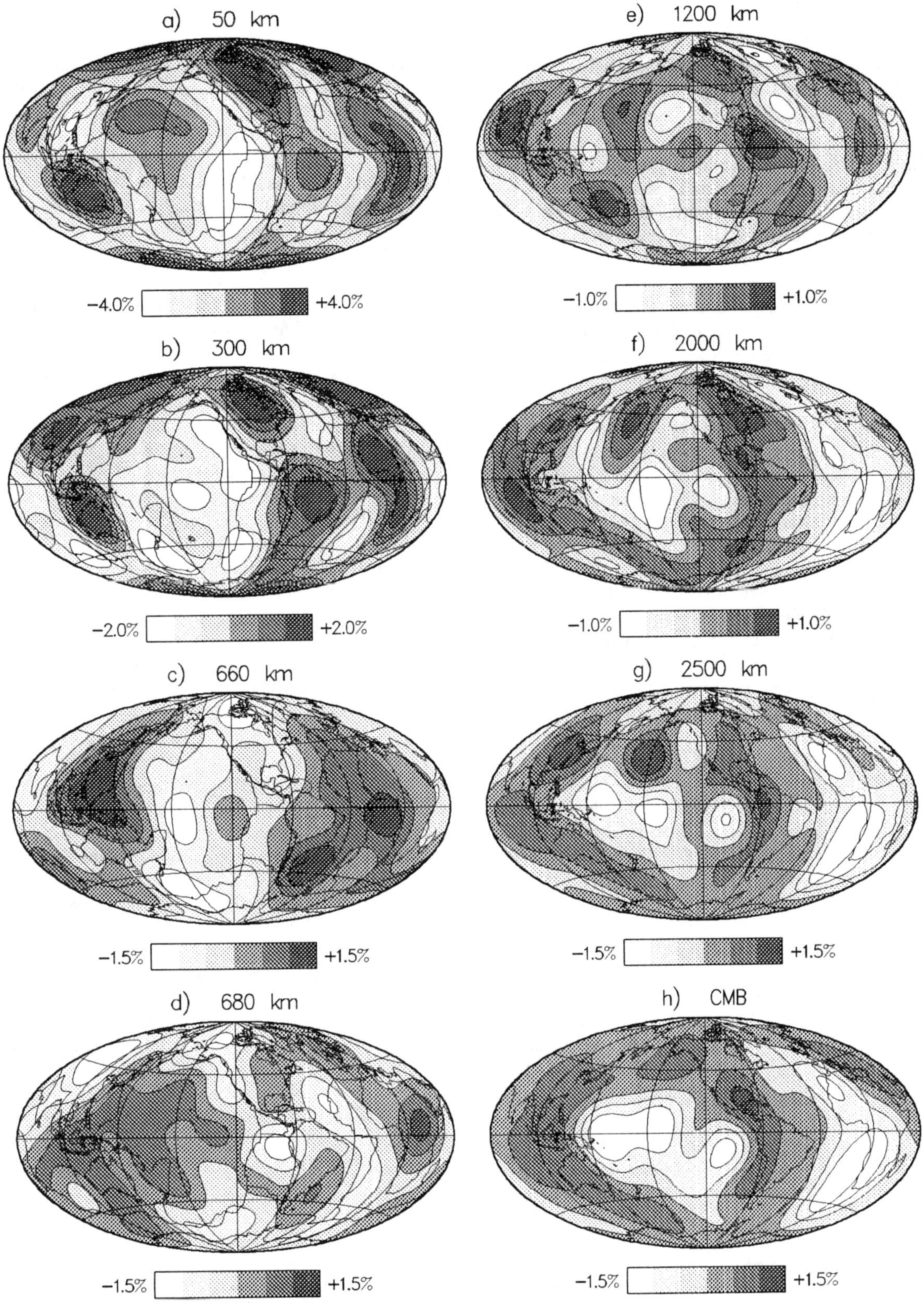

Fig. 11. Maps of relative deviations from the average shear wave speed at eight depths in the mantle obtained by the synthesis of the coefficients of model SH8/U4L8 listed in Table 1. The depth and minimum and maximum scale values are shown with each map. The scale bar is shown at the bottom and the minimum (slowest) and maximum (fastest) values of each map are represented by the white and darkest gray shades, respectively.

ity [*Dziewonski*, 1984; *Inoue et al.*, 1990] as well as in other studies of shear velocity *Woodhouse and Dziewonski*, 1989; *Tanimoto*, 1990]. The slow features under the Pacific are easily traced up to the spreading center in the east Pacific and similarly, the slow region under Africa my be traced to spreading centers in the Atlantic and Indian Oceans.

It is interesting to compare Figures 11f-h with those of both observed and predicted $ScS - S$ travel time residuals (Figures 8a and b). These $ScS - S$ observations directly reflect the lowermost mantle structure because the ScS leg has a long path through the lowermost 1000 km of the mantle, which is characterized by spatially consistent, strong velocity anomalies, whereas the S leg bottoms in the middle mantle where the amplitude of heterogeneity is relatively low and the spectrum is nearly flat.

Cross-sections and a Discontinuity Between Upper and Lower Mantle. Figures 12a–c show great-circle cross-sections through the model SH8/WM13. Perhaps the most striking feature seen in these images is the structure near 670 km. One can easily discern the location of the 670 km discontinuity in the cross-sections of Figure 12 by the changes in the character of heterogeneity at this depth. Clearly there are features which extend vertically through the 670 km discontinuity, such as the slow features beneath the east Pacific, Indian Ocean, and mid-Atlantic spreading centers. The slow upwelling material in the lower mantle can be traced to these features and is feeding the spreading centers. There are also fast features which extend through the 670 discontinuity, such as beneath Australia (Figure 12b) or Indonesia (Figure 12c). However, not all structures are continuous through the 670 km discontinuity. There are a number of structures which seem to be blocked or deflected at this depth: under North Pacific, Europe and North Africa in Figure 12a, Indian Ocean and Tasman Sea in Figure 12b, the Pacific and Atlantic oceans in Figure 12c.

Figure 12d and e show cross-sections through the model SH8/U4L8, which has separate upper and lower mantle parameterizations. Comparison of Figures 12a and 12b to figures 12d and 12e shows that the models we obtain are essentially independent of the parameterization used. This is also clear from Figure 10c, which shows the cross-correlation between these models as a function of depth. With the exception of some differences in the immediate vicinity of the 670 km discontinuity, these two models are virtually the same.

Figure 12f is a composite cross-section through the upper mantle of model M84C [*Woodhouse and Dziewonski*, 1984] and the lower mantle of model $L02.56$ [*Dziewonski*, 1984]; in the latter, the P-velocity anomalies are re-scaled by a factor of 2 [*Li et al.*,1991b]. The details of $L02.56$ near the 670 km discontinuity should be ignored, since some of the upper mantle heterogeneity might have been mapped into this part of the model. The upper mantle images in Figures 12c and f are difficult to tell apart. Parameterization of model $L02.56$ is much coarser than that of $SH8/WM13$, hence Figure 12c contains more detail. But the similarity of large scale features is unmistakable. It seems, therefore, that the principal features of mantle heterogeneity have been properly identified nearly a decade ago, using different types of data.

Planetary-scale Structures. The map at 2500 km depth (Figure 11g) perhaps best summarizes several features which dominate the heterogeneity in the mantle. Four distinct features can be seen in this map: two slow, 'plume like' upwelling regions under the Pacific and Africa, and two linear 'trough like' regions of seismically fast, descending flow beneath the Indian Ocean and beneath North and South America. Cross-sections through these features are shown in Figure 13. Recognizing that these features dominate the heterogeneity in the lower mantle, we suggest names for them to aid future discussion.

The 'Great African Plume', shown in Figure 13a, is the structure of largest amplitude. The African continent seems to possess what is effectively a 'super-root,' being anomalously fast to 1300 km depth. The large-scale upwelling originating at the CMB is split by this root, and the low velocities (high temperatures) can be traced up to the mid-Indian Rise and mid-Atlantic Ridge. Figures 13a and c provide other cross-sections through this structure.

The 'Pangea Trough' (Figure 13b) is a narrow region of anomalously high velocities. It follows, roughly, the western edge of this historical supercontinent. It is centered approximately on 75°W and extends in latitude beyond both polar circles. With minute exceptions, the velocities are higher than normal from the Moho to the CMB; fragments of this structure have been recognized earlier [*Jordan and Lynn*, 1974; *Grand*, 1987; *van der Hilst and Engdahl* , 1991); see Figures 12b and c for nearly orthogonal sections. The virtual geomagnetic pole paths appear to bunch along this structure [*Laj et al.*, 1991]. *Richards and Engebretson* [1992] obtain good correlation of the gravest terms of the spherical harmonic expansion of the geoid, lateral variation of seismic velocities in the lower mantle and subduction history integrated over the last 200 My. It is of interest to note that the

Fig. 12. Great-circle cross-sections through various 3-D velocity models. Each cross-section is made along a particular great circle (the heavy line in the inset map) and passes through the center of the Earth. The outermost ring is closest to the Earth's surface, the innermost corresponds to the CMB. In the continuous model SH8/WM13 (panels a–c) the depth of the 670 km discontinuity is indicated by a dashed line. For the discontinuous models (panels d–f) the 670 km discontinuity is shown by a solid black line. The scale of shades ranges from -1.5% (white) to +1.5% (darkest gray). Significant saturation of the scale is possible in the upper mantle.

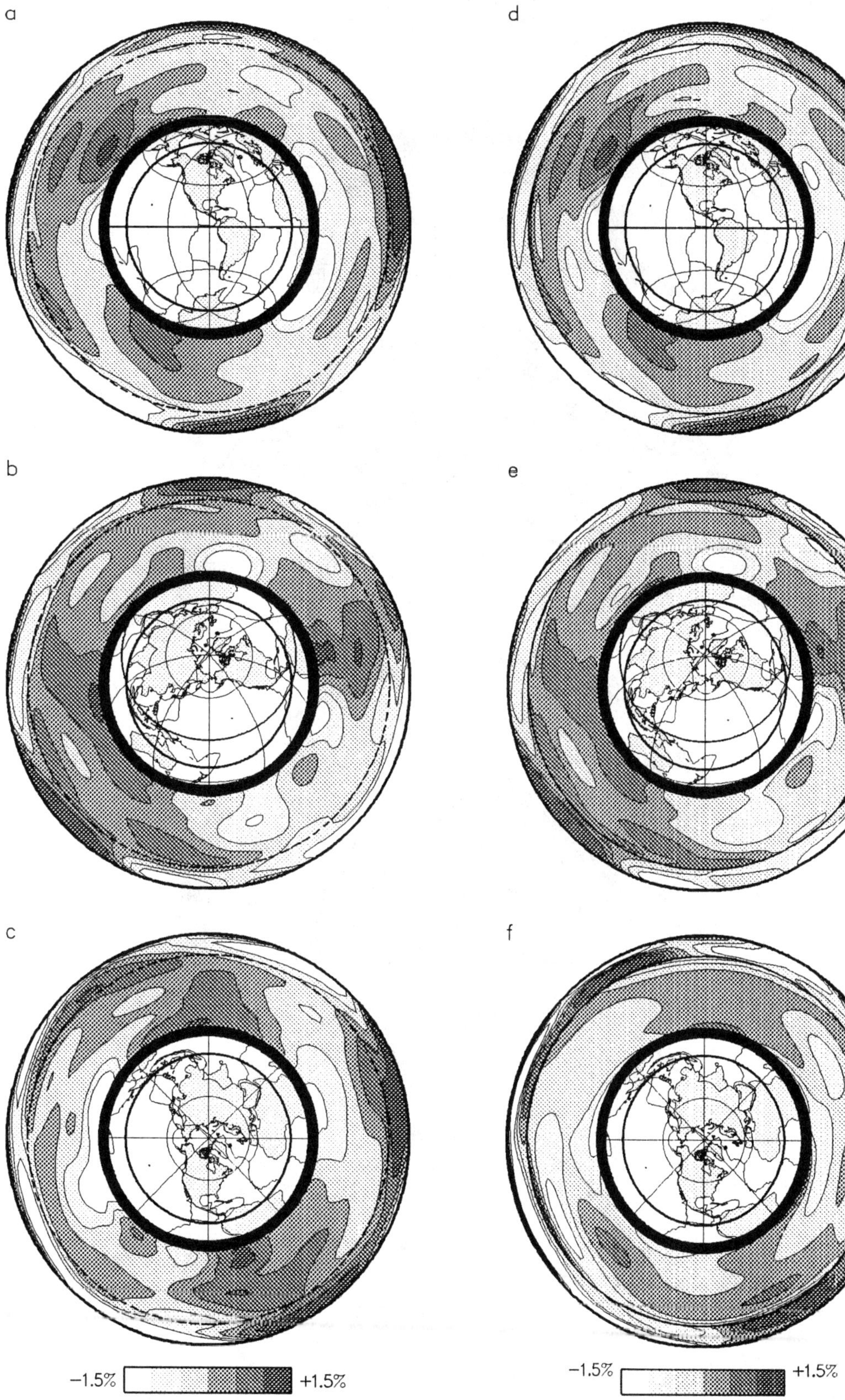
a
d
b
e
c
f
−1.5%
+1.5%
−1.5%
+1.5%

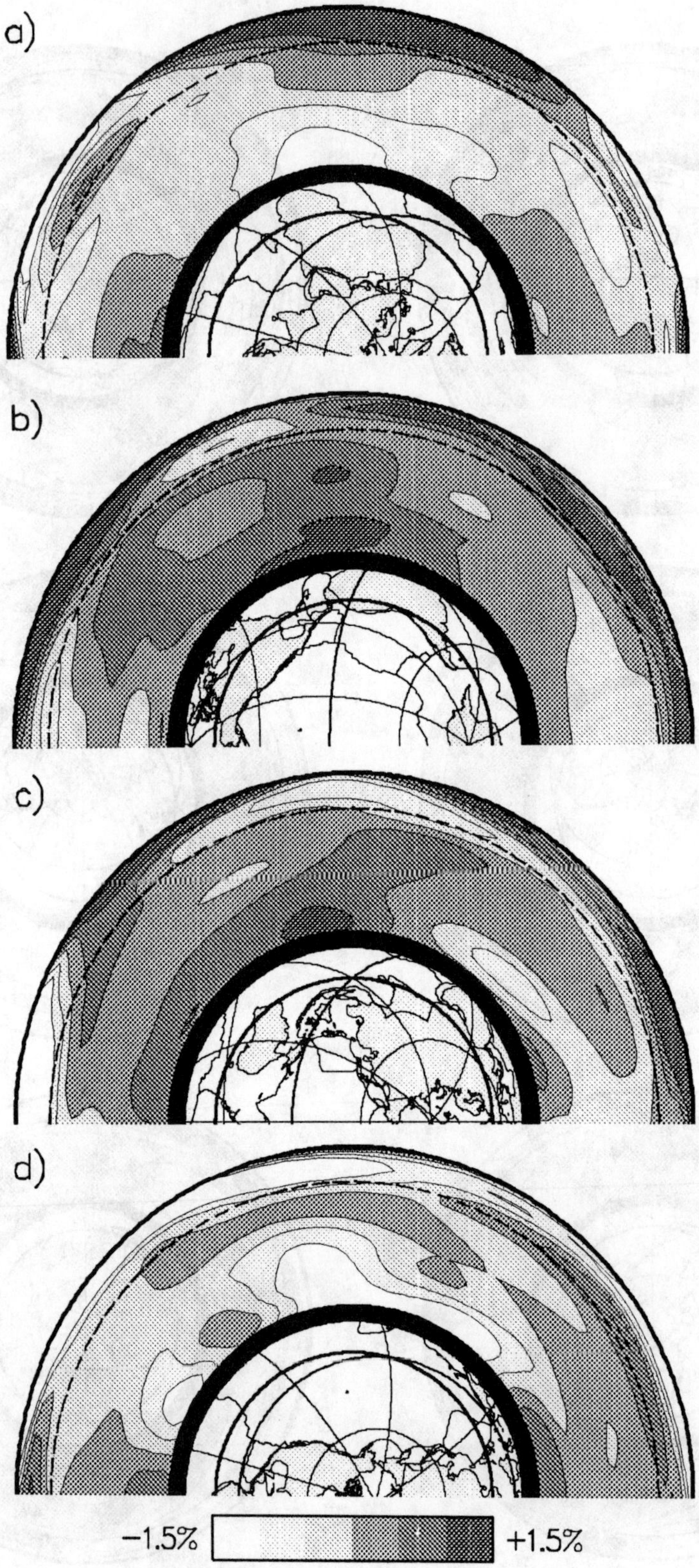

Fig. 13. Four cross-sections through the mantle of the 3-D shear velocity model SH8/WM13, showing large-scale structures named: a) 'Great African plume', b) 'Pangea Trough', c) 'Tethys Trough', and 'Equatorial Pacific Plume Group'.

present day location of the Pangea trough corresponds to the position of a subduction zone some 100 My [see Figure 2 in *Richards and Engebretson*, 1992].

Figure 13c shows the 'Tethys Trough', a zone of high velocities which follows the axis of the Tethys Sea convergence. This structure is continuous only in the lower mantle; in the upper mantle it is interrupted by what are, perhaps, more recent features. An exception is the Australian continent, under which the velocities are systematically high from the surface to the CMB.

The 'Equatorial Pacific Plume Group' is shown well in Figures 11g and h and the easternmost plume of this group (15°S, 110°W) is shown in Fig. 13d. This feature appears to be connected with the development of the East Pacific Rise; see Figure 13b. The central plume (10°S, 150°W) affects the thermal structure of the entire South and Central Pacific, including the upper mantle; Figure 13a shows a north-south section of the western edge of this plume. The westernmost plume (5°N, 160°E) has large amplitude at the CMB, but it decreases rapidly at shallower depths; suggesting that it may be a relatively young feature. Additional important but less extensive features are the 'China High' and 'North Pacific High'; these are clearly visible in the maps at 2000 and 2500 km depth in Figure 11.

General Remarks . We note that the cross-sections of the model shown in Figures 12 and 13 bear a strong resemblance to the results of recent convection calculations. On a very large scale, one sees relatively narrow upwellings in the lower mantle which mushroom out into much larger heads in the upper mantle. Return flow is perhaps less distinctive, and primarily consists of large features which descend straight down (e.g. under Australia and the Canadian shield) or with some deflection. Converting the velocity perturbations to density perturbations allows the calculation of mantle flow. Such calculations will be discussed in the next section. We also note that the pattern of convection indicated by our model may have important bearing on large-scale patterns of isotopic signatures in oceanic basalts. The similarity in the pattern of the Dupal anomaly [*Dupre and Allegre*, 1983; *Hart*, 1984] and the pattern of low velocities in the lower mantle has been noted by *Castillo* [1988]. From the cross-sections of Figure 13c we see that the upwelling beneath Africa is apparently feeding the spreading centers in both the South Atlantic and Indian Ocean, implying that the source of the Dupal anomaly may, in fact, be in the lower mantle.

We have obtained improved resolution of structure at all depths by combining two major classes of seismic data in a joint inversion. We observe a number of what are, by now, relatively familiar features in the lower-most mantle. What is striking is that these features can be traced through the lower mantle and many of them continue on through the upper mantle as well. The continuity of such features through the 670 km discontinuity implies that convective transport of material between the upper and lower mantle may be occurring. However, there are other regions where material is either blocked or deflected at depths of 600 to 800 km or more. In general, significant changes in the character of heterogeneity occur between depths from 600 to 800 km. The amplitude of heterogeneity begins to increase above this depth, and both the spectrum and pattern of heterogeneity change.

TOMOGRAPHIC MODELS AND VISCOUS FLOW IN THE MANTLE

This account of the geodynamic implications of the recent seismic models of mantle heterogeneity provides an updated discussion of the results presented in *Forte et al.* [1993a]. In the following discussion we shall employ a new radial viscosity profile of the mantle which differs from that used in *Forte et al.* [1993a]; the principal difference lies in the introduction of a high-viscosity lithospheric layer and a region of increasing viscosity in the bottom half of the lower mantle. Comparison of the results obtained using two different viscosity models will allow the reader to appreciate the stability and reliability of the geodynamic predictions we present below.

In addition, we also provide an account of recent work [*Forte et al.*, 1993b] which demonstrates that the dynamic topography calculated on the basis of our predicted buoyancy-induced flow agrees rather well with the observed dynamic surface topography. This last result is an important new verification of the validity of three-dimensional mantle flow predicted on the basis of the seismic tomographic models.

The well-known ability of the mantle to creep indefinitely over geological time scales is understood in terms of the existence of atomic-scale defects in the lattice of crystal grains [e.g. *Nicolas and Poirier*, 1976]. If the ambient mantle temperature is sufficiently high, the imposition of nonhydrostatic stresses causes the lattice defects to propagate and thus allows the mantle rocks to creep or 'flow' slowly. This process may be characterized in terms of a single parameter, namely an effective viscosity [e.g. *Stocker and Ashby*, 1973; *Weertman and Weertman*, 1975]. The effective viscosity of the mantle depends strongly on the pressure, and the temperature. We therefore expect that the viscosity of the mantle will vary significantly in three dimensions as a result of the lateral and depth variation of these thermodynamic state variables.

In this study we employ a simplified treatment of mantle flow in which we assume that the mantle viscosity varies only with radius. The radial viscosity profile we will propose, in the course of fitting the geoid and plate motion data, may thus reflect the effective horizontal average of the viscosity distribution at any given depth.

The characterization of the creep properties of the mantle in terms of an effective viscosity allows us to model the slow flow of the mantle with the conventional hydrodynamic field equations which express the principles of mass and momentum conservation. The previous treatments of mantle-flow in spherical shells with radially-varying viscosity [e.g. *Ricard et al.*, 1984; *Richards and Hager*, 1984; *Forte and Peltier*, 1987b] have assumed that the mantle is incompressible. In recent studies of mantle flow in spherical geometry by *Forte* [1989] and *Forte and Peltier* [1991b] it was shown that the effects of the finite compressibility of the mantle are not negligible. In particular, these studies demonstrated that the very-long wavelength nonhydrostatic geoid is quite sensitive to the effects of mantle compressibility. In the viscous-flow predictions of the geoid, plate motions, dynamic surface topography, and CMB topography presented below we shall therefore employ the compressible-flow theory described in *Forte and Peltier* [1991b].

The principal assumption we make in modeling the mantle flow is that the 670 km seismic discontinuity is not a barrier to radial flow. The cross-section views of the $\delta v_S/v_S$-heterogeneity in model SH8/U4L8, shown in Figures 12 and 13, provide examples of plume-like features extending across the 670 km seismic discontinuity. In addition we have found [*Forte et al.*, 1993a] that all our attempts to model the geoid and plate motions with a single layered-flow model, which explicitly includes a barrier at 670 km depth, result in rather unsatisfactory fits to the observations. Layered-flow models which provide fair fits to the nonhydrostatic geoid provide poor fits to the plate motions and *vice versa.* In contrast, as we show below, the assumption of whole-mantle flow allows us to derive a single model which provides very good fits to both the nonhydrostatic geoid and the plate motions.

Nonhydrostatic Geoid

The solution of the hydrodynamic field equations in spherical geometry [e.g. *Richards and Hager*, 1984; *Ricard et al.*, 1984; *Forte and Peltier*, 1987b,1991b] expresses the three-dimensional flow velocity **u** excited by the internal buoyancy forces due to the density perturbations $\delta\rho(r,\vartheta,\varphi)$. On the basis of these theoretical flow predictions it is possible to obtain the so-called 'kernels' which relate, in the spherical-harmonic spectral domain, the various flow-related geophysical surface observables (e.g. the nonhydrostatic geoid) to density perturbations in the mantle. The nonhydrostatic geoid kernels $G_\ell(r)$ relate the spherical harmonic coefficients of the geoid δN_ℓ^m to the radially-varying spherical harmonic coefficients of the perturbed density $\delta\rho_\ell^m(r)$ as follows :

$$\delta N_\ell^m = \frac{3}{(2\ell+1)\bar{\rho}} \int_b^a G_\ell(r)\delta\rho_\ell^m(r)dr, \qquad (28)$$

in which $\bar{\rho}$ is the average density of the Earth, b km is the radius of the CMB, and a km is the radius of the solid surface. The geoid kernels $G_\ell(r)$ have been shown [e.g. *Richards and Hager*, 1984] to be rather sensitive to the radial variation of mantle viscosity. It is also important to remember that these kernels are only sensitive to relative viscosity variations $\eta(r)/\eta_0$ (where $\eta(r)$ is the actual viscosity profile and η_0 is a reference viscosity) and therefore the nonhydrostatic geoid data cannot be used to independently constrain the absolute value of mantle viscosity [*Forte and Peltier*, 1987b,1991b].

We exploit the viscosity-sensitivity of the predicted nonhydrostatic geoid, using the flow calculated on the basis of the $\delta v_S/v_S$-heterogeneity in model *SH8/U4L8*, to infer the viscosity profile $\eta(r)/\eta_0$ shown in Figure 14. This viscosity profile was derived essentially by trial-and-error forward modeling, in which we exploit the sensitivity of the geoid data provided by the viscosity Fréchet kernels shown in *Forte et al.* [1993a].

The most notable features in the inferred viscosity profile in Figure 14 are the high-viscosity lithosphere, the zone of increased viscosity in the lower mantle, and the thin (70 km thick) layer of very low viscosity at the base of the upper mantle. The viscosity increase within the lower mantle is dictated by the apparent requirement (based on the viscosity Fréchet kernels for the geoid) for a zone of increased strength in the lower mantle. The low-viscosity zone at the base of the upper mantle is modeled with the following exponential variation in the depth-interval 600-670 km:

$$\eta(r)/\eta_0 = (\eta_{670}/\eta_0)\exp\left[\frac{r-r_{670}}{h}\right], \qquad (29)$$

in which η_{670}/η_0 is the normalized viscosity at the bottom of the upper mantle, and h is the scale height for the viscos-

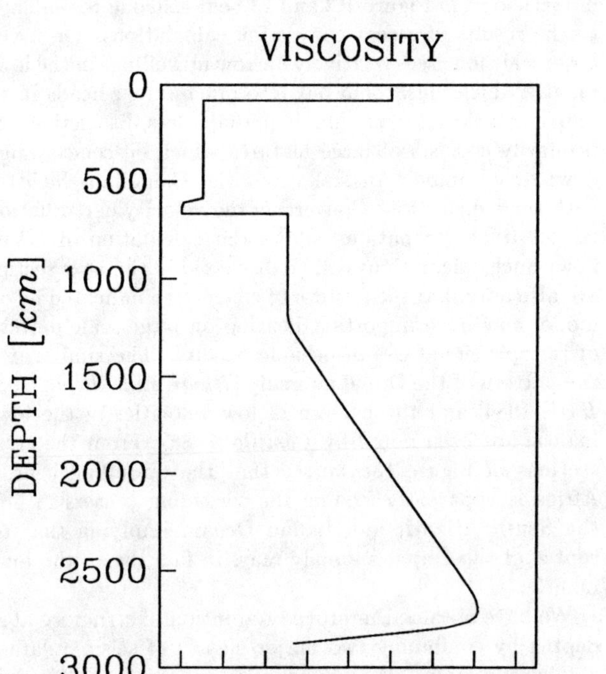

Fig. 14. The relative viscosity profile $\eta(r)/\eta_0$ in which $\eta(r)$ is the actual mantle viscosity and η_0 is a reference mantle viscosity. This viscosity profile is inferred on the basis of the nonhydrostatic geoid data (see text for details).

ity variation. The quantities η_{670}/η_0 and h are selected so that the viscosity at 600 km depth is 100 times greater than the viscosity at the bottom of the upper mantle. The non-hydrostatic geoid data seem to favor such steep decreases of viscosity at the base of the upper mantle. In fact, if we increase the scale-height h so that $\eta_{670}/\eta_{600} = 1/10$ (rather than 1/100) we find that our fit to the longest wavelength (*i.e.* ℓ=2 and 3) geoid is completely lost.

A low-viscosity zone at the base of the upper-mantle might be interpreted as a thermal boundary layer effect, in analogy to the hypothesized low-viscosity zone at the base of the lower mantle [e.g. *Yuen and Peltier*, 1980; *Stacey and Loper*, 1983]. We emphasize however that we are assuming a whole-mantle flow and therefore the existence of a thermal boundary layer at the base of the upper mantle, which would be expected if there were separate upper- and lower-mantle circulations, is somewhat puzzling. A possible explanation may be the local distortion of the geotherm arising from the latent-heat release accompanying the continual flux of hot upwelling mantle across the spinel - post-spinel phase-change boundary (this is the so-called "Verhoogen effect" described by *Verhoogen* [1965] and *Turcotte and Schubert*, 1971]. An alternative explanation may lie with the long-term thermal effect of the extensive 'pooling' of hotter-than-average mantle in the vicinity of the 670 km seismic discontinuity which is quite apparent in the cross-section views of the $\delta v_S/v_S$-heterogeneity in model $SH8/U4L8$. This 'pooling' is indicative of an apparent barrier to radial flow across some portions of the 670 km discontinuity which is evident in Figures 12 and 13. The horizontally-averaged effect of these hotter-than-average 'pools' on the strongly temperature-sensitive mantle viscosity might lead to an effective low-viscosity layer at these depths. Such hypotheses are of course only speculative but it is worth noting that in the PREM radial Earth model [*Dziewonski and Anderson*, 1981] the region from 600 to 670 km depth is characterized by a noticeable decrease in the radial gradients of density, P-velocity, and S-velocity. Such reduced gradients would be expected if there were an increased (superadiabatic) temperature gradient at the base of the upper mantle.

A new interpretation of the geoid-inferred low-viscosity zone at the bottom of the upper mantle derives from the recent study of creep in perovskite by *Karato and Li* [1992]. *Karato and Li* argue that the grain-size reduction associated with the spinel to perovskite + magnesiowustite phase change will lead to a considerable softening of material descending below the 670 km discontinuity. They also argue that an additional effect, which may be as important, is the softening in perovskite due to structural phase transformations which are expected to occur in the shallow part of the lower mantle. In either case *Karato and Li* expect that the top of the lower mantle may possess a much lower effective viscosity than the deeper portions of the mantle. Since our viscous flow models assume a whole-mantle style of flow, the geoid data cannot distinguish whether we place the low-viscosity zone at the bottom of the upper mantle or at the top of the lower mantle.

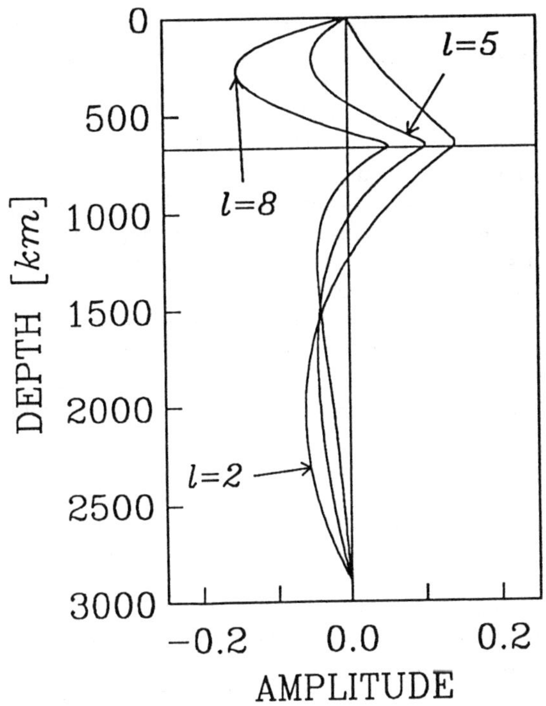

Fig. 15. The nonhydrostatic geoid kernels $G_l(r)$, defined in (28), calculated for the viscosity profile shown in Figure 14. The location of the 670 km seismic discontinuity is indicated by the heavy horizontal line.

In Figure 15 we now show a selection of geoid kernels, defined in (28), calculated for the viscosity profile $\eta(r)/\eta_0$ in Figure 14. The most important feature of the geoid kernels shown in Figure 15 is their peak amplitude found at 670 km depth. This behavior is a direct consequence of the low viscosity zone at the base of the upper mantle and is in accord with the peak correlations between the geoid and $\delta v_S/v_S$-heterogeneity near 670 km depth. The depth-variation of the sign of the geoid kernels also matches the overall depth-variation of the $\delta v_S/v_S$-geoid correlations discussed in detail by *Forte et al.* [1993a] and this explains why the viscosity profile in Figure 14 provides a good match between predicted and observed geoid, as we show below.

An issue of great importance in the modeling of mantle flow with seismically-inferred heterogeneity concerns the choice of appropriate density-velocity proportionality factors $\partial ln\rho/\partial ln v_S$ which are required to translate the $\delta v_S/v_S$-heterogeneity in model $SH8/U4L8$ into an equivalent density heterogeneity field:

$$\delta\rho_l^m(r) = \rho_o(r)\left(\frac{\partial ln\rho}{\partial ln v_S}\right)(r)\left(\frac{\delta v_S}{v_S}\right)_l^m(r), \qquad (30)$$

in which $\rho_o(r)$ is the PREM radial density profile. We have assumed here that the $\partial ln\rho/\partial ln v_S$ is only a function of radius and thus we are explicitly ignoring the possibility of substantial lateral variations in this quantity. Such lateral variations, if they are indeed important [e.g. *Forte et al.*, 1991b], may be greatest in the top 400 km of the mantle in which the chemically distinct "roots" of continents hypothesized by *Jordan* [1975] may possess $\partial ln\rho/\partial lnv_S$ values which are quite different from the ambient mantle [see *Jordan*, 1979, for a detailed discussion]. We will follow the example of *Hager and Clayton* [1989] and infer $(\partial ln\rho/\partial lnv_S)(r)$ directly from the nonhydrostatic geoid data itself. The procedure is quite straightforward and involves substituting (30) into (28), using the $\delta v_S/v_S$-heterogeneity model *SH8/U4L8*, and using the geoid kernels shown in Figure 15. If we further assume for simplicity that $\partial ln\rho/\partial lnv_S$ is constant in the depth range 0-400 km, 400-670 km, 670-1000 km, and 1000-2891 km, the problem reduces to a simple linear least-squares inversion of the nonhydrostatic geoid data. The $\partial ln\rho/\partial lnv_S$ values that are inferred in this manner are:

$$\partial ln\rho/\partial lnv_S = \begin{cases} +0.10, & \text{0-400 km} \\ +0.35, & \text{400-670 km} \\ +0.36, & \text{670-1000 km} \\ +0.15, & \text{1000-2891 km} \end{cases} \quad (31)$$

The $\partial ln\rho/\partial lnv_S$ value in the transition zone and down to 1000 km is rather similar to the value $\partial ln\rho/\partial lnv_S \approx 0.4$ expected on the basis of laboratory measurements [e.g. *Anderson et al.*, 1968; *Isaak et al.*, 1989]. There is, however, a marked decrease in the value of $\partial ln\rho/\partial lnv_S$ from the upper mantle to the bottom of the lower mantle. This decrease may reflect the important effect of increasing pressure with depth and it appears to be consistent with the recent high-pressure data of *Chopelas* [1988, 1990] which has been used by *Yuen et al.* [1991] to show a marked decrease, with increasing pressure, of the $\partial ln\rho/\partial lnv_S$ values for MgO. Since the temperature derivatives of the seismic wave speeds are expected to decrease with increasing depth [e.g. *Anderson*, 1987], it is clear from (31) that the coefficient of thermal expansion must decrease even more rapidly. The recent measurements by *Chopelas and Boehler* [1989] in fact show that the coefficient of thermal expansion will decrease significantly (perhaps by as much as an order of magnitude) across the depth-range of the mantle. This important result has also been discussed in *Chopelas* [1990], *Anderson et al.* [1990], and Reynard and Price [1990].

The magnitude of the $\partial ln\rho/\partial lnv_S$ value inferred for the top 400 km of the mantle is much smaller than typical laboratory values [e.g. *Anderson et al.*, 1968]. One possible explanation for this anomalous value is that we are inferring an effective horizontal average of a laterally-varying $\partial ln\rho/\partial lnv_S$ which may reflect the effects of partial-melting below mid-ocean ridges and/or the effects of the continent-ocean differences envisaged by *Jordan* [1975]. When considering such explanations it is important to remember that the $\delta v_S/v_S$ heterogeneity is poorly correlated to the geoid in the top 400 km of the mantle (see Figure 9 in *Forte et al.*, 1993a] and therefore our least-squares inversion of the geoid data has penalized this portion of the mantle by assigning it a very small $\partial ln\rho/\partial lnv_S$ value. It also appears that the $\partial ln\rho/\partial lnv_S$ value inferred for the top 400 km of the mantle is sensitive to the details of the assumed radial viscosity profile and is therefore rather uncertain [*Forte et al.*, 1993a].

On the basis of the $\partial ln\rho/\partial lnv_S$ values in (31), and the geoid kernels shown in Figure 15, we can now calculate the nonhydrostatic geoid expected on the basis of the $\delta v_S/v_S$ heterogeneity in model *SH8/U4L8*. The results of this calculation are shown in Figure 16 in which we compare the predicted geoid, in Figure 16b, with the GEM-T2 nonhydrostatic geoid [*Marsh et al.*, 1990] in Figure 16a. The predicted and observed geoids evidently agree rather well and this agreement may be quantified in terms of the 68% variance reduction obtained with our geoid prediction. The greatest source of misfit between the predicted and observed geoids is the poorly predicted N_2^0 geoid coefficient (predicted N_2^0 = -12.5 meters) which is much smaller than the corresponding GEM-T2 value (nonhydrostatic GEM-T2 N_2^0 = -27.6 meters). If we adjust the predicted N_2^0 coefficient to equal the GEM-T2 nonhydrostatic value the resulting geoid, shown in Figure 16c, now agrees more closely with the GEM-T2 geoid and, in fact, the variance reduction we obtain with this adjustment is 82%. The importance of the mantle heterogeneity in the vicinity of the 670 km seismic discontinuity is confirmed by the observation that the seismically-inferred density contrasts (derived from model *SH8/U4L8*) in the depth-range 400-1000 km account for 60% of the root-mean-square amplitude in the geoid prediction shown in Figure 16b.

The difficulty in matching the excess ellipticity of the observed nonhydrostatic geoid has been reported previously on the basis of earlier seismic heterogeneity models [e.g. *Forte*, 1989; *Forte and Peltier*, 1991b]. This difficulty does not, however, seem to arise in the geoid modeling studies of *Hager and Clayton* [1989] and *Hager and Richards* [1989] in which the *Clayton and Comer* [1983] model of lower-mantle P-velocity heterogeneity is employed. As discussed in detail by *Forte and Peltier* [1991b], it appears that the *Clayton and Comer* model is characterized by an anomalous Y_2^0-heterogeneity which is substantially larger than the Y_2^0-heterogeneity in *Dziewonski*'s [1984] P-velocity model *L02.56*. It is apparently because of this anomaly in the *Clayton and Comer* model that *Hager and Richards* [1989] are able to obtain such excellent ($\approx$ 80-90%) variance reductions with their geoid predictions. Indeed, if we employ the mantle viscosity model '*WL*' of *Hager and Richards* [1989] to predict the geoid, using the $\delta v_S/v_S$ heterogeneity in model *SH8/U4L8* (and also using the optimal $\partial ln\rho/\partial lnv_S$ values inferred from the geoid using the model '*WL*' geoid kernels), we obtain a poor fit (variance reduction = 38%) to the GEM-T2 nonhydrostatic geoid. In this latter case we find that the misfit is again mostly due to the poorly predicted N_2^0 geoid coefficient. *Forte et al.* [1993a] suggest that the consideration by *Clayton and Comer* [1984] of the differences between the ellipticity corrections of *Dziewonski and Gilbert* [1986] and those used by the ISC, as the data to be inverted as velocity anomalies, might have artificially increased the size of the Y_2^0-heterogeneity in their model.

It is not clear at this time whether there is a fundamental discrepancy between the 'seismic' and 'geodetic' non-

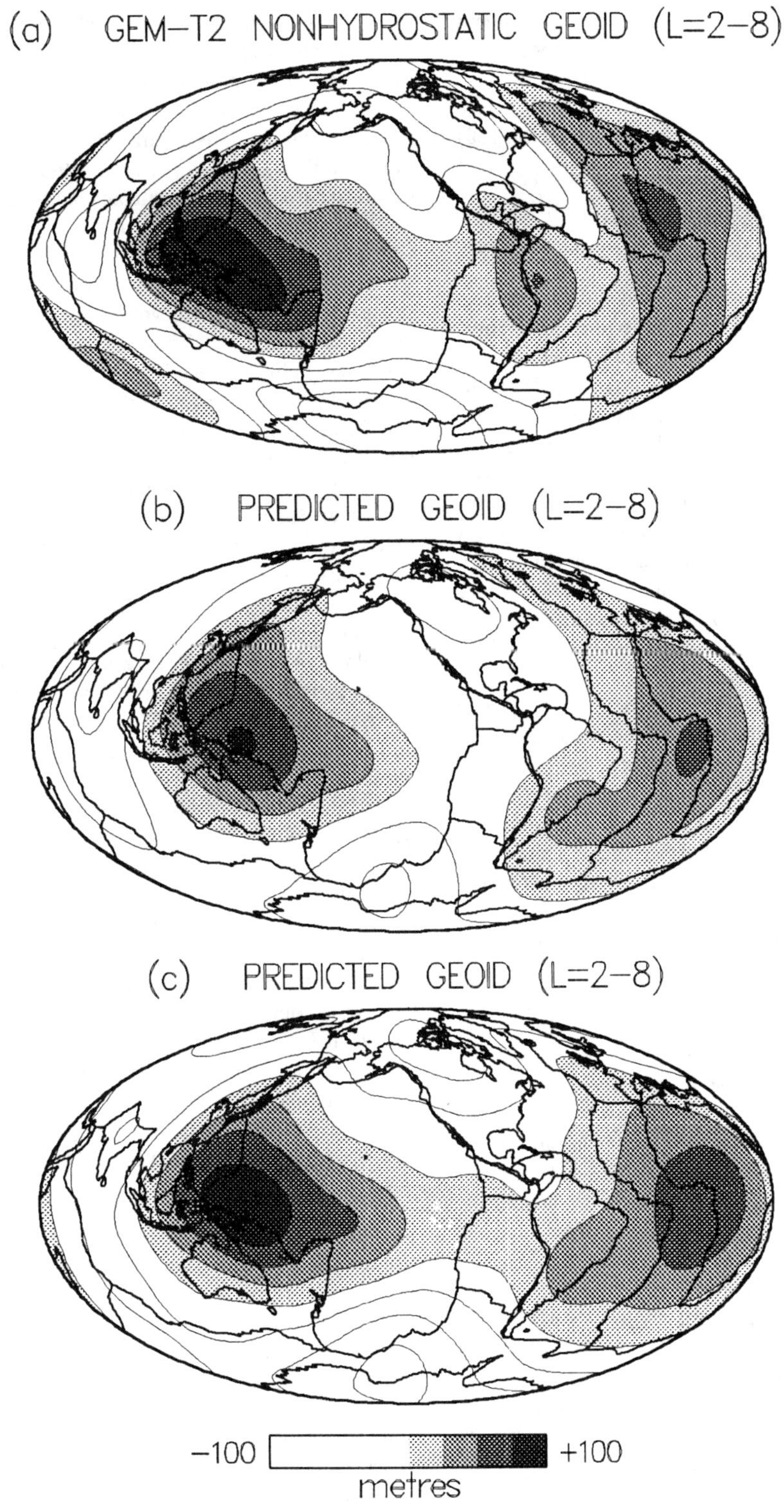

Fig. 16. (a) The GEM-T2 geoid (Marsh et al., 1990), filtered by removal of the hydrostatic flattening (e.g. Jeffreys, 1963; Nakiboglu, 1982), in the degree range $\ell = 2 - 8$. (b) The nonhydrostatic geoid, in the degree range $\ell = 2 - 8$, predicted with model $SH8/U4L8$ using the viscosity profile in Figure 14 and the $\partial ln\rho/\partial lnv_S$ values in (31). (c) The nonhydrostatic geoid of (b) which has been adjusted by setting its Y_2^0 coefficient equal to the value of the observed Y_2^0 coefficient in (a). The contour interval in all cases is 25 meters.

hydrostatic flattening coefficients. Nonhydrostatic geoid data may be incorporated into the inversion process described above in section 3. In addition to the waveform data and travel time data we can add, with an appropriate weight λ_g^2, the geodynamic data. In this way, equation (26) is modified:

$$\left(A_{k\ell m;k'\ell' m'} + \lambda_g^2 G_{k\ell m;k'\ell' m'} + \eta^2(k,\ell)\mathbf{I}\right) \cdot \delta_{k'} c_{\ell'}^{m'} + \eta^2(k,\ell) \cdot {}_k c_\ell^m = b_{k\ell m} + \lambda_g^2 b_{k\ell m}^g \qquad (32)$$

where G and b^g involve the coefficients and data related to the geodynamic observables. The initial experiments have been performed using the geoid coefficients [*Forte et al.*, 1992a] and the results up to date are encouraging. It seems possible to fit the geoid data nearly exactly (with variance reduction of 95%) with a relatively small degradation of the fit to the differential travel time data. It remains, however, to be seen whether there is a particular pattern to this misfit, which may indicate that there is indeed a discrepancy between the two kinds of data. Yet another experiment, after a model consistent with the geodynamic data is derived, is to declare it the 'target' model and then examine the ensuing model perturbation in a subsequent inversion in which the geodynamic data are withdrawn.

It is useful at this stage to consider how our analysis of the geoid data, in terms of the mantle flow models, differs from the previous detailed studies by *Hager and Richards* [1989] and *Forte and Peltier* [1991b]. It is important to appreciate, at the outset, that the geoid-derived inferences of mantle viscosity and $\partial ln\rho/\partial lnv$ are strongly conditioned by the particular seismic model which is employed to represent the lateral heterogeneity in the mantle. We noted above the striking correlation between the $\delta v_S/v_S$-heterogeneity in model $SH8/U4L8$ and the geoid (especially at $\ell = 2$) in the vicinity of the 670 km seismic discontinuity. This important observation led us to propose the existence of a thin layer of very low viscosity at the base of the upper mantle in order to obtain geoid kernels whose amplitudes peak at this depth. The possibility of low viscosity zones in the upper mantle has been considered by *Hager and Clayton* [1989] and *Hager and Richards* [1989] although in these studies the lowest viscosities were assigned to a thick 'asthenospheric' channel in the depth range 100 - 400 km. This channel causes the geoid kernels to peak at about 400 km depth and at 670 km depth the degree-2 geoid kernels have negligible amplitude [see Figure 6 in *Forte and Peltier*, 1991b]. The degree–2 $\delta v_S/v_S$-heterogeneity in model $SH8/U4L8$ does not favor such a depth variation for the geoid kernels.

An additional complexity which we have not considered arises from the possibility that some features in the 3-D models of the upper mantle may be due to frequency-dependent anelastic effects [*Romanowicz*, 1990]

The Predicted Plate Motions. The most obvious manifestation of the mantle convective circulation is of course provided by the observed 'drift' of the continents as the tectonic plates move relative to each other. It was clear from the earliest seismic tomographic models [e.g. *Masters et al.*. 1982; *Woodhouse and Dziewonski*, 1984] that the large scale movement of the Earth's plates was driven by the deep-seated buoyancy forces associated with the seismically-inferred lateral density variations [*Peltier and Forte*, 1984; *Forte and Peltier*, 1987a]. This realization led to the detailed mantle-flow studies by *Forte and Peltier* [1987b, 1991a] which demonstrated that the surface flow excited by seismically-inferred density perturbations was in good agreement with the large-scale divergence and convergence of plate velocities at ridges and trenches. From a rheological perspective the plate velocities are very important because, unlike the geoid, they are sensitive to the absolute value of mantle viscosity. This sensitivity has been exploited by *Forte and Peltier* [1987b, 1991b,c] and *Forte et al.* [1991a] to infer the absolute mantle viscosity by matching the observed plate motions to the surface mantle flow predicted on the basis of the seismic tomographic models of heterogeneity.

The vector nature of the plate-velocity field may be completely characterized in terms of two complimentary scalar functions. In *Hager and O'Connell* [1981] the plate velocities were summarized in terms of the so-called poloidal and toroidal generating scalars. In *Forte and Peltier* [1987a,b] the plate velocities $\mathbf{v}(\theta,\varphi)$ were also shown to be completely characterized by their horizontal divergence $\nabla_H \cdot \mathbf{v}$ and by their radial vorticity $\hat{\mathbf{r}} \cdot \nabla \times \mathbf{v}$. The horizontal divergence and radial vorticity scalars are convenient because they are amenable to direct physical interpretation. The field $\nabla_H \cdot \mathbf{v}$ describes the rate of plate divergence at ridges and trenches while the field $\hat{\mathbf{r}} \cdot \nabla \times \mathbf{v}$ describes plate motions at transform plate boundaries [*Forte and Peltier*, 1987a].

As in the case of the nonhydrostatic geoid, it is possible to derive the kernel functions relating the plate-velocity scalars to the internal density perturbations in the mantle. The horizontal divergence kernels $D_\ell(r)$ relate the spherical harmonic coefficients of the horizontal divergence of the predicted surface flow , $(\nabla_H \cdot \mathbf{u})_\ell^m(r=a)$, to the radially-varying spherical harmonic coefficients of the density perturbation $\delta\rho_\ell^m(r)$ as follows :

$$(\nabla_H \cdot \mathbf{u})_\ell^m(r=a) = \frac{g_0}{\eta_0}\int_b^a D_\ell(r)\delta\rho_\ell^m(r)dr, \qquad (33)$$

in which $g_0 = 10\ m/s^2$ is the gravitational acceleration in the mantle and η_0 is the reference viscosity which normalizes the viscosity profile $\eta(r)/\eta_0$ used in the geoid modeling (see previous section). In Figure 17 we show a selection of horizontal divergence kernels $D_\ell(r)$, calculated for the viscosity profile $\eta(r)/\eta_0$ in Figure 14. As in the case of the geoid kernels in Figure 15, the peak amplitude in the longest wavelength horizontal divergence kernels is found in the vicinity of the 670 km seismic discontinuity and this is due to the low-viscosity zone at the base of the upper mantle. Density heterogeneities near the bottom of the upper mantle are therefore most effective in generating surface flow.

The mantle flow predicted by models with a spherically symmetric viscosity, such as in Figure 14, is entirely poloidal and cannot therefore account for the strong toroidal flows (or

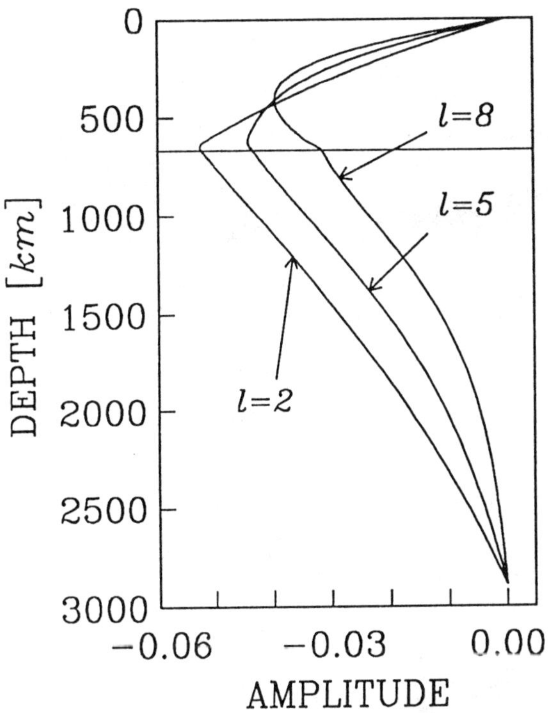

Fig. 17. The horizontal divergence kernels $D_l(r)$, defined in (33), calculated for the viscosity profile shown in Figure 14. The location of the 670 km seismic discontinuity is indicated by the heavy horizontal line.

equivalently, the radial vorticity at transform faults) which characterize the actual plate motions. The toroidal component of the plate velocities may be explained in terms of a lithosphere possessing lateral viscosity variations [e. g. *Ribe*, 1992; *Forte*, 1992]. The tectonic plates themselves are generally regarded as the most extreme manifestation of lateral rheology variations in the mantle [e.g. *Hager and O'Connell*, 1981] and this assumption has motivated several recent studies of mantle flow which have explicitly accounted for the presence of rigid plates. Two approaches have been taken in these studies which may be distinguished by the specific assumptions made regarding the nature of the constraints posed by the presumed rigidity of the tectonic plates. One approach, advocated by *Hager and O'Connell* [1981], involves matching the average stresses acting on the the plates and this is the technique employed by *Ricard and Vigny* [1989] and *Gable et al.* [1991]. The second approach, taken by *Forte and Peltier* [1991b,c], does not require the evaluation of flow-induced stresses on the plates and instead involves an explicit consideration of the limited class of surface flows which are permitted by the requirement that plate motions may only occur by rigid-body rotations around Euler rotation poles. In this second approach it may be shown [*Forte and Peltier*, 1991c] that the rigid-plate constraint on predicted surface flows may ultimately be summarized by the following expressions :

$$(\nabla_H \cdot \mathbf{v})_\ell^m = \sum_{s,t} P_{\ell s}^{mt} (\nabla_H \cdot \mathbf{u})_s^t, \tag{34}$$

$$(\hat{\mathbf{r}} \cdot \nabla \times \mathbf{v})_\ell^m = \sum_{s,t} Q_{\ell s}^{mt} (\nabla_H \cdot \mathbf{u})_s^t, \tag{35}$$

in which $P_{\ell s}^{mt}$ and $Q_{\ell s}^{mt}$ are matrix operators (which depend only on the observed plate geometries) describing the effects of the surface plates on the buoyancy-induced flow in (33), and $(\nabla_H \cdot \mathbf{v})_\ell^m$ and $(\hat{\mathbf{r}} \cdot \nabla \times \mathbf{v})_\ell^m$ are the spherical harmonic coefficients of the predicted plate-like horizontal divergence and radial vorticity, respectively. These predicted plate-like surface flows are sensitive to the absolute mantle viscosity via the reference viscosity η_0 in (33).

On the basis of expressions (33)-(35) we can now calculate the plate-like surface flows expected from the $\delta v_S/v_S$ heterogeneity in model $SH8/U4L8$. In these calculations we shall employ the $\partial ln\rho/\partial ln v_S$ values listed in (31) and the divergence kernels, shown in Figure 17, obtained with the same viscosity profile used in the geoid modeling. The results of these calculations are shown in Figure 18 in which we present the observed horizontal divergence and radial vorticity fields along with the corresponding predicted fields. Although model $SH8/U4L8$ describes lateral heterogeneity only up to spherical harmonic degree $\ell = 8$, the matrix elements in (34) and (35) were calculated up to degree $\ell = 15$ and hence the predictions in Figure 18 also include harmonics up to degree 15. The degree-8 truncation level in model $SH8/U4L8$ is not a serious limitation from our point of view since, as shown by *Su and Dziewonski* [1991] in a recent analysis of global seismic travel-time residuals, the lateral heterogeneity in the upper mantle is dominated by long wavelength features with little power in the degree range $\ell \geq 7$. Also, as shown by *Ricard and Vigny* [1989] and *Forte and Peltier* [1991b,c], the interaction of very-long wavelength mantle flow with the tectonic plates induces shorter wavelength plate motions which agree closely with the corresponding observed plate motions. The agreement between the predicted and observed divergence fields in Figure 18 is evidently very good (global correlation coefficient for the degree range ℓ = 1–15 is +0.85 and the variance reduction is 73%) while the agreement between the predicted and observed radial vorticity fields is not as good (global correlation coefficient in the range ℓ = 1-15 is +0.61 and the variance reduction is 31%). We find that the radial vorticity predictions are especially sensitive to any 'misalignment' between the observed plate boundaries and the corresponding pattern of seismically-inferred density heterogeneities in the mantle.

We have exploited the sensitivity of the predicted plate-motions to the absolute value of mantle viscosity by determining the value of η_0 in (33) which minimizes the least-squares misfit between the observed and predicted plate divergence. In this manner we find $\eta_0 = 1.1 \times 10^{21}$ $Pa\ s$ and we thus infer, on the basis of the $\eta(r)/\eta_0$ profile in Figure 14, that the upper-mantle viscosity is (with the exception of the thin low-viscosity zone at the bottom of the upper-mantle) 1.1×10^{21} $Pa\ s$. Such an inference for the upper-mantle viscosity is encouraging because it agrees well with the 'Haskell'-value of 1×10^{21} $Pa\ s$ which has been

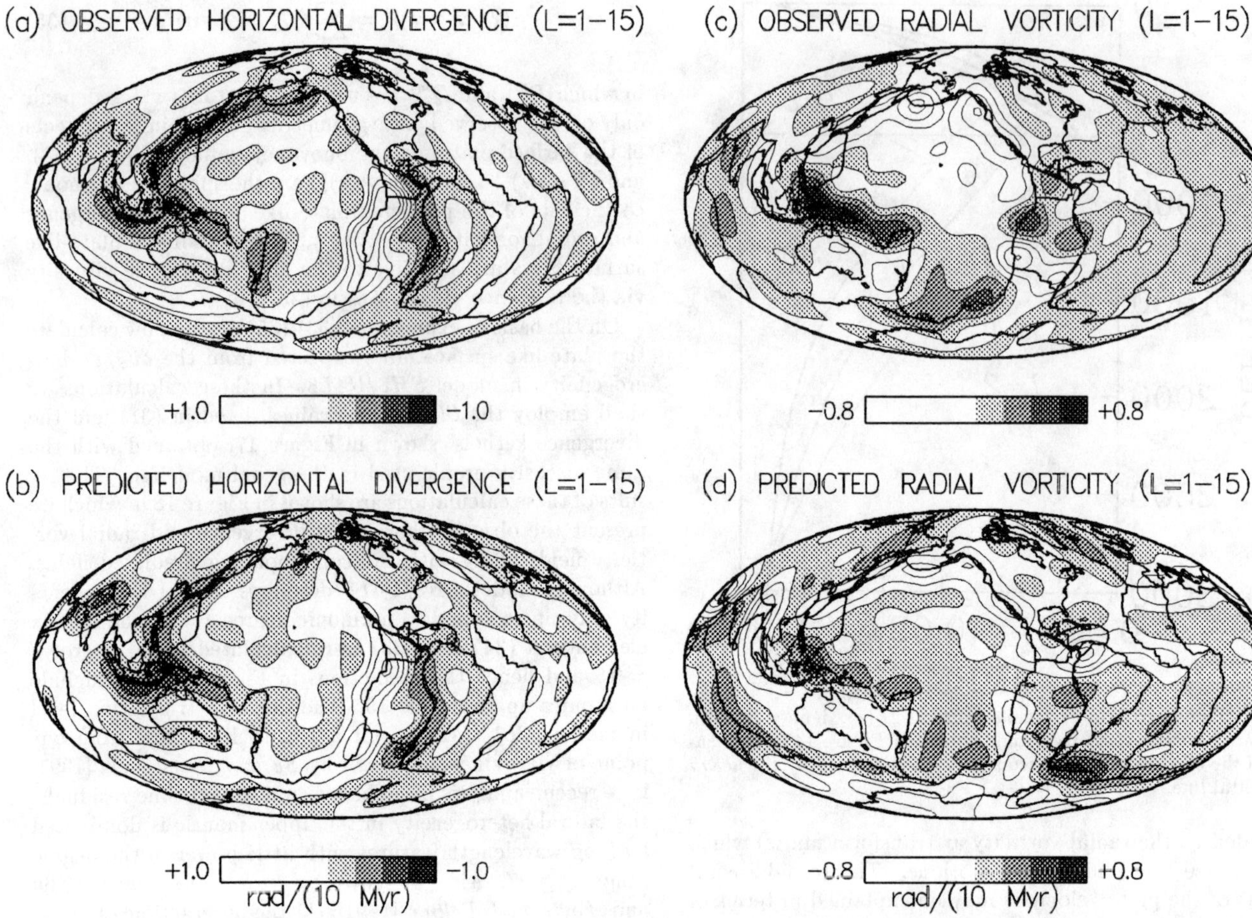

Fig. 18. (a) The horizontal divergence of the Minster and Jordan (1978) tectonic plate velocities (Forte and Peltier, 1987b) in the degree range $\ell = 1-15$. The contour interval is 0.25 $\times 10^{-7}$ *rad/yr*. (b) The horizontal divergence, in the degree range $\ell = 1-15$, predicted with model $SH8/U4L8$ using the mantle viscosity profile in Figure 14 and the $\partial ln\rho/\partial lnv_S$ values in (31). The contour interval is 0.25×10^{-7} *rad/yr*. (c) The radial vorticity of the Minster and Jordan (1978) tectonic plate velocities (Forte and Peltier, 1987b) in the degree range $\ell = 1-15$. The contour interval is 0.2×10^{-7} *rad/yr*. (d) The radial vorticity, in the degree range $\ell = 1-15$, predicted with model $SH8/U4L8$ using the mantle viscosity profile in Figure 14 and the $\partial ln\rho/\partial lnv_S$ values in (31). The contour interval is 0.2×10^{-7} *rad/yr*. In the calculation of the predicted plate motions in (b) and (d), according to (33)-(35), the reference viscosity $\eta_0 = 1.1 \times 10^{21}$ *Pa s* was employed (see text for details).

traditionally inferred from the analysis of glacial isostatic adjustment data [e.g. *Haskell*, 1935, 1936, 1937; *Cathles*, 1971; *Peltier and Andrews*, 1976; *Fjeldskaar and Cathles*, 1991].

Viscous Flow Calculation of the Dynamic CMB Topography. The previous subsections demonstrated that the mantle flow predicted on the basis of the $\delta v_S/v_S$ heterogeneity in model $SH8/U4L8$ provides good matches to both the observed nonhydrostatic geoid and the observed plate motions. In particular we wish to emphasize that the inferences of $\partial ln\rho/\partial lnv_S(r)$ and $\eta(r)$ obtained from the geoid and plate-motion data are in good agreement with independent laboratory data and also the independent viscosity inferences derived from postglacial rebound data. Such a degree of consistency suggests to us that both the geometry and the magnitude of the mantle flow calculated using model $SH8/U4L8$ is realistic. We believe therefore that our predictions of flow-induced CMB topography, presented below, will be equally realistic.

The CMB kernel functions $B_l(r)$ relate the spherical harmonic coefficients of the flow-induced CMB topography, δb_l^m, to the radially-varying coefficients of the driving density contrasts $\delta\rho_l^m(r)$ as follows :

$$\delta b_l^m = \frac{1}{\Delta\rho_{cm}} \int_b^a B_l(r)\delta\rho_l^m(r)dr, \qquad (36)$$

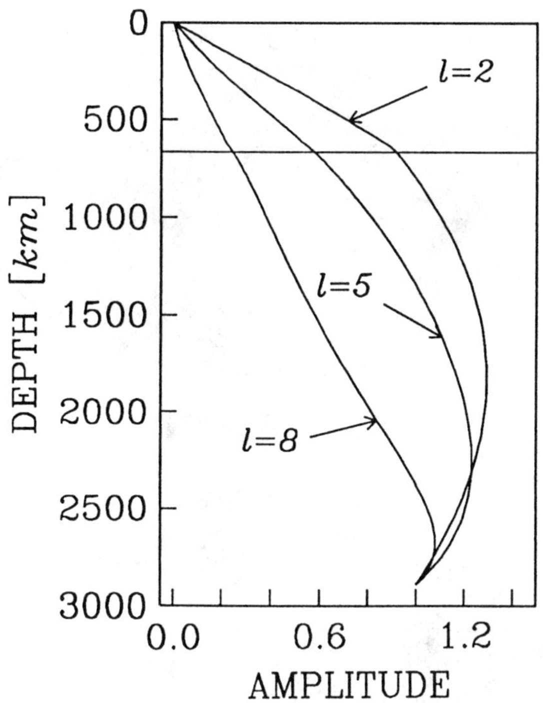

Fig. 19. The CMB topography kernels $B_\ell(r)$, defined in (36), calculated for the viscosity profile shown in Figure 14. The location of the 670 km seismic discontinuity is indicated by the heavy horizontal line.

in which $\Delta\rho_{cm}$ = -4.434 Mg/m^3 is the density jump across the CMB. A selection of CMB topography kernels $B_l(r)$, calculated for the radial viscosity profile in Figure 14, is shown in Figure 19. The predicted CMB topography is, like the geoid, sensitive only to the relative viscosity variations $\eta(r)/\eta_0$ and cannot therefore be used to constrain the absolute mantle viscosity.

The $\partial ln\rho/\partial ln v_S$ values in (31), inferred from the geoid data, and the kernels in Figure 19 will now be employed to calculate the flow-induced CMB topography expected on the basis of the $\delta v_S/v_S$ heterogeneity in model $SH8/U4L8$. In Figure 20a we show the resulting CMB topography prediction and, for comparison, we also show the seismically inferred CMB topography of *Morelli and Dziewonski* [1987; referred to as MD] in Figure 20b. Although the MD model only includes spherical harmonics up to degree $\ell = 4$ it is possible that the least-squares inversion upon which this model is based may have aliased into it the effect of higher degree CMB topography. It is for this reason that we show predicted CMB topography for the degree range $\ell = 1$–8. It is evident that, apart from the mismatch beneath southern Africa and beneath the Indian ocean, the CMB topography prediction is in good agreement with the MD model as both show the well-defined ring of depressed CMB below the circum-Pacific region and the elevated CMB below the central Pacific and Atlantic oceans. It is also rather evident that the major difference between the CMB topography prediction and the MD model is the approximately factor of 1.5 difference in the overall amplitude of the CMB undulations.

An important motivation for calculating the dynamic CMB topography is that the seismic determination of the CMB topography is at least one order of magnitude more difficult than, for example, the mapping of velocity anomalies in the mantle. Consequently, there has been little agreement among papers addressing this issue. In Figure 20c we show model 'SAF' of *Li et al.* [1991; referred to as LGW] retrieved from normal-mode multiplet splitting data, and the 'Model 6' of *Doornbos and Hilton* [1989; referred to as DH] retrieved from PKP and $PKKP$ travel-time residuals is shown in Figure 20d. There is clearly a very good agreement in the patterns of the CMB topography in Figures 20a-c, all of which display the characteristic circum-Pacific depression.

Each of the models of CMB topography shown in Figure 20 show topographic undulations with amplitudes of several kilometers. Such amplitudes are considerably in excess of the ≈ 0.5 km amplitudes advocated by *Hide* [1989] on the basis of estimates of the topographic torque between the mantle and core needed to explain decade-scale length-of-day variations. This difficulty has led some authors [e.g. *Hager and Richards*, 1989] to argue for the existence of a thick (≈ 300 km) layer of very low viscosity at the base of the mantle in order to reduce the amplitude of the flow-induced CMB topography. If, however, we model the possible viscosity decrease across the seismic D" layer in a physically realistic manner then the amplitude of the dynamic CMB topography is not significantly reduced [*Forte and Peltier*, 1991b; *Forte et al.*, 1993a]. The necessity for advocating a small-amplitude CMB topography has been put into serious doubt by the recent studies of *Jault and Le Mouël* [1990] and *Bloxham* [1991] who show that core-mantle torque is very sensitive to the alignment between the CMB undulations and the flow at the top of the core. *Bloxham* [1991] shows that, with small alteration, the core flows inferred from magnetic secular variation data produce acceptable core-mantle torques when they interact with the dynamic CMB undulations shown in Figure 20a.

Dynamic Surface Topography. The mantle convective circulation gives rise to viscous normal stresses which act upon the upper bounding surface of the mantle to deflect it from its hydrostatic reference position. This dynamic surface topography, like the CMB topography, is an observable property of the mantle convective flow. The issue is somewhat confused by the different definitions of this variable. In one recently published paper [*Cazenave and Lago*, 1991], the authors considered only the oceans and removed the topographic signature of the cooling oceanic lithosphere as a function of the seafloor age. Clearly, this topography is a part of the dynamic process [e.g. *Jarvis and Peltier*, 1982] and it is reflected in the tomographic models by the age dependence of seismic velocities in the lithosphere: see Figure 2 in *Su et al* [1992], for example. The following is a brief account of the development presented by *Forte et al.* [1993b].

As in the case of the dynamic CMB topography, we may

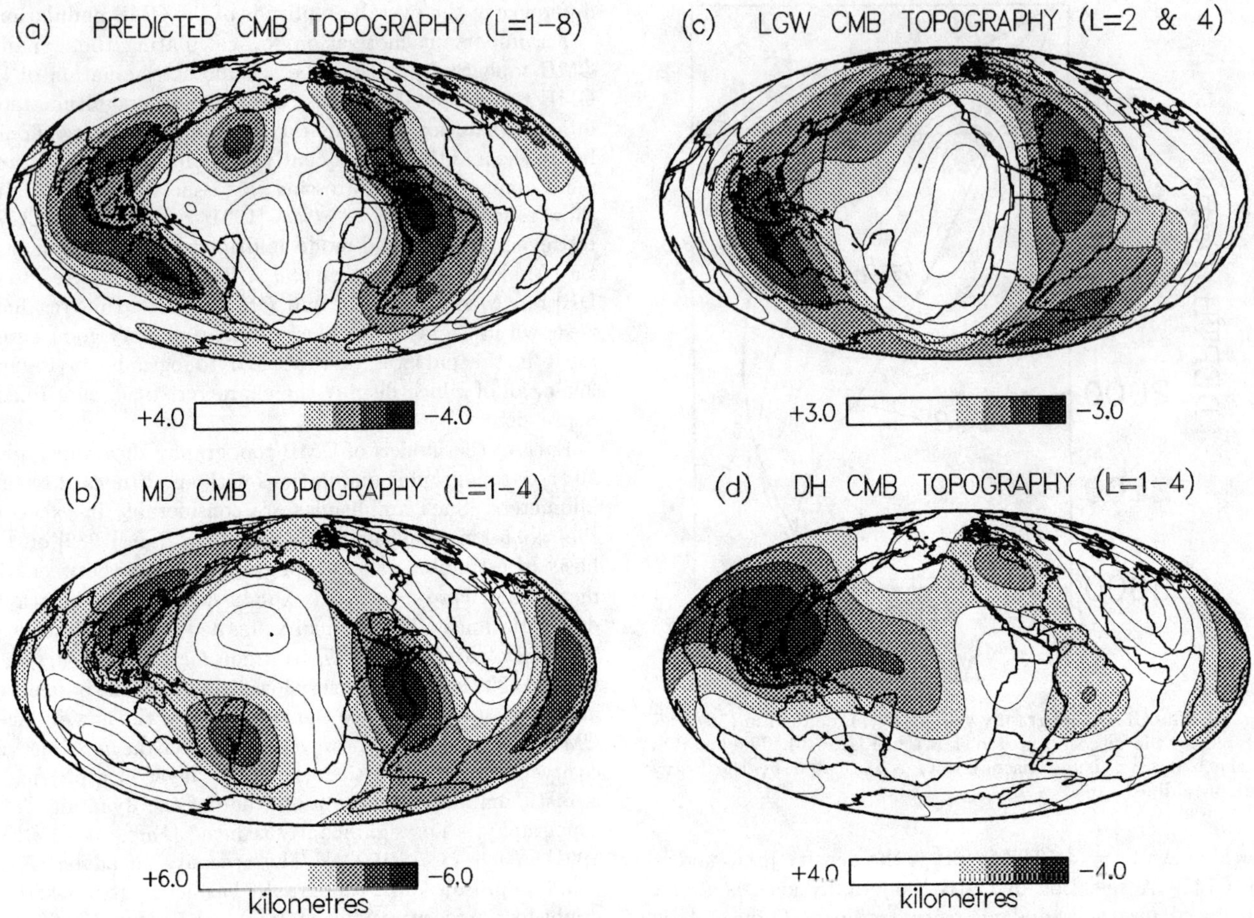

Fig. 20. (a) The flow-induced CMB topography, in the degree range $\ell = 1 - 8$, predicted with model $SH8/U4L8$ using the viscosity profile in Figure 14 and the $\partial ln\rho/\partial ln v_S$ values in (31). The contour interval is 1.0 kilometers. (b) The seismically-inferred CMB topography model of Morelli and Dziewonski (1987) in the degree range $\ell = 1 - 4$. The contour interval is 1.5 kilometers. (c) The seismically-inferred CMB topography model "SAF" of Li *et al.* (1991) for harmonic degrees $\ell = 2$ and 4. The contour interval is 0.75 kilometers. (d) The seismically-inferred CMB topography model "Model 6" of Doornbos and Hilton (1989) in the degree range $\ell = 1 - 4$. The contour interval is 1 kilometer.

calculate the relationship between the spherical harmonic coefficients of the flow-induced surface topography δa_ℓ^m and the radially varying coefficients of the driving density contrasts $\delta\rho_\ell^m(r)$ as follows:

$$\delta a_\ell^m = \frac{1}{\Delta_{mo}} \int_b^a A_\ell(r)\, \delta\rho_\ell^m(r)\, dr; \qquad (37)$$

in which $\Delta_{mo} = 2.2 Mg/m^3$ is the density jump across the mantle-ocean boundary. A selection of surface topography kernels $A_\ell(r)$, calculated for the radial viscosity profile in Figure 14 is shown in Figure 21. The surface topography kernels attain a value identically equal to 1 at the top surface and, according to (37), this shows that any density contrasts placed on the bounding surface itself are isostatically compensated by the ensuing deformation of this boundary. The kernels $A_\ell(r)$ are, as in the case of the CMB topography kernels, sensitive only to relative viscosity variations $\eta(r)/\eta_0$.

A comparison between the actual dynamic topography at the Earth's surface and the surface topography predicted using (37) is not straightforward, because the global topographic signal arising from mantle convection is apparently overwhelmed by the global surface topography associated with the isostatic compensation of continent-ocean crustal density differences. In Figure 22a we show the long wavelength (ℓ = 1–8) component of the actual topography at the Earth's surface which has been obtained from the ETOPO5 bathymetry-topography data base [*National Geophysical Data Center*, 1986]. In this figure it is quite evident that the observed topography is almost completely

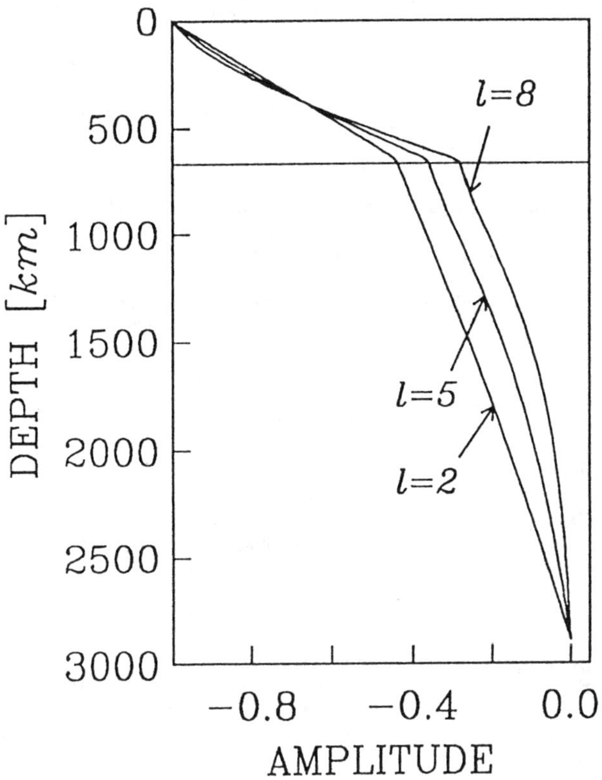

Fig. 21. The dynamic surface topography kernels $A_l(r)$, defined in (37), calculated for the viscosity profile shown in Figure 14. The location of the 670 km seismic discontinuity is indicated by the heavy horizontal line.

characterized by elevated continents and depressed ocean basins, and this is of course a well known manifestation of the isostatic compensation of the density contrast between the thick (light) continental crust and thin (heavy) oceanic crust [e.g. *Love*, 1911; *Jeffreys*, 1970]. To determine the dynamic (*i.e.* nonhydrostatic) contribution to the observed surface topography it is clear that we must first remove the

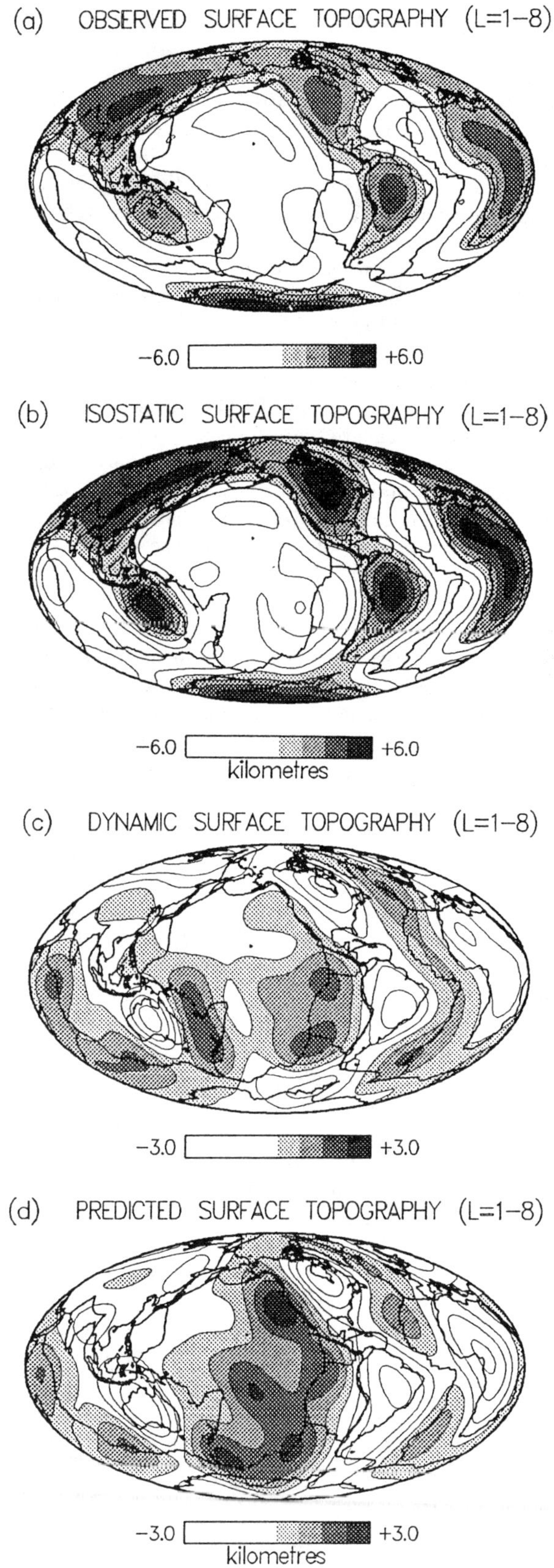

Fig. 22. (a) The solid surface topography obtained from the ETOPO5 data base (National Geophysical Data Center, 1986) calculated for the degree range $\ell = 1 - 8$. The contour interval is 1.5 km. (b) The isostatic solid surface topography described in the text, calculated for the degree range $\ell = 1 - 8$. The contour interval is 1.6 km. (c) The estimated dynamic component of the solid surface topography obtained by subtracting the field in (b) from the field in (a). The contour interval is 0.875 km. (d) The flow-induced solid surface topography, in the degree range $\ell = 1 - 8$, predicted with model $SH8/U4L8$ using the viscosity profile in Figure 14 and the $\partial ln\rho/\partial ln v_S$ values in (31). The contour interval is 0.875 km.

topography due to the isostatic compensation of continents and oceans.

Although various specifications of the isostasy principle have been formulated [*Love*, 1911; *Garland, 1977*; *Dahlen*, 1982], they all yield, to the first order accuracy, the well known local condition that the total mass in any given vertical column is constant to some depth (the so-called 'depth of compensation'). We have determined the isostatic solid-surface topography by separately determining the radial locations of the solid surface, the midr - crust boundary, and the Mohorovicic discontinuity in the continental and the oceanic portions of the crust, so that the mass in any vertical column is constant.

In this calculation we employ the continent-ocean crustal density contrasts given in the PEM models [*Dziewonski et al.*, 1975] and we ensure that all density perturbations and boundary perturbations average to zero on the sphere, so as to preserve the total mass and the scalar moment of inertia of the Earth. The calculated difference between the isostatic radial location of the solid surface in the oceans and the continents is shown in Figure 22b, from which we clearly see, as expected, that the observed topography in Figure 22a is dominated by the isostatic compensation of the laterally heterogeneous crust. We may now obtain an estimate of the actual dynamic component of the solid-surface topography by subtracting the isostatic topography in Figure 22b from the actual topography in Figure 22a. This difference – which we call the observed dynamic surface topography – is shown in Figure 22c.

The dynamic surface topography calculated using the kernels in Figure 21, the velocity-density proportionality factors in (31) inferred from the geoid data and the $\delta v_S/v_S$ heterogeneity in $SH8/U4L8$ is shown in Figure 22d. The inferred dynamic topography in Figure 22c and our predictions agree quite well, both in the spatial pattern and in amplitude.

This agreement provides important confirmation of the overall validity of our viscous flow calculations, which have now been shown to provide good fits to a diverse collection of global observables such as the geoid, plate motions. CMB topography and, now, solid surface topography. An important common characteristic of the inferred and predicted dynamic topography is the significant depression of the surface of the continents and elevation of the mid-oceanic ridges. The continental depressions are evidently a manifestation of the dynamic 'suction' which develops as a consequence of the flow driven by the deep high-velocity anomalies which in model $SH8/U4L8$ extend right to the core-mantle boundary below North America, Australia and Antarctica.

Acknowledgements. The authors wish to thank Barbara Romanowicz for many helpful and insightful comments in her review. The waveform portion of the data set from which models $SH8/U4L8$ and $SH8/WM13$ were derived and several programs used in this study were developed in collaboration with John Woodhouse now at Oxford University. We thank the staffs of the seismographic networks GDSN and IDA for their efforts in collecting data used in this study. We are grateful to Yu-Szen Zhang for sending us a copy of his thesis and numerical details of his upper mantle S-velocity model. This work has been supported by grant EAR90-05013 from the National Science Foundation and by a Canadian NSERC postdoctoral fellowship awarded to A.M.F.

REFERENCES

Abramowitz, M., and I. A. Stegun. *Handbook of Mathematical Functions*, Dover, New York, 1965.

Agnew, D., J. Berger, R. Buland, W. Farrell, and F. Gilbert, International deployment of accelerometers: A network of very long period seismology. *EOS Trans. AGU, 57*, 180-188, 1976.

Anderson, D.L., A seismic equation of state II. Shear properties and thermodynamics of the lower mantle, *Phys. Earth Planet. Inter., 45*, 307-323, 1987.

Anderson, O.L., E. Schreiber, R.C. Lieberman, and N. Soga, Some elastic constant data on minerals relevant to geophysics, *Rev. Geophys. Space Phys., 6*, 491-524, 1968.

Anderson, O.L., A. Chopelas, and R. Boehler, Thermal expansion vs. pressure at constant temperature: a re-examination, *Geophys. Res. Lett., 17*, 685-688, 1990.

Bloxham, J., Length-of-day variations, topographic core-mantle coupling, and the steady and time-dependent components of core-flow, *EOS Trans. AGU,72*,451,1991.

Castillo, P.R., The Dupal anomaly–low velocity regions of the lower mantle correlation: implications for mantle convection, *EOS Trans. AGU, 69*, 490-491, 1988.

Cathles, L.M., The Viscosity of the Earth's Mantle, Ph.D. dissertation, Princeton University, Princeton, New Jersey, 1971.

Cazenave, A., and B. Lago, Long wavelength topography, seafloor subsidence and flattening, *Geophys. Res. Lett., 18*, 1257-1260, 1991.

Chopelas, A., New accurate sound velocity measurements of lower mantle materials at very high pressures, *EOS Trans. AGU, 69*, 1460, 1988.

Chopelas, A., Thermal expansion, heat capacity, and entropy of MgO at mantle pressures, *Phys. Chem. Minerals, 17*, 249-257, 1990.

Chopelas, A., and R. Boehler, Thermal expansion measurements at very high pressure, systematics, and a case for a chemically homogeneous mantle, *Geophys. Res. Lett., 16*, 1347-1350, 1989.

Clayton, R.W., and R.P. Comer, A tomographic analysis of mantle heterogeneities from body wave travel time data, *EOS Trans. AGU, 64*, 776, 1983.

Clayton, R. W. and R. P. Comer, A tomographic analysis of mantle heterogeneities, *Terra Cognita*, **4**, 282-283, 1984.

Dahlen, F.A., The normal modes of a rotating, elliptical earth, *Geophys. J. R. Astron. Soc., 16* 329–367, 1968.

Dahlen, F.A., Models of the lateral heterogeneity of the earth consistent with eigenfrequency splitting data, *Geophys. J. R. Astron. Soc., 44*, 77–105, 1976.

Dahlen, F. A., Isostatic geoid anomalies on a sphere, *J. Geophys. Res., 87*,, 3943–3947, 1982.

Doornbos, D.J., and T. Hilton, Models of the core-mantle boundary and the travel times of internally reflected core phases, *J. Geophys. Res., 94*, 15,741-15,751, 1989.

Dupre, B. and C.J. Allegre, Pb-Sr isotope variation in Indian Ocean basalts and mixing phenomena, *Nature, 303*, 142–146, 1983.

Dziewonski, A.M., Resolution of large scale velocity anomalies in the mantle, *EOS Trans. AGU, 56*, 395, 1975.

Dziewonski, A.M., Mapping the lower mantle, *EOS Trans. AGU, 63*, 1035, 1982.

Dziewonski, A.M., Mapping the lower mantle: Determination of lateral heterogeneity in P velocity up to degree and order 6, *J. Geophys. Res., 89*, 5929-5952, 1984.

Dziewonski, A.M., and F. Gilbert, The effect of small, aspherical perturbations on travel times and a re-examination of the corrections for ellipticity, *Geophys. J. R. Astr. Soc.*, *44*, 7-18, 1976.

Dziewonski, A.M., and D.L. Anderson, Preliminary reference Earth model, *Phys. Earth Planet. Inter.*, *25*, 297-356, 1981.

Dziewonski, A.M., and J. Steim, Dispersion and attenuation of mantle waves through wave-form inversion, *Geophys. J. R. Astr. Soc.*, *70*, 503-527, 1982.

Dziewonski, A.M., and J.H. Woodhouse, Global images of the Earth's interior, *Science*, *236*, 37-48, 1987.

Dziewonski, A.M., and R.L. Woodward, Acoustic imaging at the planetary scale, in *Acoustical Imaging*, *19*, H. Ermert and H.-P. Harjes, eds., Plenum Press, New York, 785-797, 1992.

Dziewonski, A. M., T.-A. Chou and J. H. Woodhouse, Determination of earthquake source parameters from waveform data for studies of global and regional seismicity, *J. Geophys. Res.*, *86*, 2825-2952, 1981.

Dziewonski, A. M., A. L. Hales and E. R. Lapwood, Parametrically simple Earth models consistent with geophysical data, *Phys. Earth Planet. Int.*, *10*, 12-48, 1975.

Dziewonski, A. M., B.H. Hager, and R.J. O'Connell, Large scale heterogeneity in the lower mantle, *J. Geophys. Res.*, *82*, 239-255, 1977.

Ekström, G., A.M. Dziewonski, and J. Ibañez, Deep earthquakes outside slabs, *EOS Trans. AGU, 71* 1462, 1990.

Fjeldskaar, W., and L.M. Cathles, Rheology of mantle and lithosphere inferred from post-glacial uplift in Fennoscandia, in *Glacial Isostasy, Sea-Level and Mantle Rheology*, R. Sabadini, K. Lambeck, and E. Boschi, eds., NATO ASI Series *334*, 1-19, 1991.

Forte, A.M., Mantle Convection and Global Geophysical Observables, Ph.D. dissertation, University of Toronto, Toronto, 1989.

Forte, A. M., The kinematics and dynamics of poloidal–toroidal coupling of mantle flow, *EOS Trans. AGU, 73*, 1992 Spring Meeting Suppl., 273, 1992.

Forte, A.M., and W.R. Peltier, Surface plate kinematics and mantle convection, in *The Composition, Structure, and Dynamics of the Lithosphere-Asthenosphere System*, K. Fuchs and C. Froidevaux, eds., AGU Geodynamic Series *16*, 125-136, 1987a.

Forte, A. M., and W.R. Peltier, Plate tectonics and aspherical Earth structure: The importance of poloidal-toroidal coupling, *J. Geophys. Res..*, *92*, 3645-3679, 1987b.

Forte, A.M., and W.R. Peltier, Core-mantle boundary topography and whole-mantle convection, *Geophys. Res. Lett.*, *16*, 621-624, 1989.

Forte, A.M., and W.R. Peltier, Mantle convection and core-mantle boundary topography: Explanations and implications, in *Silver Anniversary of Plate Tectonics*, T.W.C. Hilde and R.L. Carlson, eds., *Tectonophysics*, *187*, 91-116, 1991a.

Forte, A.M., and W.R. Peltier, Viscous flow models of global geophysical observables. I. Forward problems, *J. Geophys. Res.*, *96*, 20,131-20,159, 1991b.

Forte, A.M., and W.R. Peltier, Gross Earth data and mantle convection: New inferences of mantle viscosity, in *Glacial Isostasy, Sea-Level and Mantle Rheology*, R. Sabadini, K. Lambeck, and E. Boschi, eds., NATO ASI Series *334*, 425-444, 1991c.

Forte, A.M., W.R. Peltier, and A.M. Dziewonski, Inferences of mantle viscosity from tectonic plate velocities, *Geophys. Res. Lett.*, *18*, 1747-1750, 1991a.

Forte, A.M., A.M. Dziewonski, and R.L. Woodward, Geodynamic implications of seismically inferred heterogeneity, *EOS Trans. AGU, 72*, 1991 Fall Meeting Suppl., 451, 1991b.

Forte, A. M., R. L. Woodward, A. M. Dziewonski and W. R. Peltier, 3-D models of mantle heterogeneity derived from joint inversion of seismic and geodynamic data, *EOS, Trans. AGU, 73,*, 1992 Spring Meeting Suppl., 200, 1992a.

Forte, A.M., A.M. Dziewonski, and R.L. Woodward, Aspherical structure of the mantle, tectonic plate motions, nonhydrostatic geoid, and topography of the core-mantle boundary, in *Proceedings of IUGG Symposium 6: Dynamics of the Earth's Deep Interior and Earth Rotation*, edited by J.-L. Le Mouël, AGU , in press, 1993a.

Forte, A. M., W. R. Peltier, A. M. Dziewonski and R. L. Woodward, Dynamic surface topography: a new interpretation based upon mantle flow models derived from seismic topography, *Geophys. Res. Lett.*, in press, 1993b.

Fukao, Y., M. Obayashi, H. Inoue, and M. Nenbai, Subducting slabs stagnant in the mantle transition zone, *J. Geophys. Res.* it 97,4809-4822, 1992.

Gable, C.W., R.J. O'Connell, and B.J. Travis, Convection in three dimensions with surface plates, *J. Geophys. Res.*, *96*, 8391-8405, 1991.

Garland, G. D., *The Earth's Shape and Gravity*, Pergamon Press (Oxford), 1977.

Giardini, D., X.-D. Li, and J.H. Woodhouse, Three dimensional structure of the Earth from splitting in free oscillation spectra, *Nature*, *325*, 405-411, 1987.

Giardini, D., X.-D. Li, and J.H. Woodhouse, The splitting functions of long period normal modes of the Earth, *J. Geophys. Res.*, *93*, 13,716-13,742, 1988.

Gilbert, F., Ranking and winnowing gross earth data for inversion and resolution, *Geophys. J. R. Astron. Soc.*, *23*, 125–128, 1971.

Gilbert, F., and A.M. Dziewonski, An application of normal mode theory to the retrieval of structural parameters and source mechanisms from seismic spectra, *Phil. Trans. R. Soc. London, A278*, 187–269,1975.

Grand, S. P., Tomographic inversions for shear structure beneath the North American plate, *J. Geophys. Res.*, *92*, 14065-14090, 1987.

Gudmundsson, O., J. H. Davies and R. W. Clayton, Stochastic analysis of global travel time data: mantle heterogeneity and random errors in the ISC data, *Geophys. J. Int.*, *102*, 25-43, 1990.

Gurnis, M. and S.-J. Zhong, Generation of long wavelength heterogeneity in the mantle by the dynamic interaction between plates and mantle convection. *Geophy. Res. Lett.*, *18*, 581-584, 1990.

Hager, B.H., and R.J. O'Connell, A simple global model of plate dynamics and mantle convection, *J. Geophys. Res.*, *86*, 4843-4867, 1981.

Hager, B.H., R.W. Clayton, M.A. Richards, R.P. Comer, and A.M. Dziewonski, Lower mantle heterogeneity, dynamic topography and the geoid, *Nature*, *313*, 541-545, 1985.

Hager, B.H., and R.W. Clayton, Constraints on the structure of mantle convection using seismic observations, flow models, and the geoid, in *Mantle Convection*, W.R. Peltier, ed., Gordon and Breach Science Publishers, 657-763, 1989.

Hager, B.H., and M.A. Richards, Long-wavelength variations in the Earth's geoid: physical models and dynamical implications, *Phil. Trans. R. Soc. Lond. A*, *328*, 309-327, 1989.

Hart, S.R., A large-scale isotope anomaly in the Southern Hemisphere mantle, *Nature*, *309*, 753–757, 1984.

Haskell, N.A., The motion of a viscous fluid under a surface load, *Physics*, *6*, 265-269, 1935.

Haskell, N.A., The motion of a viscous fluid under a surface load, Part II., *Physics*, *7*, 56-61, 1936.

Haskell, N.A., The viscosity of the asthenosphere, *Amer. J. Sci.*, *33*, 22-28, 1937.

Hide, R., Fluctuations in the Earth's rotation and the topography of the core-mantle interface, *Phil. Trans. R. Soc. Lond. A*, *328*, 351-363, 1989.

Inoue, H., Y. Fukao, K. Tanabe, and Y. Ogata, Whole mantle P-wave travel time tomography, *Phys. Earth Planet. Inter.*, *59*, 294-328, 1990.

Isaak, D.G., O.L. Anderson, T. Goto, and I. Suzuki, Elasticity of single-crystal forsterite measured up to 1700 K, *J. Geophys. Res., 94*, 10,637-10,646, 1989.

Jarvis, G.T., and W.R. Peltier, Mantle convection as a boundary layer phenomenon, *Geophys. J. R. Astr. Soc., 68*, 385-424, 1982.

Jault, D., and J.-L. Le Mouël, Core-mantle boundary shape: constraints inferred from the pressure torque acting between the core and mantle, *Geophys. J. Int., 101*, 233-241, 1990.

Jeffreys, H., On the hydrostatic theory of the figure of the Earth, *Geophys. J. R. Astr. Soc., 8*, 196-202, 1963.

Jeffreys, H., *The Earth*, 5th ed., Cambridge University Press (New York), 1970.

Jordan, T.H., The continental tectosphere, *Rev. Geophys. Space Phys., 13*, 1-12, 1975.

Jordan, T. H., A procedure for estimating lateral variations from low-frequency eigenspectra data, *Geophys. J. R. Astron. Soc., 52*, 441-455, 1978a.

Jordan, T.H., Composition and development of the continental tectosphere, *Nature, 274*, 544–548, 1978b.

Jordan, T.H., Mineralogies, densities, and seismic velocities of garnet lherzolites and their geophysical implications, in *The Mantle Sample: Inclusions in Kimberlites and Other Volcanics*, F.R. Boyd and H.O.A. Meyer, eds., AGU Publications, 1-14, 1979.

Jordan, T. and W. S. Lynn, A velocity anomaly in the lower mantle, *J. Geophys. Res., 79*, 2679-2685, 1974.

Karato, S.-I., and P. Li, Diffusion creep in perovskite: implications for the rheology of the lower mantle, *Science, 255*, 1238-1240, 1992.

Laj, C., A. Mazaud, R. Weeks, M. Fuller and E. Herrero-Bervera, Geomagnetic reversal paths, *Nature*, **351**, 447 (1991).

Lancsoz, C., *Linear Differential Operators*, D. Van Nostrand, London, 1961.

Lognonne, P., and B. Romanowicz, Modeling of coupled modes of the earth: the spectral method, *Geophys. J. Int.,102*, 365-395, 1990.

Li, X.-D. and T. Tanimoto, Waveforms of long-period body waves in a slightly aspherical Earth model, *Geophys. J. Int.*, in press, 1992.

Li, X.-D., D. Giardini, and J.H. Woodhouse, Large-scale three-dimensional even-degree structure of the Earth from splitting of long-period normal modes, *J. Geophys. Res., 96*, 551-577, 1991a.

Li, X.-D., D. Giardini, and J.H. Woodhouse, The relative amplitudes of mantle heterogeneity in P velocity, S velocity and density from free oscillation data, *Geophys. J. Int., 105*, 649–657, 1991.

Love, A. E. H., *Some Problems of Geodynamics*, Cambridge University Press, 1911.

Marsh, J.G., F.J. Lerch, B.H. Putney, T.L. Felsentreger, B.V. Sanchez, S.M. Klosko, G.B. Patel, J.W. Robbins, R.G. Williamson, T.L. Engelis, W.F. Eddy, N.L. Chandler, D.S. Chinn, S. Kapoor, K.E. Rachlin, L.E. Braatz, and E.C. Pavlis, The GEM-T2 gravitational model, *J. Geophys. Res., 95*, 22,043-22,071, 1990.

Masters, G., and F. Gilbert, Structure of the inner core inferred from observations of its spheroidal shear modes, *Geophys. Res. Lett., 8*, 569-571, 1981.

Masters, G., T.H. Jordan, P.G. Silver, and F. Gilbert, Aspherical earth structure from fundamental spheroidal mode data, *Nature, 298*, 609-613, 1982.

Minster, J.B., and T.H. Jordan, Present-day plate motions, *J. Geophys. Res., 83*, 5331-5354, 1978.

Montagner, J.P., and T. Tanimoto, Global anisotropy in the upper mantle inferred from the regionalization of phase velocities, *J. Geophys. Res., 95*, 4797-4819, 1990.

Montagner, J.P., and T. Tanimoto, Global upper mantle tomography of seismic velocities and anisotropies, it J. Geophys. Res., 96, 20337–20351, 1991.

Morelli, A., and A.M. Dziewonski, Topography of the core-mantle boundary and lateral homogeneity of the liquid core, *Nature, 325*, 678-683, 1987.

Morelli, A., and A.M. Dziewonski, Joint determination of lateral heterogeneity and earthquake location, In *Glacial Isostasy, Sea-Level and Mantle Rheology*, R. Sabadini, K. Lambeck, and E. Boschi, eds., NATO ASI Series *334*, 515-534, 1991.

Nakanishi, I., and D.L. Anderson, Worldwide distribution of group velocity of mantle Rayleigh waves as determined by spherical harmonic inversion, *Bull. Seism. Soc. Am., 72*, 1185-1194, 1982.

Nakiboglu, S.M., Hydrostatic theory of the Earth and its mechanical implications, *Phys. Earth Planet. Int., 28*, 302-311, 1982.

Nataf, H.-C., I. Nakanishi, and D.L. Anderson, Anisotropy and shear velocity heterogeneities in the upper mantle, *Geophys. Res. Lett., 11*, 109-112, 1984.

Nataf, H.-C., I. Nakanishi, and D.L. Anderson, Measurement of mantle wave velocities and inversion for lateral heterogeneity and anisotropy, III. Inversion, *J. Geophys. Res., 91*, 7261-7307, 1986.

Nicolas, A., and J.P. Poirier, *Crystalline Plasticity and Solid State Flow in Metamorphic Rocks*, John Wiley and Sons, 444pp., 1976.

Nolet, G., Partitioned waveform inversion and two-dimensional structure under the Network of Autonomously Recording Seismographs, *J. Geophys. Res., 95*, 8499-8512, 1990.

Olson, P., P.G. Silver, and R.W. Carlson, The large-scale structure of convection in the Earth's mantle, *Nature, 344* 209–215, 1990.

Peltier, W.R., and J.T. Andrews, Glacial-isostatic adjustment-I. The forward problem, *Geophys. J. R. Astr. Soc., 46*, 605-646, 1976.

Peltier, W.R., and A.M. Forte, The gravitational signature of plate tectonics, *Terra Cognita, 4*, 251, 1984.

Peterson, J., M. Howell, L. Butler, G. Holcomb, and C. R. Hutt, The Seismic Research Observatory, *Bull. Seism. Soc. Am., 66*, 2049-2068, 1976.

Reynard, B., and G.D. Price, Thermal expansion of mantle minerals at high pressure - a theoretical study, *Geophys. Res. Lett., 17*, 689-692, 1990.

Ribe, N. M., A thin-shell model for the interaction of surface plates and convection, *EOS Trans. AGU, 71*, 1992 Spring Meeting Suppl., 273, 1992.

Ricard, Y., L. Fleitout, and C. Froidevaux, Geoid heights and lithospheric stresses for a dynamic Earth, *Ann. Geophys., 2*, 267-286, 1984.

Ricard, Y., and C. Vigny, Mantle dynamics with induced plate tectonics, *J. Geophys. Res., 94*, 17,543-17,559 , 1989.

Ricard, Y., C. Doglioni, and R. Sabadini, Differential rotation between lithosphere and mantle: A consequence of lateral mantle viscosity variations, *J. Geophys. Res., 96*, 8407-8415, 1991.

Richards, M.A., and B.H. Hager, Geoid anomalies in a dynamic Earth, *J. Geophys. Res., 89*, 5987-6002, 1984.

Richards, M. A. and D. C. Engebretson, Large-scale mantle convection and the history of subduction, *Nature, 355*, 437-440, 1992.

Ritzwoller, M., G. Masters, and F. Gilbert, Observations of anomalous splitting and their interpretation in terms of aspherical structure, *J. Geophys. Res., 91*, 10,203-10,228. 1986.

Ritzwoller, M., G. Masters, and F. Gilbert, Constraining aspherical Earth structure with low-degree interaction coefficients: application to uncoupled multiplets, *J. Geophys. Res., 93*, 6369-6396, 1988.

Romanowicz, B., Multiplet–multiplet coupling due to lateral heterogeneity: asymptotic effects on the amplitude and frequency of the earth's normal modes, *Geophys. J. R. Astron. Soc., 90*, 75-100, 1987.

Romanowicz, B., The upper mantle degree two: constraints and inferences on attenuation tomography from global mantle wave measurements, *J. Geophys. Res., 95*, 11051–11071, 1990.

Romanowicz, B., Seismic tomography of the Earth's mantle, *Annu. Rev. Earth Planet. Sci., 19*, 77-99, 1991.

Romanowicz, B., G. Roult and T. Kohl, The upper mantle degree two pattern: constraints from Geoscope fundamental spheroidal mode eigenfrequency and attenuation measurements, *Geophys. Res. Let., 14*, 1219-22, 1987.

Roult, G., B. Romanowicz and J.P. Montagner, 3D upper mantle shear velocity and attenuation from fundamental mode free oscillation data, *Geophys. J. Int., 101*, 61-80, 1990.

Sheehan, A.F. and S.C. Solomon, Joint inversion of shear wave travel time residuals and geoid and depth anomalies for long-wavelength variations in upper mantle temperature and composition along the Mid-Atlantic Ridge, *J. Geophys. Res., 96* 19981–20009, 1991.

Shure, J., R. L. Parker and G. E. Backus, Harmonic splines for geomagnetic modeling, *Phys. Earth Planet. Inter., 28*, 215-229, 1982.

Snieder, R., J. Beckers and F. Neele, The effect of small-scale structure on normal mode frequencies and global inversions. *J. Geophys. Res., 96*, 501-515, 1991.

Solheim, L.P., and W.R. Peltier, Heat transfer and the onset of chaos in a spherical, axisymmetric anelastic model of whole mantle convection, *Geophys. Astrophys. Fluid Dyn., 53*, 205-255, 1990.

Stacey, F.D., and D.E. Loper, The thermal boundary-layer interpretation of D" and its role as a plume source, *Phys. Earth Planet. Int., 33*, 45-55, 1983.

Stocker, R.L., and M.F. Ashby, On the rheology of the upper mantle, *Rev. Geophys. Space Phys., 11*, 391-426, 1973.

Su, W.-J., and A.M. Dziewonski, Predominance of long-wavelength heterogeneity in the mantle, *Nature, 352*, 121-126, 1991.

Su, W.-J. and A. M. Dziewonski, On the scale of mantle heterogeneity, *Phys. Earth Planet. Inter.*, it 74, 29–54, 1992.

Su, W.-J., R. L. Woodward and A. M. Dziewonski, Deep origin of mid-oceanic ridge velocity anomalies, *Nature, 360*, 149-152, 1992.

Tanimoto, T., The three dimensional shear wave structure in the mantle by overtone waveform inversion I. Radial seismogram inversion, *Geophys. J. R. Astr. Soc., 89*, 713-740, 1987.

Tanimoto, T., Long-wavelength S-wave velocity structure throughout the mantle, *Geophys. J. Int., 100*, 327-336, 1990.

Tanimoto, T., Predominance of large-scale heterogeneity and the shift of velocity anomalies between the upper and lower mantle, *J. Phys. Earth, 38*, 493–509, 1991.

Turcotte, D.L., and G. Schubert, Structure of the olivine-spinel phase boundary in the descending lithosphere, *J. Geophys. Res., 76*, 7980-7987, 1971.

van der Hilst, R. D. and E. R. Engdahl, On ISC *PP* and *pP* data and their use in delay-time tomography of the Caribbean region, *Geophys. J. Int.*, **106**, 169-188 (1991).

van der Hilst, R., R. Engdahl, W. Spakman, and G. Nolet, Tomographic imaging of subducted lithosphere below northwest Pacific island arcs, *Nature, 353* 37–43, 1991.

Verhoogen, J., Phase changes and convection in the Earth's mantle, *Phil. Trans. R. Soc. Lond. A, 258*, 276-283, 1965.

Weertman, J., and J.R. Weertman, High temperature creep of rock and mantle viscosity, *Annual Rev. Earth Planet. Sci., 3*, 293-315, 1975.

Widmer, R., The large-scale structure of the deep earth as constrained by free oscillation observations, *Ph.D. thesis*, Univ. of California, San Diego, Scripps Inst. of Oceanography, 1991.

Woodhouse, J.H., and A.M. Dziewonski, Mapping the upper mantle: Three dimensional modeling of Earth structure by inversion of seismic waveforms, *J. Geophys. Res., 89*, 5953-5986, 1984.

Woodhouse, J.H., and A.M. Dziewonski, Three dimensional mantle models based on mantle wave and long period body wave data, *EOS Trans. AGU, 67*, 307, 1986.

Woodhouse, J.H., and A.M. Dziewonski, Seismic modeling of the Earth's large-scale three dimensional structure, *Philos. Trans. R. Soc. Lond. A, 328*, 291-308, 1989.

Woodhouse, J.H., and D. Giardini, Inversion for the splitting function of isolated low order normal mode multiplets, *EOS Trans. AGU, 66*, 301, 1985.

Woodhouse, J. H., D. Giardini and X.-D. Li, Evidence for inner core anisotropy from free oscillations, *Geophys. Res. Lett.* **13**, 1549-1552, 1986.

Woodward, R. L., Structure of the Earth's upper mantle from long-period seismic data, *Ph.D. thesis*, Univ. of California, San Diego, Scripps Inst. of Oceanography, 1989.

Woodward, R.L., and G. Masters, Global upper mantle structure from long-period differential travel times, *J. Geophys. Res., 96* 6351–6377, 1991a.

Woodward, R.L., and G. Masters, Lower-mantle structure from $ScS - S$ differential travel-times, *Nature, 352*, 231-233, 1991b.

Woodward, R. L., A. M. Forte, W.-J. Su, and A. M. Dziewonski, Constraints on the large-scale structure of the Earth's mantle, in *Proc. Union Symposium 12, IUGG*, Takahashi *et al.*, eds., Am. Geophys. Un., Washington, D.C.,in press, 1993.

Yuen, D.A., and W.R. Peltier, Mantle plumes and the thermal stability of the D" layer, *Geophys. Res. Lett., 7*, 625-628, 1980.

Yuen, D.A., A.M. Leitch, and U. Hansen, Dynamical influences of pressure-dependent thermal expansivity on mantle convection, in *Glacial Isostasy, Sea-Level, and Mantle Rheology*, R. Sabadini, K. Lambeck, and E. Boschi, eds., NATO ASI Series vol. 334, 663-701, 1991.

Zhang, Y.-S., Three-dimensional modeling of upper mantle structure and its significance to tectonics, *Ph. D. Thesis*, California Institute of Technology, Pasadena, Ca., 1991.

Zhang, Y.-S. and T. Tanimoto, Global Love wave phase velocity variation and its significance to plate tectonics, *Phys. Earth Planet. Int., 66*, 160-202, 1991.

Zhang, Y.-S. and T. Tanimoto, Ridges, hotspots and their interpretation as observed in seismic velocity maps, *Nature, 355*, 45-49, 1992.

A. M. Dziewonski, A. M. Forte, W.-J. Su and R. L. Woodward, Department of Earth and Planetary Sciences, Harvard University, Cambridge, MA 02138.

Topographic Core-Mantle Coupling and Fluctuations in the Earth's Rotation

R. HIDE

Robert Hooke Institute, The Observatory
Clarendon Laboratory, Parks Road
Oxford OX1 3PU, England, U. K.

R. W. CLAYTON

Seismological Laboratory, California Institute of Technology
Pasadena, CA 91125. U. S. A.

B. H. HAGER

Department of Earth, Atmospheric and Planetary Sciences
Massachusetts Institute of Technology
Cambridge, MA 02139, U. S. A.

M. A. SPIETH

Jet Propulsion Laboratory, California Institute of Technology
Pasadena, CA 91109, U. S. A.

C. V. VOORHIES

Geodynamics Branch Code 921, Goddard Space Flight Center
Greenbelt, MD 20771, U. S. A.

Astronomically-determined irregular fluctuations in the Earth's rotation vector on decadal time scales can be used to estimate the fluctuating torque on the lower surface of the Earth's mantle produced by magnetohydrodynamic flow in the underlying liquid metallic core. A method has been proposed for testing the hypothesis that the torque is due primarily to fluctuating dynamic pressure forces acting on irregular topographic features of the core-mantle boundary and also on the equatorial bulge. The method exploits (a) geostrophically-constrained models of fluid motions in the upper reaches of the core based on geomagnetic secular variation data, and (b) patterns of the topography of the CMB based on the mantle flow models constrained by data from seismic tomography, determinations of long wavelength anomalies of the Earth's gravitational field and other geophysical and geodetic data. According to the present study, the magnitude of the axial component of the torque implied by determinations of irregular changes in the length of the day is compatible with models of the Earth's deep interior characterized by the presence of irregular CMB topography of effective "height" no more than about 0.5km (about 6% of the equatorial bulge) and strong horizontal variations in the properties of the D" layer at the base of the mantle. The investigation is now being extended to cover a wider range of epochs and also the case of polar motion on decadal time scales produced by fluctuations in the equatorial components of the torque.

1. Introduction

Electric currents generated in the Earth's liquid metallic core are responsible for the main geomagnetic field and its secular changes [see *Jacobs*, 1987*ab*; *Melchior*, 1986; *Moffatt*, 1978*a*]. The currents are produced by dynamo action involving irregular magnetohydrodynamic flow in the core. Concomitant dynamical stresses acting on the overlying mantle are invoked in the interpretation of the so-called "decadal" fluctuations in the rotation of the "solid Earth" (mantle, crust and cryosphere). Studies of these rotational manifestations of core motions bear directly on investigations of the structure, composition and dynamics of the Earth's deep interior [*Aldridge*, 1990; *Anufriev and Braginsky*, 1977; *Benton*, 1979;

Relating Geophysical Structures and Processes: The Jeffreys Volume
Geophysical Monograph 76, IUGG Volume 16

Eltayeb and Hassan, 1979; *Hide*, 1969; 1970; 1977; 1986; 1989; *Hide and Dickey*, 1991; *Hinderer et al.*, 1990; *Jault and Le Mouël*, 1989; 1990; *Lambeck*, 1980; 1988; *Moffatt*, 1977*b*; *Morrison*, 1979; *Paulus and Stix*, 1989; *Roberts*, 1972; *Rochester*, 1984; *Spieth et al.*, 1986; *Voorhies*, 1991*a*; *Wahr*, 1988].

Consider a set of body-fixed axes x_i, $i = 1, 2, 3$, aligned with the principal axes of the solid Earth and rotating about the center of mass of the whole Earth with angular velocity

$$\hat{\omega}_i = \hat{\omega}_i(t) = (\hat{\omega}_1, \hat{\omega}_2, \hat{\omega}_3) = \Omega\,(\hat{m}_1, \hat{m}_2, 1+\hat{m}_3) \qquad (1.1)$$

Here t denotes time and Ω the mean speed of rotation of the solid Earth in recent times, 0.7292115 x 10^{-4} radians per second [*Cazenave*, 1986; *Lambeck*, 1980; *Moritz and Mueller*, 1987; *Munk and MacDonald*, 1960; *Rochester*, 1984]. Over time scales short compared with those characteristic of geological processes, the rotation of the Earth departs only slightly from steady rotation about the polar axis of figure, so that $\hat{m}_1$, $\hat{m}_2$, and $\hat{m}_3$ are all very much less than unity and $|\dot{m}_i| << \Omega$, where $\dot{\hat{m}}_i \equiv d\hat{m}_i / dt$. Periodic variations in $\hat{\omega}_i$ on time scales less than a few years are caused by periodic lunar and solar tidal torques and related changes in the moment of inertia of the solid Earth. Irregular variations on these time scales are produced by atmospheric and oceanic torques due to tangential stresses in surface boundary layers and normal pressure forces acting on surface topography, and they are largely associated with seasonal, intraseasonal and interannual fluctuations in the total angular momentum of the atmosphere. When these rapid variations (including the Chandlerian wobble of the figure axis relative to the rotation axis) have been removed from the observational data, the smoothed time series

$$\omega_i = \Omega\,(m_1, m_2, 1 + m_3) \qquad (1.2)$$

that remains reveals (within the errors involved) the decadal variations, the axial component of which is illustrated in Fig. 1. Geophysicists have long argued that these decadal variations in m_3 are largely manifestations of angular momentum exchange between the core and mantle [*Jacobs*, 1987*ab*; *Jault and Le Mouël*, 1991; *Stoyko*, 1951; *Vestine*, 1952].

Expressions needed in the study of the variable rotation of the non-rigid solid Earth due to core-mantle coupling are readily obtained by standard methods based on Euler's dynamical equations (see e. g., *Munk and MacDonald*, [1960]). If $L_i^*(t)$, $i = 1, 2, 3$, is the fluctuating torque exerted by the core on the mantle then the axial component L_3^* satisfies [*Hide*, 1989]:

$$\dot{m}_3 = -\Lambda_0\, d(\Delta\Lambda)/dt = L_3^*/\Omega\, C + \alpha_3 \qquad (1.3)$$

where C is the principal moment of inertia of the solid Earth about the polar axis, $\Lambda_0 \equiv 2\pi/\Omega$ is the average length of the day (LOD), $\Lambda(t) \equiv 2\pi/\omega_3$ and $\Delta\Lambda \equiv \Lambda(t) - \Lambda_0$. The quantity α_3, which can be calculated explicitly [*Hide*, 1989], represents secondary effects due to a variety of causes, such as fluctuations in the inertia tensor of the solid Earth (including changes associated with stresses responsible for L_i^* and to comparatively weak torques applied at the Earth's surface by the atmosphere and oceans on the relevant time scales).

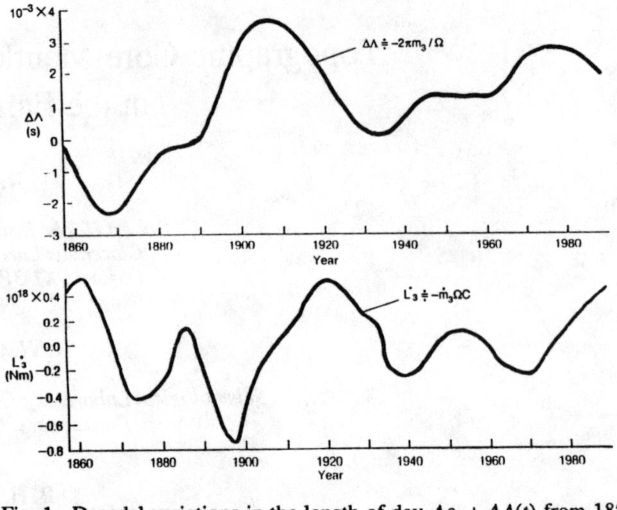

Fig. 1. Decadal variations in the length of day $\Lambda_0 + \Delta\Lambda(t)$ from 1855 to 1985 (upper curve) and corresponding variations in the axial component $L_3^*(t)$ of the torque L^* acting upon the solid Earth (lower diagram); see equation (1.3) and *Dickey et al.* [1989] or *Hide and Dickey* [1991]. More rapid variations (of tidal and atmospheric origin) were removed from the original time series by taking a ten-year running average.

2. The Torque Exerted by the Core on the Mantle

The fluctuating torque $L_i^*(t)$ exerted by the core on the overlying mantle is largely a consequence of (a) tangential stresses at the CMB associated with shearing motions in the thin (less than 1 m) viscous boundary layer just below the CMB, (b) normal dynamical pressure forces acting on the equatorial bulge and other (possibility) smaller and more irregular departures from spherical of the shape of the CMB, (c) Lorentz forces due to the flow of electric currents in the weakly-conducting lower mantle generated by the electromotive forces associated with the geodynamo processes in the core and (d) gravitational forces associated with horizontal density variations in the core and mantle and especially at the CMB [see *Hide*, 1969; *Jacobs*, 1987*ab*; *Jault and Le Mouël*, 1991; *Melchoir*, 1986; *Rochester*, 1984; *Voorhies*, 1991*a*]. Denote by $L_{i(+)}^*$ and $L_{i(-)}^*$ the respective contributions to L_i^* from regions where tractions act in the positive sense and from regions where tractions act in the negative sense and introduce the "canceling factor"

$$\gamma_i^* = \gamma_i^*(t) \equiv L_i^*/[L_{i(+)}^* - L_{i(-)}^*] \qquad (2.1)$$

By definition, each γ_i^* fluctuates about an average of zero and attains magnitudes less than or equal to unity. The accuracy with which L_i^* can be determined obviously depends on the value of $|\gamma_i^*|$, the most favorable situation being when $|\gamma_i^*| \sim 1$.

Rough dynamical arguments show that the pressure coupling associated with CMB topography might predominate over other effects [*Hide*, 1969]. The contribution made by viscous stresses is negligible on all but the most extreme assumptions about the viscosity of the core, but electromagnetic coupling, according to detailed studies by a number of investigators, might be adequate if the (unknown) electrical conductivity of the lower mantle were sufficiently high and the toroidal magnetic field in the core concentrated in a boundary layer just below the CMB [*Bullard et al.*, 1950; *Paulus and Stix*, 1989; *Roberts*, 1972; *Rochester*, 1984]. So it is of interest to investigate the topographic contribution $L_i(t)$ (say) to $L_i^*(t)$ using available geophysical data, in order to establish the extent to which decadal variations in the Earth's rotation might reasonably be attributed to topographic core-mantle coupling.

If p_s is the dynamical pressure associated with core motions $u = u_s$ in the free stream just below the viscous boundary layer at the CMB (u being the Eulerian flow velocity relative to a reference frame fixed to the solid Earth) and the CMB is the locus of points where the distance from the Earth's center of mass is $r = c + h(\theta, \phi)$, (c being the mean radius of the CMB and (θ, ϕ) the co-latitude and longitude of a general point), then [see *Hide*, 1989; *Voorhies*, 1991*a*].

$$L_i(t) = -c^2 \int_0^{2\pi} \int_0^{\pi} (r \times p_s \nabla h)_i \sin\theta \, d\theta \, d\phi \tag{2.2}$$

if $|h| << c$ and $(\nabla_s h) << 1$. Here $\nabla_s \equiv c^{-1}(\hat{\theta}\, \partial/\partial\theta, \hat{\phi}\, cosec\theta\, \partial/\partial\phi)$, $\hat{\theta}$ and $\hat{\phi}$ are unit vectors in the directions of increasing θ and ϕ respectively, and r is the vector distance from the Earth's center of mass. An important step in the analysis is the recognition of the fact that the surface integral of $r \times \nabla_s(h\, p_s)$ over the whole CMB is equal to zero, which leads to a more useful expression for $L_i(t)$, namely

$$L_i(t) = c^2 \int_0^{2\pi} \int_0^{\pi} (r \times h \nabla p_s)_i \sin\theta \, d\theta \, d\phi \tag{2.3}$$

Nearly everywhere within the core, owing to the presence there of electric currents, Lorentz forces may be comparable in magnitude with the Coriolis forces due to the Earth's rotation acting on core flow. But in the upper reaches of the core, within tens of kilometers of the CMB, Lorentz forces should be about 10^{-2} times the Coriolis forces or less, unless geodynamo action is confined to a thin boundary layer just below the CMB or metallic electrical conductivities are attained at the base of the "solid" mantle. In regions where Lorentz forces are negligible, geostrophic balance between the horizontal components of the pressure gradient and Coriolis forces should obtain to sufficient accuracy [*Backus and Le Mouël*, 1986; 1987; *Bloxham and Jackson*, 1991; *Gire and Le Mouël*, 1990; *Hide*, 1986; 1989; *Hills*, 1979; *Le Mouël*, 1984; *Voorhies*, 1991*a*]. Whence

$$2\bar{\rho}_s \Omega \cos\theta\, (-w_s, v_s) = -c^{-1}(\partial p_s/\partial\theta, \operatorname{cosec}\theta\, \partial p_s/\partial\phi) \tag{2.4}$$

where $\bar{\rho}_s$ is the horizontally averaged value of the density ρ in the upper-reaches of the core. Here (v_s, w_s) are the (θ, ϕ) components of u_s, which are typically much greater in magnitude than u_r, the r-component of u_s. It follows from equations (2.3) and (2.4) that on the time scales of interest here, over which u_s may change significantly but h does not, that

$$L_3(t) = -2\bar{\rho}_s \Omega c^3 \int_0^{2\pi} \int_0^{\pi} h(\theta,\phi)\, v_s(\theta,\phi,t) \sin^2\theta \cos\theta \, d\theta \, d\phi \tag{2.5}$$

The basic theoretical relationships needed are given by equation (1.3) with $\alpha_3 = 0$ and the working hypothesis that $L_3^* = L_3$, together with equations (2.2) to (2.5). The integral on the right-hand side of equation (2.5) involves CMB topography $h(\theta, \phi)$. When dealing with the equatorial components of the torque and the polar motion they produce, the dominant contribution to h is the equatorial ϕ-independent bulge of the CMB, which corresponds to a 9 km difference between the equatorial and polar radii of the core. [e.g., *Herring et al.*, 1991]. But the equatorial bulge makes no contribution to the axial component L_3, which changes the LOD $\Lambda(t)$ (see equation (1.3)), so when dealing with such changes it is necessary to look in detail at features of h that depend on ϕ as well as θ. Over the past-twenty years various attempts have been made to infer $h(\theta, \phi)$ from seismology and from the pattern of long-wavelength gravity anomalies under various assumptions about the structure and rheology of the solid Earth, with more recent studies making use of results from seismic tomography; several hypothetical fields of $h(\theta, \phi)$ are now available, as discussed in Section 4 below.

The other quantity needed in the evaluation of $L_i(t)$ is either the field of pressure $p_s(\theta,\phi,t)$ or the field of horizontal flow $(v_s(\theta, \phi, t), w_s(\theta, \phi, t))$, which is related to p_s through equation (2.4). Geomagnetic secular variation data have been used by various workers to infer (v_s, w_s) by a method that invokes the geostrophic relationship (equation (2.4)) in combination with the equations of electrodynamics appropriate to the case when the mantle can be treated as a perfect electrical insulator of uniform magnetic permeability and the core as a perfect conductor, as we shall now discuss.

3. VELOCITY AND PRESSURE FIELDS IN THE CORE

Denote by $B(r, \theta, \phi, t)$ the value of the main geomagnetic field at a general point (r, θ, ϕ) at time t, and by $\dot{B} \equiv \partial B/\partial t$ the so-called geomagnetic secular variation. Determinations of B made at and near the Earth's surface at various epochs can be used to infer u_s, the Eulerian flow velocity just below the CMB [*Backus and Le Mouël*, 1986; 1987; *Benton*, 1981*ab; Benton and Celaya*, 1991; *Bloxham*, 1988; 1989; *Bloxham and Jackson*, 1991; *Courtillot and Le Mouël*, 1988; *Gire et al.*, 1986; *Gire and Le Mouël* 1990; *Gubbins*, 1982; *Lloyd and Gubbins*, 1990; *Voorhies*, 1986*ab;* 1987; 1991; 1992; *Voorhies and Backus*, 1985; *Whaler*, 1986; 1990; 1991; *Whaler and Clarke*, 1988]. From flows that are constrained to be tangentially geostrophic (see below), an estimate of the horizontal pressure gradient just below the CMB can be deduced through equations (2.4).

The first of the three reasonable key assumptions that underlie the method used is that the electrical conductivity of the mantle and magnetic permeability gradients there are negligibly small, so that B can be written as the gradient of a potential V satisfying Laplace's equation $\nabla^2 V = 0$. This facilitates the downward extrapolation of the observed field at and near the Earth's surface in order to obtain B and $\dot{B}$ at the CMB.

The second assumption is that the electrical conductivity of the core is so high that when dealing with fluctuations in B on time scales very much less than that of the Ohmic decay of magnetic fields in the core (which is several thousand years for global-scale features), B satisfies Alfven's "frozen flux" theorem [*Backus*, 1968; *Roberts and Scott*, 1965]. This is expressed by the equation

$$\partial \mathbf{B}/\partial t = \nabla \times (\mathbf{u} \times \mathbf{B}), \tag{3.1}$$

which may be shown to imply that the lines of magnetic force emerging from the core are advected by the horizontal flow (v_s, w_s) just below the CMB. Accordingly, if $\boldsymbol{B} = (B_r, B_\theta, B_\phi)$, then B_r at the CMB satisfies

$$\frac{\partial B_r}{\partial t}+\frac{v_s}{c}\frac{\partial B_r}{\partial \theta}+\frac{w_s}{c\sin\theta}\frac{\partial B_r}{\partial \phi}=B_r\left[\frac{\partial u}{\partial r}\right]_{r=c} \quad (3.2)$$

Now equation (3.1) alone does not permit the unique determination of $\boldsymbol{u}_s$ from knowledge of $\boldsymbol{B}$ and $\dot{\boldsymbol{B}}$ at the CMB. Of the various additional considerations that have been employed to secure uniqueness, a physically plausible approach is to invoke the assumption we have already made in Section 2, namely that the flow in the upper reaches of the core is in geostrophic balance with the pressure field there [*Backus and Le Mouël*, 1986; 1987; *Hills*, 1979; *Jault et al.*, 1988; *Le Mouël*, 1984]. The corresponding radial vorticity balance can be expressed by the equation

$$\cos\theta\left[\frac{\partial u}{\partial r}\right]_{r=c}+\frac{v_s}{c}\sin\theta=0, \quad (3.3)$$

which is readily deduced by eliminating p_s from equations (2.4) and using the mass continuity equation $\nabla \cdot \boldsymbol{u} = 0$ for flow in an effectively incompressible fluid.

Various groups of geomagnetic workers have produced maps of (v_s, w_s) and investigated the errors and uncertainties encountered in practice [*Bloxham and Jackson*, 1991]. In this study, we use ten flow models, two of which have been published elsewhere and eight of which were produced for this study. Four of these models are shown in Figure 2. Figure 2a shows model IIa of *Bloxham* [1989], a geostrophic model determined assuming steady flow and using a smoothly varying model of the magnetic field at the CMB for the interval 1975 - 1980. This model is expressed in terms of spherical harmonics through degree and order 14. The model shown in Figure 2b is model GVC1E6 of *Voorhies* [1991*a*; 1992], derived, assuming steady flow, using $\boldsymbol{B}$ and $\dot{\boldsymbol{B}}$ from the "Definitive Geomagnetic Reference Field" (DGRF) at epochs starting in 1980 and moving back to 1945 in 5-year intervals. This model incorporates spherical harmonic expansions up to degree and order 16 and provides a weighted variance reduction of 98.074%. *Voorhies* [1988] presented a series of models fit to the DGRF for shorter intervals of time. Model G6070.1, shown in Figure 2c, is a relatively smooth fit to the DGRF from 1960 - 1970. Model G8070.1, shown in Figure 2d, uses the same damping parameter to fit the DGRF data from 1980 - 1970. Other models of the G6070.*m*/G8070.*m* series (not shown here) for the same time intervals were used, with increasing index *m* corresponding to increasing roughness of the model flows.

These flow models, which were used to produce hypothetical flow fields $\boldsymbol{u}_s$ and corresponding pressure fields p_s, are all similar in appearance. *Voorhies'* [1988; 1991*a*; 1992] models CVC1E6 and G8070.1 are barely distinguishable in these plots, reflecting the fact that the models were fit using a similar procedure starting with the 1980 DGRF as the initial condition. The most apparent differences between these two models and the model of Bloxham [1989], appropriate for approximately the same time span as G8070.1, are beneath southern Africa and in the northwestern Pacific. The differences between *Voorhies'* [1988] models G8070.1 and G6070.1, fit to different epochs of the DGRF using identical procedures, are also small, being most easily seen beneath northern Siberia, Alaska, and the southeastern Pacific. Although the calculation of the time-varying core surface flows needed to model variations in p_s, (hence L_i) is straightforward, the similarity among these models demonstrates the care needed to estimate such variations reliably. In particular, there is as much variation between models fit to the same epoch by different workers as there is between models by a single worker of the flow for different epochs.

4. Topography of the Core-Mantle Boundary

The topography $h(\theta, \phi)$ of the CMB (see equation (2.3)) has been investigated using various techniques. The problem is complicated by the suspected presence of the D" layer at the base of the mantle; its variable mechanical and chemical properties make it difficult to separate the effects of D" and CMB topography [*Anderson*, 1989; *Bullen*, 1963; *Doornbos*, 1980; *Gubbins*, 1989; *Gudmundsson et al.*, 1990; *Gudmundsson and Clayton*, 1991, 1992; *Haddon*, 1982; *Hager et al.*, 1985; *Hager and Richards*, 1989; *Jacobs*, 1987*a*; *Jeanloz*, 1990; *Knittle and Jeanloz*, 1991; *Loper*, 1991; *Morelli and Dziewonski*, 1987]. Determinations of the amplitude and phase of the free-core nutation lead to an estimate of 0.5 km of excess ellipticity at the CMB [*Babcock and Wilkins*, 1988; *Gwinn et al.*, 1986; *Herring et al.*, 1985; 1986; 1991; *Kinoshita and Suchay*, 1990; *Reid and Moran*, 1988; *Wahr*, 1988]. While this topographic component does not affect L_3 (and hence the LOD, see equation (1.3)), it may indicate the likely magnitude of the relief.

The most direct method (in principle) of estimating CMB topography is to examine variations in travel times of seismic waves that interact with the CMB. Core reflected phases such as PcP and phases that propagate through the core such as PKP and PKIKP are affected similarly by the presence of velocity anomalies in the mantle, which, for rays traversing the same path, affect the travel-times of both phases comparably. However, deformations of the CMB affect the two phases in opposite senses; in regions where $h < 0$, PcP travel times are increased owing to increased path length, but PKP and PKIKP travel times are reduced because a greater portion of the path is through the "faster" mantle material. Potentially, a joint investigation of data from both types of phase could lead to a unique determination of seismic properties of the CMB [*Morelli and Dziewonski*, 1987]. Unfortunately, the quality of the observations of PcP, PKP, and PKIKP travel times given in the International Seismic Centre (ISC) catalogue tends to be poor, both in terms of travel-time measurements and geographic distribution.

Morelli and Dziewonski [1987] boldly carried out inversions for CMB topography using PcP and PKPef phases. The models for these two phases showed a reasonable correlation, which they took as evidence that the inversions were successful, despite the potential pitfalls. They inverted both data sets together to determine a model of CMB topography through degree and order 4, with peak-to-peak amplitudes in excess of 10 km.

Gudmundsson and Clayton [1992] carried out a number of inversions of the ISC data for these phases, as well as the additional phases PKPab, PKPbc; and PKPde. They also investigated the

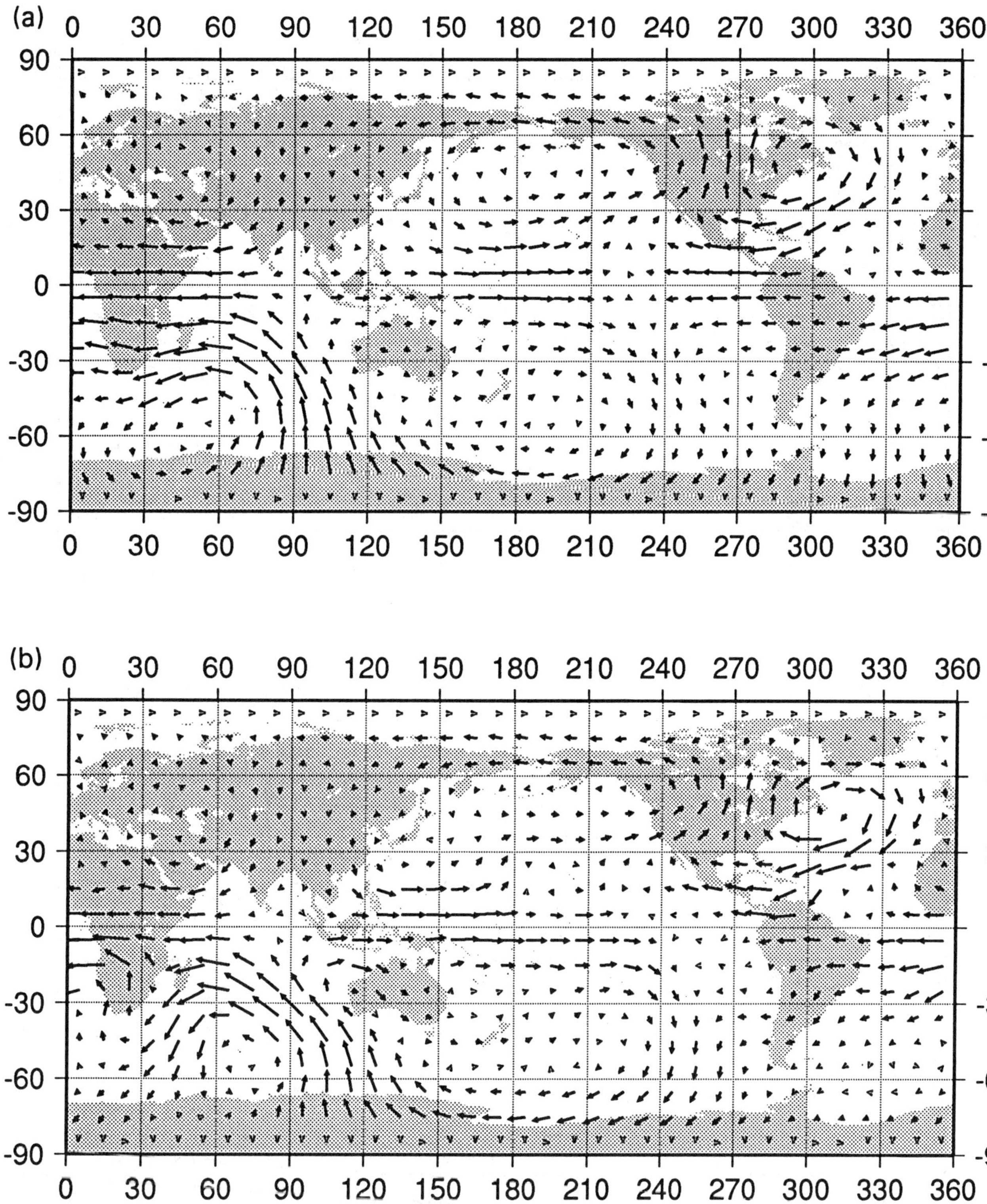

Fig. 2. Geostrophic flow fields estimated by *Bloxham* [1989] (2a), *Voorhies* [1991*a*] (2b) and *Voorhies* [1988] (2c and 2d). An arrow with a magnitude equal to the distance between grid points (10°) corresponds to a velocity of 20 km/year.

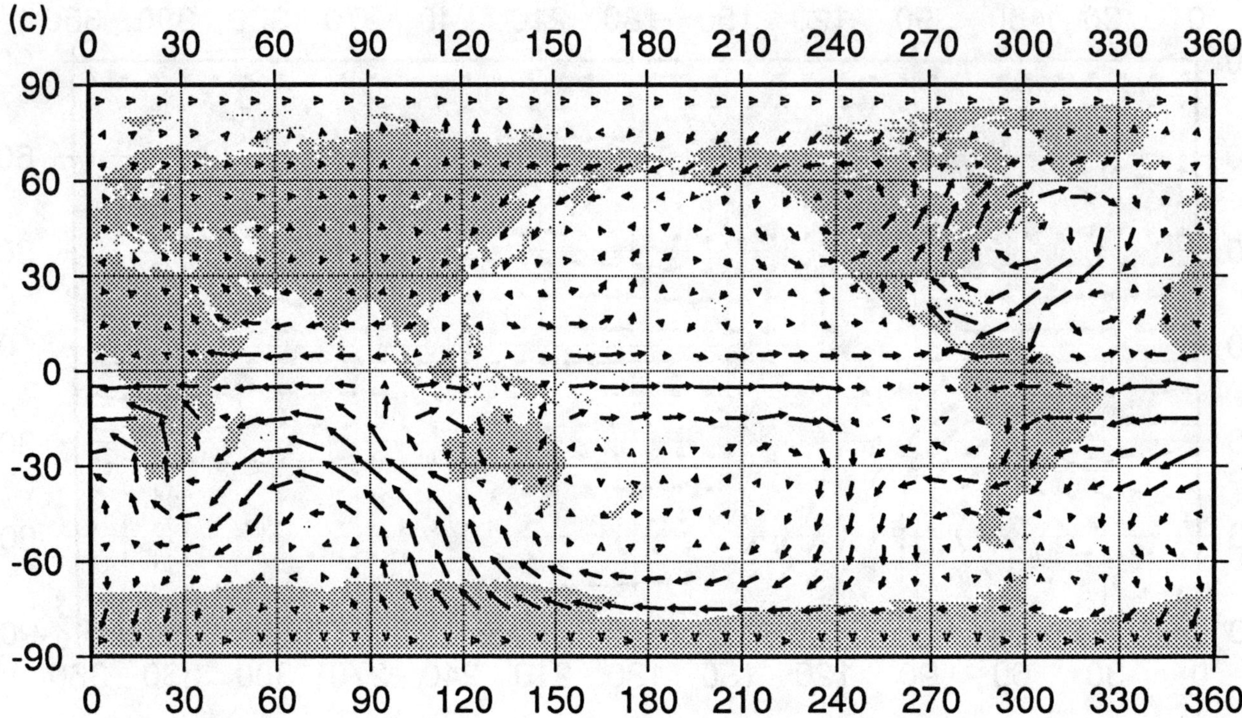

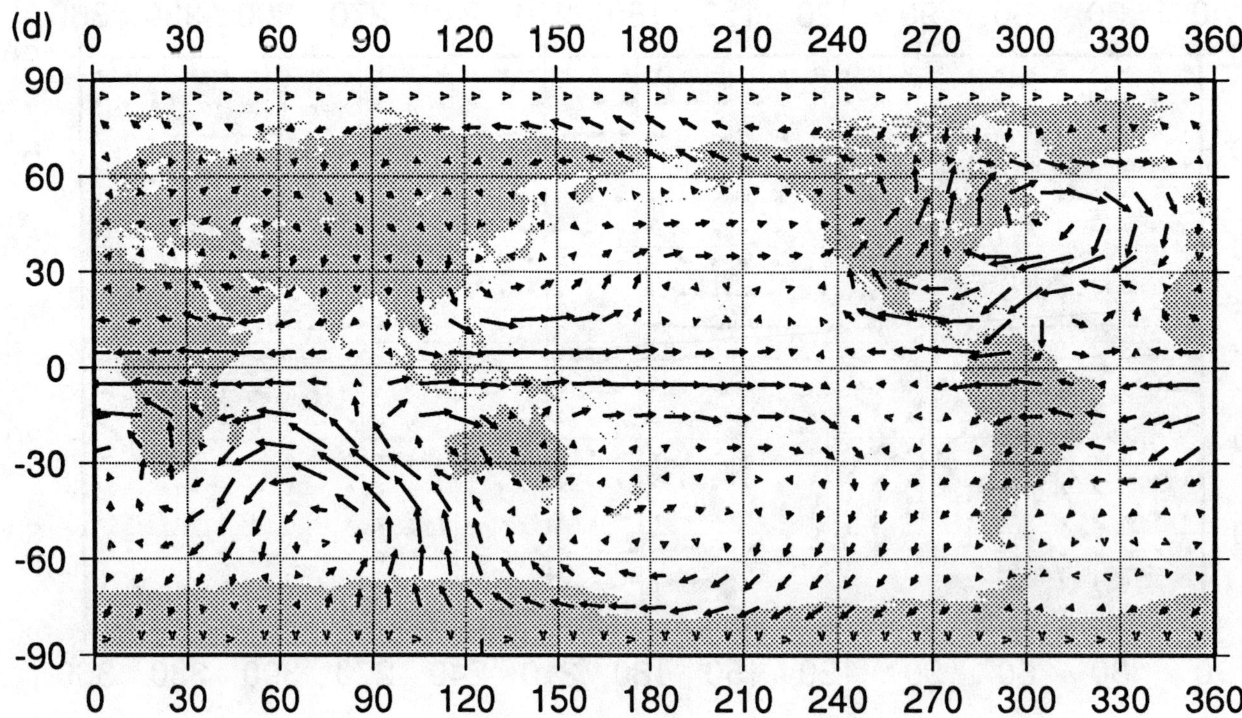

Fig. 2. (continued)

effects of including a horizontally-variable D" layer in their models. The CMB topography obtained for the different phases differs substantially. In addition, they found that by varying the "tradeoff" between the effects due to D" variations and CMB topography respectively, the amplitude of the ϕ-dependent part of h ranged from 1 km to 6 km. Their models were parameterized in terms of spherical harmonic expansions through degree and order 10, although they employed a "stochastic inverse," choosing the trade-off parameter to retain only 25 parameters (as measured by the trace of the resolution matrix) in their solution. The topography for one particular choice that includes the effects of variable velocity in the D" layer in the model parameterization (Figure 9 in *Gudmundsson and Clayton* [1992]), expanded through degree and order 4 for comparison with the model of *Morelli and Dziewonski* [1987], is illustrated in the top panel of Figure 3.

Another approach to modeling CMB topography is to use density variations inferred from lateral variations in seismic-wave velocities in the mantle or other geophysical inference to drive models of mantle flow. These models, which predict dynamically-maintained topography at the Earth's surface and at the CMB, are constrained by long-wavelength features of the geoid. This approach assumes that the seismic-velocity variations are caused by density variations (via temperature), to which they are linearly related. The mantle viscosity must also be specified, along with major chemical boundaries in the mantle, including the problematic D" layer. In a spherical harmonic expansion of $h(\theta, \phi)$ up to degree and order 9, models with no D" layer give amplitudes of the ϕ-dependent part of h up to 2.5 km (e.g., model WO of *Hager and Clayton* [1989]), nearly 25% of the equatorial bulge. On the other hand, models with a low viscosity and/or chemically distinct D" layer give much lower amplitudes. For example, the CMB topography for model WL of *Hager and Richards* [1989], a model which has a D" layer which has a low viscosity, but is not compositionally distinct from the overlying mantle, is shown in the bottom panel of Figure 3. It has an excess ellipticity compatible with the inferences from nutation studies. The model is dominated by long-wavelength variations, with a peak-to-peak amplitude of the topography of < 2 km.

Clearly, models of CMB topography differ more than do models of core flow. Thus, the comparison of the implied LOD variations with observed values on the basis of the method outlined in Section 1 has the potential to distinguish among classes of models such as those presented here.

5. PREDICTED TORQUES AND LENGTH-OF-DAY VARIATIONS

Estimates of the topographic torque exerted by the core on the mantle were produced from models of the flow fields in the outer core and the CMB topography via numerical integration of equation (2.5), and compared with values implied by the LOD determinations presented in Figure 1. Predicted LOD values obtained on the basis of equations (1.3) and (2.5) (with $\alpha_3 = 0$ and $L_3^* = L_3$) for 50 combinations of 10 different models of the flow field, (v_s, w_s), and 5 models of CMB topography, h, are given in Table 1. Four of the core flow models are those shown in Figure 2; the others represent additional models presented by *Voorhies* [1988] for the DGRF in the epochs 1980-1970 and 1960-1970. The models differ in the damping parameter used to control the roughness of the flow. For example, G6070.1 is the smoothest model for 1960 - 1970, while G6070.4 is the roughest. In addition to the two CMB topography models shown in Figure 3, we also used the model of *Morelli and Dziewonski* [1987], model WO of *Hager and Clayton* [1989], and the model of *Gudmundsson and Clayton* [1992] shown in Figure 3, but expanded through degree and order 10.

Core flow Model IIa of *Bloxham* [1989] assumes steady flow during the epoch 1975-1980, during which time the average value of L_3^* was ~ 0.1 x 10^{18} Nm; the corresponding change in LOD is about −0.5 msec/decade. Models G8070.*m* of *Voorhies* [1988] assume steady flow starting in 1980, going back to 1970. During this interval, $\Delta\Lambda$ first increased, then decreased, with negligible net torque averaged over the decade. Models G6070.*m* of *Voorhies* [1988] assume steady flow between 1960 and 1970, corresponding to L_3^* of ~ −0.2 x 10^{18} Nm and a change in LOD of about +1 msec/decade. Model GVC1E6 of *Voorhies* [1991*a*] assumes steady flow starting in 1980, going back to 1945, with the fit to the data being better at later than at earlier times in this interval. An average torque of ~ −0.1 x 10^{18} seems appropriate for the interval over which this model best fits the geomagnetic data; this would correspond to a change in LOD of 0.5 msec/decade.

By chance, all combinations of flow models and topography models presented here give positive predictions for changes in LOD. (Other models we have used, not shown, give negative predictions.) The combinations using the seismological models of h, with irregular ϕ-dependent amplitudes of several kilometers, "predict" LOD changes much larger, some by over a factor of 100, than the observed changes. Much closer agreements with the magnitude of observed LOD changes were found in the case of combinations for model WL, which has irregular CMB topography < 1 km in amplitude, although the predicted changes are still a factor of 4 - 8 to large. For the topography models used here, *Bloxham's* [1989] flow model predicts the wrong sign of LOD variation during 1975 - 1980. *Voorhies'* [1988] models predict the correct sign for the decade represented by models G6070.*m*. The models for the epoch represented by models G8070.*m* predict too large changes in LOD; the observed value for this decade is about zero. But, as discussed next, we are more confident of the order of magnitude of the torques than their sign, so the differences for these different flow models may not be diagnostic.

There is of course concern about the effect of inadequacies in the flow and topography fields on "predicted" values of the LOD variations. The problem is illustrated in Figure 4, which presents maps of the spatial distribution of contributions to the L_3 torque, $2\bar{\rho}_s\Omega_0\, c \sin\theta \cos\theta$, for the two flow models shown in Figures 3c and 3d, interacting with the topography shown in the bottom panel of Figure 2. As expected, a high degree of canceling is exhibited (see equation (2.1)). The net axial torque (and hence the LOD estimate) is the sum of these individual parts, which means it ends up being the small difference between two large numbers. For example, the total torque for the model in the top panel of Figure 4 is −1 x 10^{18} N-m (comparable to the that inferred for ~ 1900, as can be seen from Figure 1). A constant torque per unit area of −5 x 10^3 N/m

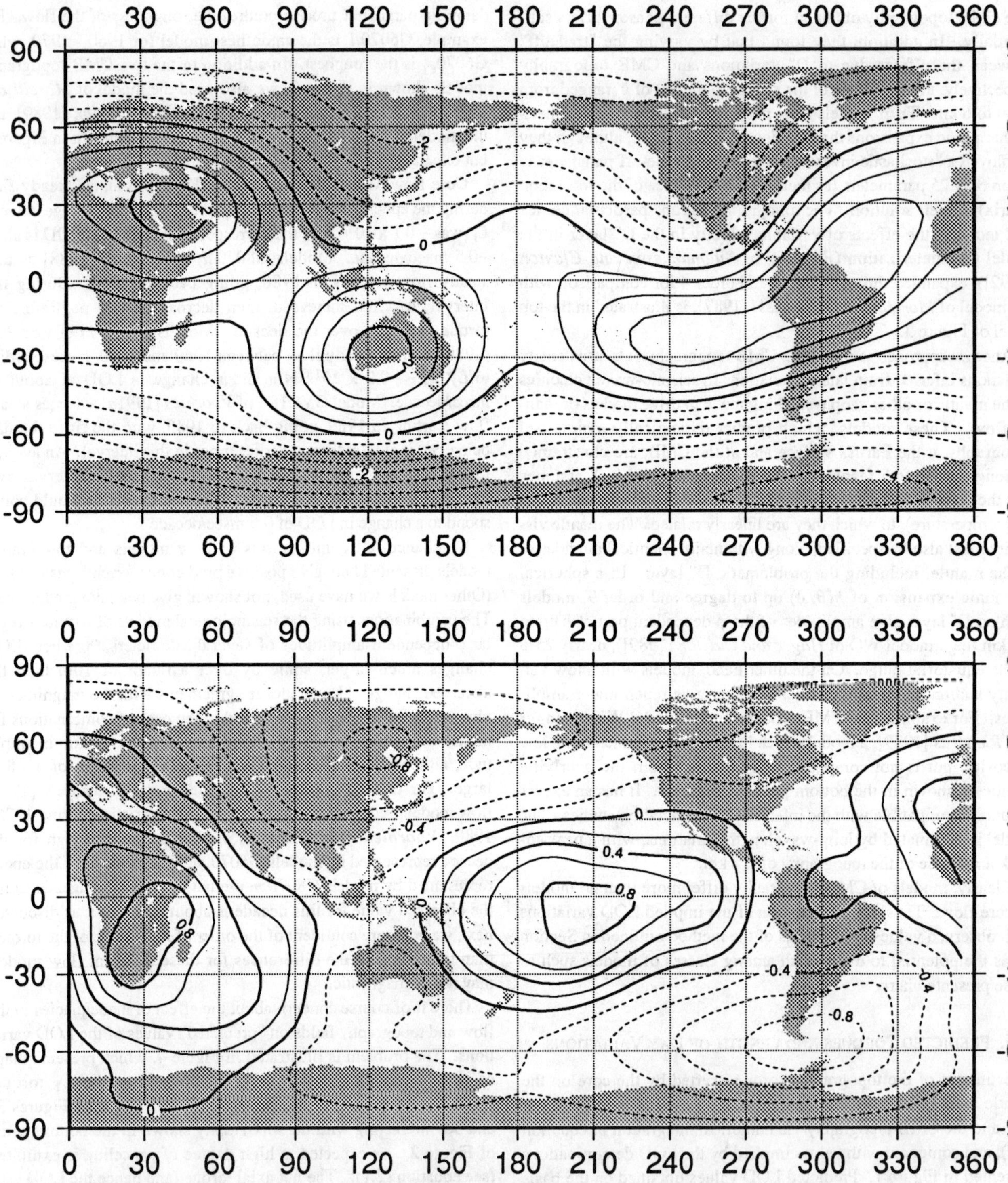

Fig. 3. Models of the topography of the CMB. The upper panel is a model derived using seismic tomography incorporating the effects of velocity variations in the D" layer above the CMB [*Gudmundsson and Clayton*, 1992]. The lower panel shows the dynamic topography inferred for model WL of *Hager and Richards* [1989]. Solid contours correspond to regions where the topography is upwarped towards the Earth's surface, while dashed contours represent regions where mantle is depressed into the core. The contour interval in the upper panel is 500 m; in the lower panel, 200 m.

TABLE 1. Predicted values of LOD variations for combinations of CMB topography models and geostrophic flow models discussed in the text.

$\Delta\Lambda$ (msec/decade)

CMB Topography \ Flow	Bloxham IIa [1989]	Voorhies GVC1E6 [1992]	Voorhies G6070.1 [1988]	Voorhies G8070.1 [1988]	Voorhies G6070.2 [1988]	Voorhies G8070.2 [1988]	Voorhies G6070.3 [1988]	Voorhies G8070.3 [1988]	Voorhies G6070.4 [1988]	Voorhies G8070.4 [1988]
Morelli & Dziewonski [1987]	143	115	110	133	115	131	97	120	58	112
Gudmundsson Clayton [1992] degree 1-4	85	71	66	75	59	81	47	80	33	78
Gudmundsson Clayton [1992] degree 1-10	88	70	64	73	57	77	46	75	34	78
Hager & Clayton WO [1989]	4.2	7.6	3.1	3.4	11	4.7	24	5.7	31	6.8
Hager & Richards WL [1989]	2.2	3.5	4.8	6.7	5.7	5.5	8.1	4.2	8.4	2.1

acting over the area of the CMB would result in a torque of this magnitude. The maximum amplitudes of the contributions to the equivalent torque integral plotted in Figure 4 are more than a factor of 100 greater than this average value.

The contributions to the torque integral for the two flow models at different epochs are similar, but there are discernable differences. For example, beneath southern Africa, the largest negative torque integral contribution for 1960 - 1970 has decreased in magnitude by 1980 - 1970, but the magnitude of the negative contribution south of India has increased. During the same interval, the contribution beneath Alaska changes sign, while the maximum over Manchuria splits into separated highs. Although these two flow models, and the patterns of torque contributions computed from them, are similar, the integrated effect of their small differences leads to calculated changes of LOD that differ by almost 2 msec/decade, comparable to the total predictions at the two epochs. Interestingly, the difference in LOD change predicted for these two models is comparable to the difference in LOD change observed for these two decades. But different flow models for the same epochs produce changes in predicted LOD comparable in magnitude, but opposite in sign. For this reason, we cannot assign very high significance to detailed LOD predictions. Indeed, it would probably be possible to construct an acceptable core flow model that exerts no net torque on the mantle [*Bloxham*, 1991; *Voorhies*, 1991*b*]. However, the order of magnitude supports the hypothesis that decadal fluctuations in LOD are largely effected by the topographic torque, even though the time-averaged torque is, of course, equal to zero. [*Hide*, 1969].

The effect of spherical harmonic degree truncation in the topography fields can be investigated by comparing the contributions in the torque integral and changes in length of day for two representations of the CMB topography model of *Gudmundsson and Clayton* [1992]. The first representation, shown in Figure 2, is expanded through degree and order 4. The second carries the expansion through degree and order 10. (Because only 25 parameters were used in their inversion, the amplitudes of the coefficients fall off fairly rapidly with harmonic degree.) As can be seen from Table 1, increasing the maximum degree and order from 4 to 10 has a very small effect on calculated LOD variations. Inspection of maps of the contributions to the torque integral calculated using *Voorhies'* flow model and these two topographic models (Figure 5) indicates that most of the features are well-represented by the smoother CMB models. The relative amplitudes of the contributions from different geographic regions vary, but the total torque remains almost constant. We take the stability of these estimates to indicate that contributions from higher harmonics (and their associated errors) are probably insignificant, owing to a high degree of cancellation at these scales (see equation (2.4)).

6. DISCUSSION AND CONCLUSIONS

The determination of the efficacy of topographic coupling is clearly fraught with practical difficulties. But the results of this paper demonstrate the usefulness of the method employed by providing independent evidence in favor of the hypothesis that topographic coupling can account for the observed decadal changes in LOD if $\tilde{h}$, the rms. value of the ϕ-dependent portion of h, the topographic relief, is typically about 1 km in amplitude or possibly slightly less.

We emphasize that this value of $\tilde{h}$ depends on the suppositions underlying the method. In particular, the velocity fields as determined from geomagnetic secular variation data (see Section 3) refer

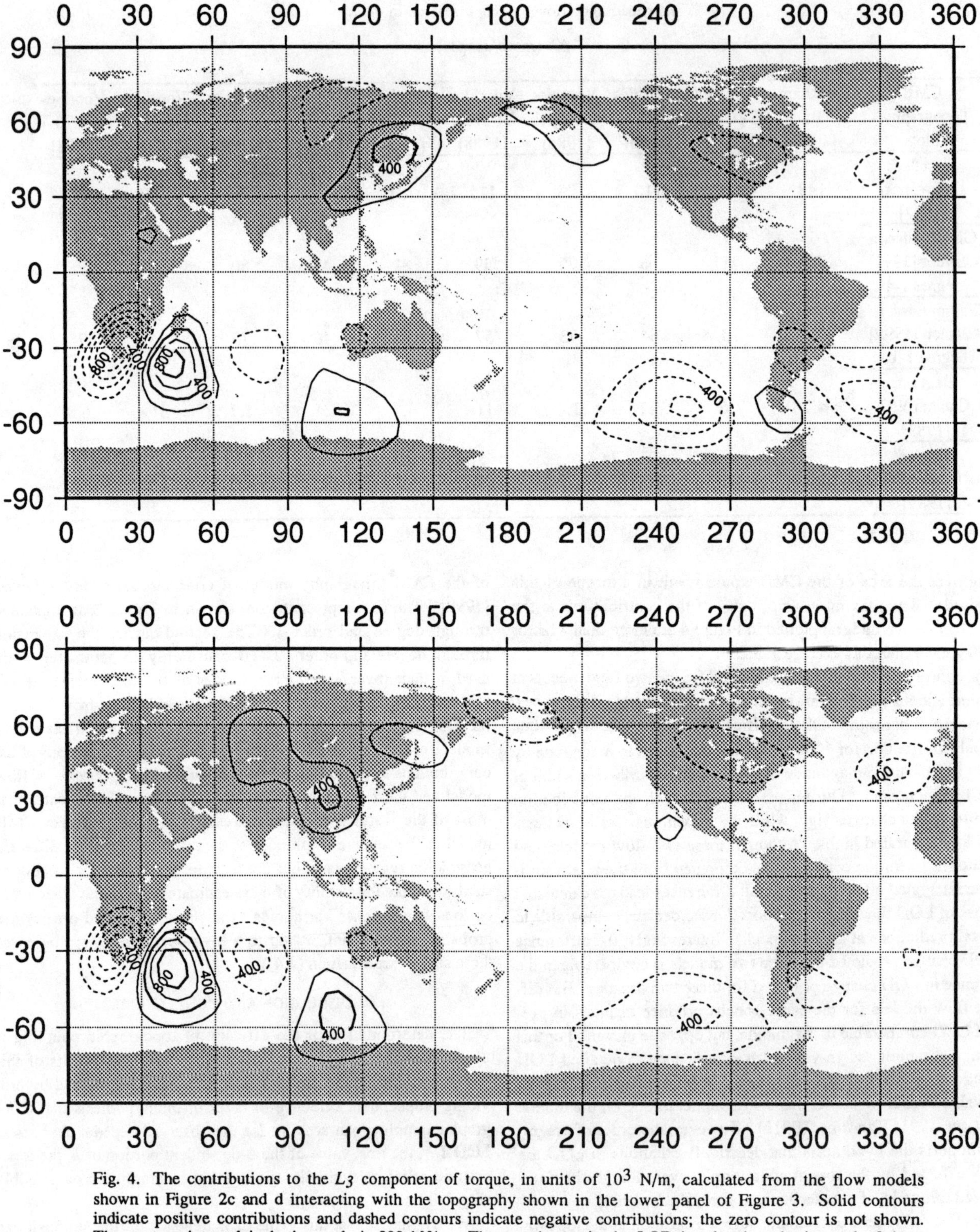

Fig. 4. The contributions to the L_3 component of torque, in units of 10^3 N/m, calculated from the flow models shown in Figure 2c and d interacting with the topography shown in the lower panel of Figure 3. Solid contours indicate positive contributions and dashed contours indicate negative contributions; the zero contour is not shown. The contour interval in both panels is 200 kN/m. The net change in the LOD is related to the integral of these contributions over the area of the CMB. A constant value of 200 kN/m, integrated over the area of the CMB, would give a torque of 0.3 x 10^{18} Nm and a variation in LOD of −1.4 msec/decade.

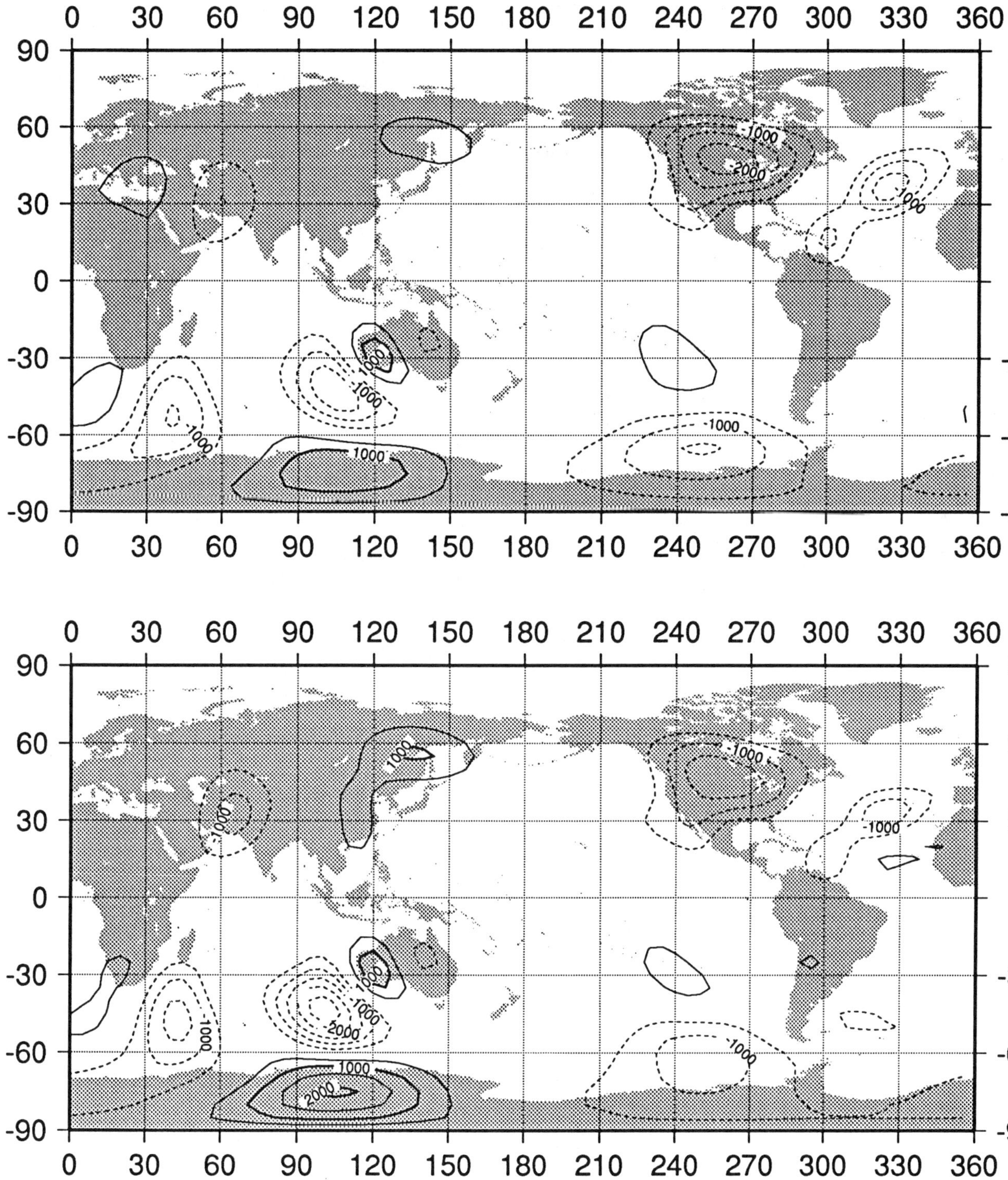

Fig. 5. The contributions to the L_3 component of torque, in units of 10^3 N/m, calculated from the flow model shown in Figure 2b interacting with the topography from the model of *Gudmundsson and Clayton* [1992] for two different degrees of truncation. The upper panel is for topography expanded through degree and order 4 (shown in the upper panel of Figure 3). The lower panel is for the model expanded through degree and order 10. Solid contours indicate positive contributions. The contour interval in both panels is 500 kN/m. The net change in the LOD is related to the integral of these contributions over the area of the CMB.

to the free stream at the top of the electrically-conducting liquid core. In calculating from these velocity fields the pressure fields acting upon topographic relief of the CMB, it has been assumed that the electrically-conducting core is in contact with the mantle over the whole of the CMB. This is probably a reasonable supposition, but it is interesting to speculate on the possibility that between the highly conducting core and the weakly conducting lower mantle there might exist a thin continuous layer or "pools" of weakly-conducting or insulating liquid which have escaped detection by seismic methods. The actual pressure field at the CMB might then differ from that given by equation (2.4) [*Hager and Richards*, 1989]. It would have to be determined from considerations of the dynamics of the hypothetical intermediate layer or pools, and the relationship between the two pressure fields might not be simple. If typical pressure gradients at the CMB were weaker than those at the top of the electrically-conducting core, then the value of $\tilde{h}$ needed to produce the necessary torques on the mantle would have to be larger. However, the idea of an intermediate layer is not supported by a parallel study to the present one [R. Hide. and A. Jackson, in preparation], in which decadal polar motion induced by topographic core-mantle coupling is investigated.

As we have seen in Section 4, inferences concerning $\tilde{h}$ can be drawn from direct seismic measurements and also from attempts to interpret the horizontal variations of the Earth's gravitational field in terms of density variations in the mantle maintained by slow convection there. In both cases, the implied value of $\tilde{h}$ depends critically on the assumptions made concerning the properties of the lower mantle. Seismic measurements give values greater than 1 km unless there are substantial (~ 1%) lateral variations in seismic velocities in the zone just above the CMB. The most likely candidate is the D" layer of 100 - 300 km thickness. Consistent with this "scenario" are the mantle convection studies, which give $\tilde{h}$ of about 1 km where a chemically-distinct or low viscosity layer in the D"-region is included, but a significantly greater value otherwise.

So far as future work on topographic core-mantle coupling is concerned, as better geophysical and geodetic data and models become available it will be necessary to repeat, refine, and extend the calculations and comparisons made in the present paper along the obvious lines already indicated. The results will have important implications not only for the nature of the stresses responsible for torques at the CMB and the structure of the lower mantle and CMB topography, but also, indirectly, for the magnetohydrodynamics of the core and the nature of the geodynamo. Dynamo "models" can be classified in terms of two characteristic features, namely (a) the average strength of the toroidal magnetic field in the core (which for "strong field" dynamos is typically much greater than that of the poloidal field, whereas for "weak field" dynamos the two field types are comparable in strength), and (b) whether or not dynamo action extends throughout most of the volume of the core or is confined to the upper reaches. Should more refined calculations either weaken the case for significant topographic coupling, or consistently indicate an excessive topographic couple, this would constitute evidence in favor of electromagnetic torques at the CMB produced by dynamo action concentrated just below the CMB.

Acknowledgements. We thank Don Anderson, Jean Dickey, Olafur Gudmundsson, David Stevenson, and Charles Yoder for helpful comments on various aspects of this work. We also thank Dr. Dickey and other members of the Space Geodesy Science and Applications Group of the Jet Propulsion Laboratory for facilitating this collaborative study. Jeremy Bloxham kindly provided coefficients for his core flow model. Partial support was provided by NASA grants NAG5-819 and NAG5-315 to RWC and BHH.

REFERENCES

Aldridge, K.D. (ed), Core-mantle interactions, *Surveys in Geophys.*, *11*, 329-353, 1990.

Anderson, D.L. *Theory of the Earth*, Oxford, Blackwell Scientific Publications, 1989.

Anufriev, A.P., and S.L. Braginsky, Influences of irregularities in the boundary of the Earth's core on fluid velocity and magnetic field, *Geomagn. Aeron.*, *17*, 492-496, 1977.

Babcock, A., and G. Wilkins, (Eds.) *The Earth's Rotation and Reference Frames*, Reidel, Dordrecht, 1988.

Backus, G.E., Kinematics of secular variation in a perfectly conducting core, *Philos. Trans. Roy. Soc.*, *A263*, 239-263, 1968.

Backus, G.E., and J-L Le Mouël, The region of the core-mantle boundary where the geostrophic velocity field can be determined from frozen flux data, *Geophys. J. Roy. Astron. Soc.*, *85*, 617-628, 1986.

Backus, G.E., and J-L Le Mouël, The region of the core-mantle boundary where the geostrophic velocity field can be determined from frozen flux data, addendum, *Geophys. J. Roy. Astron. Soc.*, *88*, 321-322, 1987.

Benton, E.R., Magnetic probing of planetary interiors, *Phys. Earth Planet. Int.*, *20*, 111-118, 1979.

Benton, E.R., A simple method for determining the vertical growth rate of vertical motion at the top of Earth's outer core, *Phys. Earth Planet. Int.*, *24*, 242-244, 1981a.

Benton, E.R., Inviscid, frozen-flux velocity components at the top of Earth's core from magnetic observations at earth's surface: Part 1. A new methodology, *Geophys. Astophys. Fluid Dyn.*, *18*, 154-174, 1981b.

Benton E. R., and M. A. Celaya, The simplest unsteady surface flow of a frozen flux core that exactly fits a geomagnetic field model, *Geophys. Res. Lett.*, *18*, 577-580, 1991.

Bloxham, J., The dynamical regime of fluid flow at the core surface, *Geophys. Res. Lett.*, *15*, 585-588, 1988.

Bloxham, J., Length-of-day variations, topographic core-mantle coupling, and the steady and time-dependent components of core flow (abstract), *Eos, Trans. AGU*, *72*, 451, 1991.

Bloxham, J., Simple models of fluid flow at the core surface derived from geomagnetic field models, *Geophys. J. Int.*, *99*, 173-182, 1989.

Bloxham, J., and A. Jackson, Fluid flow near the surface of the Earth's core, *Rev. Geophys.*, *29*, 97-120, 1991.

Bullard, E.C., C. Freedman, H. Gellman, and J. Nixon, The westward drift of the Earth's magnetic field, *Phil. Trans. Roy. Soc.*, *A243*, 67-92, 1950.

Bullen, K.E., *Introduction to the Theory of Seismology*, Cambridge University Press, Cambridge, 1963.

Cazenave, A., editor, *Earth Rotation; Solved and Unsolved Problems*, *NATO Advanced Institute Series C; Mathematical and Physical Sciences*, Boston, D. Reidel, 1986.

Courtillot, V., and J-L Le Mouël, Time variations of the Earth's magnetic field: From daily to secular, *Ann. Rev. Earth Planet. Sci.*, *16*, 389-476, 1988.

Dickey, J.O., T.M. Eubanks, and R. Hide, Interannual and decade flunctuations in the Earth's rotation, in *Variations in the Earth's Rotation*, *Geophysical Monograph Series*, edited by D.C. McCarthy, pp. 157-162, AGU, Washington, DC, 1990.

Doornbos, D.J., The effect of a rough core-mantle boundary found only on PKKP, *Phys. Earth Planet. Int.*, *21*, 351-358, 1980.

Eltayeb, I.A., and M.H.A. Hassan, On the effects of a bumpy core-mantle interface, *Phys. Earth Planet. Int., 19*, 239-254, 1979.

Gire, C., and J-L Le Mouël, Tangentially-geostrophic flow at the core-mantle boundary compatible with observed geomagnetic secular variation: The large-scale component of the flow, *Phys. Earth Planet. Int., 59*, 259-287, 1990.

Gire, C., J-L. Le Mouël, and T. Madden, Motions at the core surface derived from SV data, *Geophys. J. R. Astr. Soc., 84*, 1-29, 1986.

Gubbins, D., Finding core motions from magnetic observations, *Phil. Trans. Roy. Soc., A306*, 247-254, 1982.

Gudmundsson, O., and R.W. Clayton, A 2-D synthetic study of global traveltime tomography, *Geophys. J. Int., 106*, 53-68, 1991.

Gudmundsson, O., and R.W. Clayton, Some problems in mapping core-mantle boundary structure, *J. Geophys. Res.*, in press, 1992.

Gudmundsson, O., J.H. Davies, and R.W. Clayton, Stochastic analysis of global travel time data: Mantle heterogeneity and random errors in the ISC data, *Geophys. J. Int., 102*, 25-43, 1990.

Gwinn, C.R., T.A. Herring, and I.I. Shapiro, Geodesy by radio interferometry: Studies of the forced nutations of the Earth, 2. Interpretation, *J. Geophys. Res., 91*, 4755-4765, 1986.

Haddon, R. A. W., Evidence of inhomogeneities near the core-mantle boundary, *Phil. Trans. Roy. Soc., A306*, 61-70, 1982.

Hager, B. H., and R. W. Clayton, Constraints on the structure of mantle convection using seismic observations, flow models, and the geoid, in *Mantle Convection*, edited by W. R. Peltier, pp. 657-763, Gordon and Breach, New York, 1989.

Hager, B.H., R. W. Clayton, M. A. Richards, R. P. Comer, and A. M. Dziewonski, Lower mantle heterogeneity, dynamic topography and the geoid, *Nature, 313*, 541-545, 1985.

Hager, B.H., and M.A. Richards, Long wavelength variations in the Earth's geoid: Physical models and dynamical implications, *Philos. Trans. Roy. Soc., A328*, 309-327, 1989.

Herring, T.A., B. A. Buffett, P. M. Mathews, and I. I. Shapiro, Forced motions of the Earth: Influence of inner core dynamics: 3. Very long interferometry data analysis, *J. Geophys. Res., 96*, 8259-8273, 1991.

Herring, T.A., C.R. Gwinn, and I.I. Shapiro, Geodesy by radio interferometry: Corrections to the IAU 1980 nutation series, *Proc. Inter. Conf. Earth Rotation and Terrestrial Reference Frame, 1*, 307-328, Ohio State University, Columbus, 1985.

Herring, T.A., C.R. Gwinn, and I.I. Shapiro, Geodesy by radio interferometry: Studies of the forced nutations of the Earth 1. Data analysis, *J. Geophys. Res., 91*, 4745-4754, 1986.

Hide, R., Interaction between the Earth's liquid core and solid mantle, *Nature, 222*, 1055-1056, 1969.

Hide, R., On the Earth's core-mantle interface, *Quart. J. Roy. Meteorol. Soc., 96*, 579-590, 1970.

Hide, R., Towards a theory of irregular variations in the length of the day and core-mantle coupling, *Phil. Trans. Roy. Soc., A284*, 547-554, 1977.

Hide, R., Presidential address: The Earth's differential rotation, *Quart. J. Roy. Astron. Soc., 278*, 3-14, 1986.

Hide, R., Fluctuations in the Earth's rotation and the topography of the core-mantle interface, *Phil. Trans. Roy. Soc., A328*, 351-363, 1989.

Hide, R., and J.O. Dickey, Earth's variable rotation, *Science, 253*, 629-637, 1991.

Hide, R., and K.I. Horai, On the topography of the core-mantle interface, *Phys. Earth Planet. Int., 1*, 305-308, 1968.

Hills, R.G., Convection in the Earth's mantle due to viscous stress at the core-mantle interface and due to large scale buoyancy. Ph.D. thesis, New Mexico State University, Las Cruces, 1979.

Hinderer, J., D. Jault, and J-L Le Mouël, Core-mantle topographic torque: A spherical harmonic approach and implications for the excitation of the Earth's rotation by core motions, *Phys. Earth Planet. Int., 59*, 329-341, 1990.

Jacobs, J.A., (Ed.) *The Earth's Core*, 2nd ed., Academic Press Ltd., New York, 1987*a*.

Jacobs, J.A. *Geomagnetism*. 2 vols. Academic Press Ltd., New York, 1987b.

Jault, D., and J-L Le Mouël, The topographic torque associated with tangentially geostrophic motion at the core surface and inferences on the flow inside the core, *Geophys. Astrophys. Fluid Dyn., 48*, 273-296, 1989.

Jault, D., and J-L Le Mouël, Core-mantle boundary shape: Constraints inferred form the pressure torque acting between the core and mantle, *Geophys. J. Int., 101*, 233-241, 1990.

Jault, D., and J-L Le Mouël, Exchange of angular momentum between the core and mantle, *J. Geomag. Geoelect., 43*, 111-129, 1991.

Jault, D., C. Gire, and J-L Le Mouël, Westward drift, core motions and exchange of angular momentum between core and mantle, *Nature, 333*, 353-356, 1988.

Jeanloz, R., The nature of the Earth's core, *Ann. Rev. Earth Planet. Sci., 18*, 257-386, 1990.

Kinoshita, H., and J. Souchay, The theory of the nutation for the rigid Earth model of second order, *Celestial Mechanics, 48*, 187-265, 1990.

Knittle, E., and R. Jeanloz, Earth's core-mantle boundary: Results of experiments at high pressures and temperatures, *Science, 251*, 1438-1443, 1991.

Lambeck, K., *The Earth's Variable Rotation*, Cambridge University Press, London and New York, 1980.

Lambeck, K., *Geophysical Geodesy: The Slow Deformation of the Earth*, Oxford, Clarendon Press, 1988.

Le Mouël, J-L, Outer-core geostrophic flow and secular variation of the main geomagnetic field, *Nature, 311*, 734-735, 1984.

Lloyd, D., and D. Gubbins, Toroidal fluid motion at the top of the Earth's core, *Geophys. J. Int., 100*, 455-467, 1990.

Loper, D.E., The nature and consequences of thermal interactions twixt core and mantle, *J. Geomag. Geoelect., 43*, 79-91, 1991.

Melchior, P. *The Physics of the Earth's Core*. Pergamon Press, Oxford , 1986.

Moffatt, H.K., *Magnetic Field Generation by Fluid Motion*, Cambridge University Press, Cambridge, 1978a.

Moffatt, H.K., Topographic coupling at the core-mantle interface, *Geophys. Astrophys. Fluid Dyn., 9*, 279-288, 1978b.

Morelli, A., and A.M. Dziewonski, Topography of the core-mantle boundary and lateral homogeneity of the liquid core, *Nature, 325*, 678-683, 1987.

Moritz, H., and I.I. Mueller (Eds.), *Earth Rotation: Theory and Observation*, The Ungar Publishing Co., New York, 1987.

Morrison, L.V., Redetermination of the decade fluctuations in the rotation of the Earth in the period 1861-1978, *Geophys. J. Roy. Astron. Soc., 38*, 349-360, 1979.

Munk, W.H., and G.J.F. MacDonald, *The Rotation of the Earth*, Cambridge University Press, Cambridge, 1960.

Paulus, M., and M. Stix, Electromagnetic core-mantle coupling: The Fourier method for solving the induction equation, *Geophys. Astrophys. Fluid Dyn., 47*, 237-249, 1989.

Reid, M.J., and J.M. Moran, (Eds.) *The Impact of VLBI on Astrophysics and Geophysics*, Kluwer Academic Publishers, Dordrecht, 1988.

Roberts, P. H., Electromagnetic core-mantle coupling, *J. Geomag. Geoelect., 24*, 231-259, 1972.

Roberts, P.H., and S. Scott, On analysis of secular variation: 1 A hydromagnetic constraint, *J. Geomag. Geoelect., 17*, 137-151, 1965.

Rochester, M.G., Causes of fluctuations in the Earth's rotation, *Phil. Trans. Roy. Soc., A313*, 95-105, 1984.

Spieth, M.A. R. Hide, R. W. Clayton, B. H. Hager, and C. V. Voorhies, Topographic coupling of the core and mantle and changes in the length of the day (abstract), *Eos Trans. AGU, 67*, 908, 1986.

Stoyko, N., Sur les variations de champ magnetique et la rotation de la Terre, *C.R. Hebd. Seanc. Acad. Sci. Paris, 233*, 80-82, 1951.

Vestine, E.H., On variations of the geomagnetic field, fluid motions, and the rate of the Earth's rotation, *Proc. Nat. Acad. Sci.*, *38*, 1030-1038, 1952.

Voorhies, C. V., Steady flows at the top of Earth's core derived from geomagnetic field models, *J. Geophys. Res.*, *91*, 12, 444-12, 466, 1986a.

Voorhies, C.V., Steady surficial core motions: an alternate method. *Geophys. Res. Lett.*, *13*, 1537-1540, 1986b.

Voorhies, C.V., The time-varying geomagnetic field, *Rev. Geophys.*, *25*, 929-938, 1987.

Voorhies, C.V., Probing surface core motions with DGRF models (abstract), *Eos Trans. AGU*, *69*, 336, 1988.

Voorhies, C.V., Coupling an inviscid core to an electrically-insulating mantle, *J. Geomag. Geoelect.*, *43*, 131-156, 1991a.

Voorhies, C.V., On the joint inversion of geophysical data for models of the coupled core-mantle system, *NASA Tech Memo 104536*, 24pp., 1991b.

Voorhies, C. V., Implications of decade fluctuations in the length of the day for geomagnetic estimates of core surface flow and geodynamo experiments, in *Geophysical Monographs*, edited by J-L. Le Mouël, submitted, 1992.

Voorhies, C.V., and G.E. Backus, Steady flows at the top of the core from geomagnetic field models: the steady motions theorem. *Geophys. Astrophys. Fluid Dyn.*, *32*, 163-173, 1985.

Wahr, J. M., The Earth's rotation, *Ann. Rev. Earth Planet. Sci.*, *16*, 231-249, 1988.

Whaler, K.A., Geomagnetic evidence for fluid upwelling at the core-mantle boundary, *Geophys. Journ. Roy. Astron. Soc.*, *86*, 563-588, 1986.

Whaler, K.A., A steady velocity field at the top of the Earth's core in the frozen-flux approximation - errata and further comments, *Geophys. J. Int.*, *102*, 507-509, 1990.

Whaler, K.A., Properties of steady flows at the core-mantle boundary in the frozen-flux approximation, *Phys. Earth Planet. Int.*, *68*, 144-155, 1991.

Whaler, K.A., and S.O. Clarke, A steady velocity field at the top of the Earth's core in the frozen-flux approximation. *Geophys. J. Int.*, *94*, 143-155, 1988.

R. W. Clayton, Seismological Laboratory, California Institute of Technology, Pasadena, CA 91125. U. S. A.

B. H. Hager, Department of Earth, Atmospheric and Planetary Sciences, Massachusetts Institute of Technology, Cambridge, MA 02139, U. S. A.

R. Hide, Robert Hooke Institute, The Observatory, Clarendon Laboratory, Parks Road, Oxford 0X1 3PU, England, U. K.

M. A. Spieth, Jet Propulsion Laboratory, California Institute of Technology, Pasadena, CA 91109, U. S. A.

C. V. Voorhies, Geodynamics Branch Code 921, Goddard Space Flight Center, Greenbelt, MD 20771, U. S. A.

Chemical Reactions at the Earth's Core–Mantle Boundary: Summary of Evidence and Geomagnetic Implications

RAYMOND JEANLOZ

Department of Geology and Geophysics, University of California, Berkeley

Laboratory experiments indicate that the liquid outer core reacts chemically with the crystalline silicates of the lowermost mantle. Formation of the reaction products, a crystalline mixture of metallic alloys and relatively insulating silicates, can explain the seismologically determined ~1-10% lateral and horizontal variations in wave velocities that are observed in the D" layer at the base of the mantle. Corresponding variations of ~2-4 orders of magnitude or more in electrical conductivity are expected to be anticorrelated with the seismic heterogeneities. The conductivity heterogeneity can affect the magnetic field lines emanating from the core, resulting in local variations of the magnetic diffusion time ranging from $<10^{-1}$ to $\geq 10^{2}$ years within D". The electrical conductivity is likely to be strongly anisotropic, possibly in a manner that is correlated with anisotropy of the seismic wave velocities. Both anisotropy and heterogeneity of the electrical conductivity in D" can mix poloidal and toroidal components of the geomagnetic field.

Evidence for Mantle-Core Reactions

Laboratory studies demonstrate that liquid iron and iron alloys react chemically with crystalline oxides and silicates at the ultrahigh pressure–temperature conditions of the lowermost mantle [*Knittle and Jeanloz*, 1986, 1989, 1991a; *Goarant et al.*, 1992]. Hence, according to the experimental results, the liquid iron alloy of the outer core is expected to react vigorously with $(Mg,Fe)SiO_3$ perovskite and other mantle oxides across the core–mantle boundary.

The primary reaction that has been experimentally shown to occur at deep-mantle conditions involves perovskite-structured $(Mg,Fe)SiO_3$, which is thought to make up ~70-90% of the lower mantle, plus liquid Fe producing a mixture of metallic alloys ($Fe_xO + Fe_ySi$) and nonmetallic silicates ($MgSiO_3 + SiO_2$) [*Knittle and Jeanloz*, 1989, 1991a]. Liquid Fe alloys, such as iron sulfide, are similarly found to react with oxides at these conditions [*Williams and Jeanloz*, 1991]. Thus in simple terms, mantle rock dissolves into the core, though not uniformly: according to the experiments, oxygen and silicon are incorporated preferentially into the liquid metal relative to magnesium, for example, which appears not to enter into the metal.

The physical properties of the phases within the reacting assemblage vary considerably in magnitude, with the liquid core material (reactant) and crystalline Fe-alloy products differing from the silicates (reactant and product phases) by $\delta\rho/\rho \sim +50\text{-}80\%$, $\delta V_P/V_P \sim -30\text{-}70\%$ and $\delta V_S/V_S \sim -20\text{-}100\%$, respectively, in density, compressional velocity and shear-wave velocity [*Jeanloz*, 1990]. Therefore, chemical reactions with the core would be expected to induce strong heterogeneity in the physical properties of the lowermost mantle. Whereas heterogeneity should be observed in the deep mantle, the reaction products contaminating the outer core may be quickly spread out and stirred throughout much of the rapidly flowing liquid; heterogeneity is thus likely to be diminished or destroyed in the core [*Knittle and Jeanloz*, 1991a].

Indeed, numerous seismological investigations demonstrate that the bottom ~200 (± 200) km of the mantle, the D" layer, is anomalous, being characterized by scattering of waves and by the presence of significant (≥1-10%) velocity anomalies which are laterally variable [e.g., *Bataille and Flatté*, 1988; *Young and Lay*, 1987, 1990; *Garnero et al.*, 1988; *Baumgardt*, 1989; *Lay*, 1989; *Vinnik et al.*, 1989; *Wysession and Okal*, 1989; *Davis and Weber*, 1990; *Weber and Davis*, 1990; *Gaherty and Lay*, 1992]. In addition, there is tentative seismological evidence for $\sim 10^3$ m topography on the core–mantle boundary and for compositional anomalies in the outermost core, the latter perhaps resulting from the metal–silicate reactions [*Creager and Jordan*, 1986; *Morelli and Dziewonski*, 1987; *Souriau and Poupinet*, 1990; *Lay and Young*, 1990; cf. *Rekdal and Doornbos*, 1992].

The seismologically observed heterogeneity of D" can be explained merely by the amounts of reaction-product phases, the proportion of crystalline metal to silicate constituents,

Relating Geophysical Structures and Processes: The Jeffreys Volume
Geophysical Monograph 76, IUGG Volume 16

varying laterally or vertically by ~2-20%. Furthermore, contamination of the core by mantle components (either in liquid or crystalline form) provides a natural explanation for the fact that the composition of the Earth's outer core cannot be pure iron: even if it had initially consisted only of iron (or Fe-Ni alloy), the core would necessarily become alloyed with mantle constituents over geological history [*Jeanloz*, 1990; *Knittle and Jeanloz*, 1991a].

Variability in the D" phase assemblage could simply be due to incomplete reaction between metal and silicates resulting in a heterogeneous mixture of crystalline reactant and product phases. Alternatively, complete reaction of varying initial assemblages (prior inhomogeneities in the mantle) would lead to variable amounts of product phases. The essential point is that only modest (~ 10%) heterogeneity in the final assemblage of phases is needed to reproduce the seismological observations because the individual phases differ so greatly in properties.

In the present interpretation, the seismic velocities in D" are predicted to be anticorrelated with the amount of alloy present and hence with the density. Also, enrichment of crystalline SiO_2 (reaction product) can produce locally high V_P and V_S, although "mode softening" associated with the high-pressure dynamical instability of stishovite may locally decrease the velocities [*Jeanloz*, 1989a; *Cohen*, 1992]. Finally, in addition to the occurrence of crystalline-alloy reaction products, the presence of unreacted core material (liquid Fe or Fe alloy) — probably in small quantities at any given time — would be highly effective in reducing the velocities of the lowermost mantle.

INFILTRATION AND PARTIAL MIXING OF METALS

To create heterogeneity in the D" layer through chemical reactions with the core, the liquid alloy must infiltrate the overlying mantle. Infiltration, the penetration of liquid metal into the crystalline silicates, leads to the chemical reaction of core liquid with mantle rock that produces the crystalline assemblage of alloys and silicates within D". Through ongoing reaction, the crystalline products (alloys and silicates) accumulate with time as heterogeneities embedded in the lowermost mantle.

Infiltration can occur by at least two processes. First, liquid iron is found to wet the silicate grain boundaries extensively in the high-pressure experiments, in contrast with the non-wetting behavior typically observed at low pressures for liquid metals on silicate or oxide surfaces [*Knittle and Jeanloz*, 1989, 1991a]. The change in wetting behavior can be directly attributed to oxygen becoming a metallic alloying component at high pressures [*Knittle and Jeanloz*, 1986, 1991b; *Jeanloz*, 1989b]. Estimates of surface energies then suggest that liquid iron alloy of the outer core should infiltrate the lowermost mantle by capillary action, penetrating upward to heights of ~10^1-10^3 m above the core–mantle boundary [*Jeanloz*, 1990; *Poirier and LeMouël*, 1992].

A second mode of infiltration is analogous to aquifers in the crust, and depends on the core–mantle boundary having topography. Given such variations in the height of the boundary, liquid metal from the regions of elevated topography must tend to percolate laterally and downward into mantle rock that is depressed into the core. That is, the dense liquid flows through the less dense rock under the action of gravity. Here, capillary forces are expected to be very effective because gravity is working in favor of infiltration, rather than against it as was the case with the first mechanism. The added effects of electromagnetic pumping [e.g., *Shercliff*, 1965] are not considered here, although they could be locally significant in aiding the core liquid permeate into the crystalline assemblage of the lowermost mantle.

Geodetic evidence points to an amplitude of ~10^2-10^3 m for the topography on the core–mantle boundary [*Gwinn et al.*, 1986; *Hide and Dickey*, 1991; *Mathews and Shapiro*, 1992]. Therefore, a zone of infiltration and reaction some hundreds of meters thick is expected at the bottom of the mantle, and it should be developed on a geologically rapid time scale (<< 10^7 years) [*Jeanloz*, 1990; *Knittle and Jeanloz*, 1991a]. In fact, recent VLBI studies reveal possible evidence for such a zone of metal infiltration ~10^2 m thick being present at the base of the mantle [*Buffett et al.*, 1990; *Buffett*, 1992].

The greatest uncertainties in modelling the infiltration involve estimating the permeability of the rock and the effective widths of grain boundaries: that is, the dimensions of the infiltration paths. Reasonable assumptions of grain sizes (dimensions $\delta \sim 10^{-3}$-10^{-2} m) and boundary widths (~10^{-6}-10^{-3} δ) are compatible with a thickness of order tens-hundreds of meters for the zone of infiltration and reaction. These suggest average permeabilities and fluid fractions far less than 1-10 percent at any given time, though locally, on scales much smaller than 10^3-10^4 m, these values could be significantly higher.

One would anticipate the amount of crystalline reaction product to build up over time. This is analogous to the hydration of the sub-oceanic lithosphere, which involves the accumulation of crystalline (hydrous) product phases due to reaction of initially anhydrous silicates with infiltrating sea water [*Jeanloz*, 1990]. Heterogeneity of the magnitude suggested by the seismological observations, with up to tens of percent dense, metallic alloy being present in the rock of the deepest mantle, can easily develop in this manner.

Once formed, an infiltration–reaction layer just above the top of the core would be expected to be pulled upward and partially stirred into the mantle by the background flow of solid-state (tectonic) convection (Figure 1). The time scale for mixing is therefore expected to be much longer, $\geq 10^7$-10^8 yr, than that required to form the reaction zone. It is implausible that significant amounts of liquid alloy (e.g., unreacted core liquid) are pulled upward in this process: first, because only a small amount of liquid is likely to be present at any given time; and second, because the liquid would most likely drain downward by the aquifer-like process noted above.

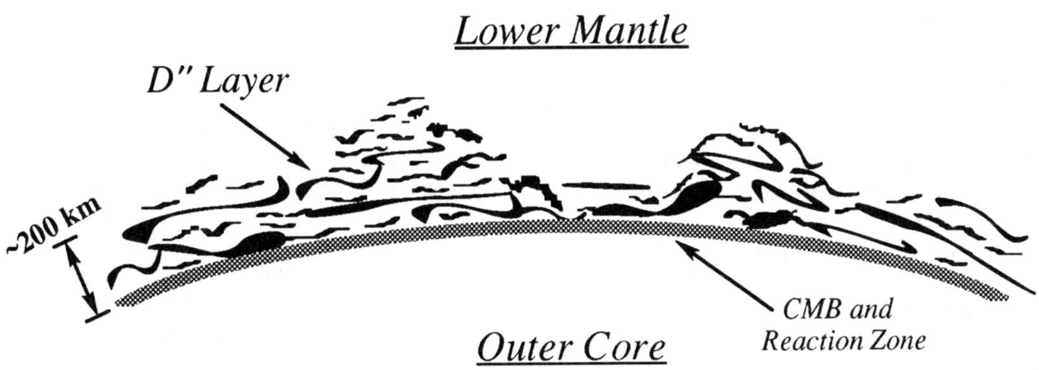

Fig. 1. Schematic cross-section of the core–mantle boundary (CMB) and overlying D" region [*Jeanloz*, 1990; *Knittle and Jeanloz*, 1991a]. Chemical reaction between the mantle and core is expected to produce a mixture of iron alloy-rich, electrically conductive regions (dark) and silicate-rich, more insulating regions (light) that are stirred by background mantle convection to form the D" layer, a heterogeneous zone of thickness varying laterally from ~0 to ~500 km. Although liquid alloy from the core is expected to be present in small quantities within the reaction zone, it is primarily crystalline alloy and silicate reaction products that are likely to be pulled upward by mantle flow to produce the heterogeneity of the D" layer. The convective streaking illustrated here emphasizes a horizontal lamination of tendrils in the lowermost mantle.

Hence, crystalline reaction products are expected to be the most important source of heterogeneity in D" [cf. *Poirier and LeMouël* , 1992]. Again, the analogy can be made with the hydrated lithosphere at the Earth's surface, by which water is incorporated into the mantle primarily through the subduction of crystalline hydrous phases rather than the liquid [*Jeanloz*, 1990].

Although quantitative models are lacking in detail, stirring of the reaction-layer "dregs" upward from the bottom of the mantle could well produce the ~10^2 km-thick zone of heterogeneity observed seismologically as the D" layer [*Davies and Gurnis*, 1986; *Sleep*, 1988; *Hansen and Yuen*, 1989; *Olson and Kincaid*, 1991; *Kellogg and King*, 1992]. Moreover, the dense, alloy-rich dregs should be piled up into drifts; that is, the thickness of D" should vary laterally, its top being elevated in regions of upwelling in the lowermost-mantle flow pattern and depressed in regions of downwelling. This expectation is matched by the seismological observations, which indicate a variable thickness or depth to the top of the D" layer; in some locations anomalous velocities begin ≥500 km above the base of the mantle, whereas in other areas there is little evidence for velocity anomalies or heterogeneity right down to the core–mantle boundary [e.g., *Lay*, 1989].

According to this model, there should be a general correlation between the seismological structure of the D" layer and that of the overlying mantle. Specifically, thick, low-velocity (i.e., alloy-rich) portions of D" should be overlain by hot, upwelling regions of the lower mantle: that is, regions having diminished seismic velocities. Similarly, locations where D" is thin should reside beneath cold, downwelling, and therefore high-velocity regions of the lower mantle. Although such correlations have not yet been proven in detail, the available data are compatible with the present expectations [*Woodward and Dziewonski*, 1992].

HETEROGENEOUS ELECTRICAL CONDUCTIVITY

In addition to producing the seismological anomalies, chemical reactions between the mantle and core are expected to create large variations in the electrical conductivity of the D" layer, both in the vertical and horizontal directions. As the electrical conductivity of the metal alloys (either liquid unreacted core material, or crystalline reaction products) is more than ~10^8-10^{10} larger than that of the $MgSiO_3$ and SiO_2 reaction products, the conductivity of the lowermost mantle is likely to vary by many orders of magnitude [*Jeanloz*, 1990].

Because the threshold for conductive percolation in a 3-dimensional random medium is ~16% by volume [*Zallen*, 1983], variations in the proportion of metal alloy between ~10 and 30% can produce local changes exceeding 2-4 orders of magnitude in the electrical conductivity of the D" layer [*Jeanloz*, 1990; *Knittle and Jeanloz*, 1991a]. Texturing, as discussed below, can affect the percolation threshold but is not likely to change the general conclusions reached here. As with the seismologically observed properties, strong heterogeneity is expected in the electrical conductivity even for relatively modest variations in the proportions of the phases present in D".

Due to the effects of magnetic induction, the movement of magnetic field lines emanating from the core ought to be strongly perturbed by the lateral variations in electrical conductivity (σ), hence magnetic diffusivity $\eta = 1/(\mu_0\sigma)$, present in the lowermost mantle (μ_0 is the permeability of free space [e.g., *Merrill and McElhinny*, 1983]). Therefore, heterogeneities in the D" layer could be manifested in the geomagnetic field observed at the Earth's surface, depending not only on the strength and dimensions of the heterogeneities but also on their geometric shapes.

For example, field lines should tend to slip around high-conductivity regions having positive curvature (~spherical heterogeneities), whereas the lines can be effectively trapped and the field intensified within conductive regions of negative curvature: e.g., field lines moving into the heterogeneity from within low-conductivity embayments curving into the edge of the high-conductivity region [*Jeanloz*, 1990]. Thus, *Poirier and LeMouël* [1992] have considered one of the least plausible geometries for investigating this effect.

The expectation, based on the known properties of the mantle–core reaction products, is that laterally extended regions of slow magnetic diffusion — conductive, alloy-rich heterogeneities — are correlated with anomalously low seismic velocities. That is, the heterogeneities are such that large-scale conductivity and velocity variations are likely to be anticorrelated. Combining the seismologically inferred dimensions of heterogeneities in the lowermost mantle ($\sim 10^3$-10^5 m) with the experimentally measured electrical conductivities of the reaction products leads to the prediction that magnetic-diffusion times vary from $<10^{-1}$ to $\geq 10^2$ years in the region above the core–mantle boundary [*Jeanloz*, 1990].

Although there is at present no independent evidence to conclusively show that the electrical conductivity of the D" layer is heterogeneous, a variety of geomagnetic observations might be explained by the presence of such variability. Thus, rapid changes in the field are allowed by the existence of low-conductivity (high diffusivity) regions in the deep mantle. Such abrupt changes or "jerks" seem to be evident both in recent observatory measurements of the normal field, and in detailed paleomagnetic recordings of the field undergoing reversal [*Courtillot and LeMöuel*, 1984; *Coe and Prévot*, 1989; cf. *Bogue and Merrill*, 1992].

In contrast, static zero-flux patches in the geomagnetic secular variation (regions of low magnetic flux moving little with time, in the field downward continued to the core–mantle boundary [*Bloxham*, 1986]) appear correlated with the seismic velocities in D". Specifically, the static zero-flux patch under Easter Island, and possibly the patches beneath the northern Pacific and the Indian Ocean south of Africa (anomalies 9, 7 and 8 of *Gubbins* [1989]) roughly coincide with regions of anomalously low velocities near the bottom of the mantle, according to *Inoue, et al.* [1990]. This suggests the possibility that the core magnetic field is partially screened by high-conductivity regions in the lowermost mantle.

Similarly, high-flux regions that are stationary (static flux bundles [*Bloxham*, 1986]) beneath the central Pacific, and possibly the two under Antarctica, that beneath Canada and — very tentatively — that beneath Siberia (anomalies 5, 4, 3, 2 and 1, respectively, of *Gubbins* [1989]), may be correlated with the edges of low-velocity regions within D" [*Inoue et al.*, 1990]. However, although the existing velocity models for the lowermost mantle are broadly in agreement with each other [*Dziewonski and Woodhouse*, 1987; *Woodhouse and Dziewonski*, 1989; *Inoue et al.*, 1990; *Woodward and Dziewonski*, 1992], the reliability and spatial resolution of both the secular-variation and the seismological observations are still too uncertain to know whether these tentative correlations are spurious or not.

Another observation is that the virtual geomagnetic pole appears to have followed similar paths during successive reversals of the field, even to the point of exhibiting nearly identical path complexities over the past ~1-5 Myr [*Laj et al.*, 1991; *Hoffman*, 1991; *Herrero-Bervera and Khan*, 1992; *Constable*, 1992]. One explanation for this regularity is that the transitional field (i.e., the geomagnetic field during a reversal) observed at the Earth's surface is controlled by the overlying mantle. In the present interpretation, the rough correlation between reversal paths [*Laj et al.*, 1991; *Constable*, 1992] and anomalously high seismic velocities in D" [*Dziewonski and Woodhouse*, 1987; *Inoue et al.*, 1990; *Woodward and Dziewonski*, 1992] would be ascribed to the transitional field recorded at the surface being screened by the locally high electrical conductivity of the lowermost mantle; the reversing field can be affected by mantle-conductivity variations through other mechanisms, but these all lead to a correlation being anticipated between conductivity, hence field configuration, and seismic wave velocities in D" [cf. *Runcorn*, 1992; *Bogue and Merrill*, 1992].

Presumably, the pattern of reversal paths would change over geological time periods ($\geq 10^7$-10^8 yr), as the D" layer evolves. Indeed, long-term changes in the conductivity distribution of the lowermost mantle may even explain variations in the field strength over hundred-million year periods [*Prévot and Perrin*, 1992].

Significant lateral heterogeneity in the magnetic diffusivity of the D" region is particularly interesting in light of *Busse and Wicht's* [1992] demonstration that such heterogeneity can itself contribute to the geodynamo (see also *Busse* [1992]). Specifically, they show that heterogeneous magnetic diffusivity in the lowermost mantle can produce dynamo action when the high- and low-conductivity regions are oriented perpendicular to the flow direction of the underlying core fluid.

Physically, the effect of the conductivity heterogeneity is to couple the poloidal (spheroidal) and toroidal (non-radial) components of the magnetic field, resulting in a growth of the field strength. Such coupling seems to be generally required for producing magnetic dynamos [*Busse and Wicht*, 1992]. Note that the finite and locally high electrical conductivity of D" allows a significant toroidal field to be maintained in the lowermost mantle [*Merrill and McElhinny*, 1983].

Two other consequences of the conductivity-heterogeneity at the base of the mantle are that a non-axisymmetric, standing poloidal field can be produced and that complex, time-dependent torques can arise between the mantle and core [*Busse*, 1992]. The latter may help explain complexities observed in the decade fluctuations of the length of the day [*Morrison*, 1979; *Hide and Dickey*, 1991].

Finally, to the degree that alloy-rich, high-conductivity

regions near the base of the mantle are interconnected with the core and distributed in a globally asymmetric pattern, the heterogeneity in D" could provide an explanation for recent observations that the normal and reversed configurations of the Earth's magnetic field have been distinguishably different, at least for the past few million years [*Schneider and Kent*, 1988; *Merrill et al.*, 1990]. Other explanations, such as thermoelectric effects in the lowermost mantle, have been suggested to explain these apparently systematic differences in the ratio of quadrupole to dipole fields during normal and reversed polarities [*Merrill and McElhinny*, 1983; *Merrill et al.*, 1990]. Though highly tentative, the present model does not conflict with such hypotheses, but suggests alternative mechanisms involving chemical reactions at the core–mantle boundary for explaining the differences between normal and reversed configurations of the geomagnetic field [cf. *Busse*, 1992].

ANISOTROPY IN D"

A likely consequence of the alloy-rich and silicate-rich zones being convectively mixed at the bottom of the mantle is that anisotropy should be developed in the physical properties of the D" layer. In particular, chaotic mixing at the high Rayleigh numbers pertaining to the Earth's deep interior would be expected to generate tendrils, sheets and other structures that impose a texture to the mantle–core region [*Ottino*, 1989; *Kellogg and Turcotte*, 1990; *Kellogg and Stewart*, 1991].

The result of such convective streaking is to create a region of strong but laterally variable anisotropy in the mantle boundary layer (Figure 1). This is analogous to the anisotropy that is thought to characterize the upper mantle [*Dziewonski and Anderson*, 1981; *Anderson*, 1989]. In fact, seismological evidence for anisotropy of this kind has been obtained in at least some regions just above the core–mantle boundary [*Vinnik et al.*, 1989; *Lay and Young*, 1991]. To prove that anisotropy is a general characteristic of the D" layer as expected here, rather than a local feature, requires further investigations, however.

Assuming that electrically conducting (alloy-rich) and resistive (oxide-rich) heterogeneities are mixed in this manner, the conductivity of D" should be highly anisotropic. Hence, the tensor quality of the electrical conductivity must be taken into account, with the conductivity in the direction of streaking being larger than that in the direction perpendicular to the layered or lineated texture.

The difference in conductivity in the two directions can be estimated by considering the former as a circuit containing resistive elements in parallel, whereas the latter is in series. Thus, as with the estimate of heterogeneity, the magnitude of the directional difference in conductivity is expected to exceed 10^2-10^4. Unlike the case of heterogeneity, however, the corresponding anisotropy in seismic velocities is expected to be correlated (rather than anticorrelated) with the conductivity: both the velocity and the conductivity parallel to the streaks are larger than in the direction perpendicular to the streaks. That is, the fast, silicate-rich layers and the conductive, alloy-rich layers efficiently transmit the seismic waves and electrical currents, respectively.

A consequence of this anisotropy is that there can be interchange between the toroidal and poloidal components of the magnetic field. Consider the diffusive term, $\partial \mathbf{B}'/\partial t = \nabla\times\mathbf{C}$, of the magnetic induction equation

$$\partial \mathbf{B}/\partial t = \nabla\times(\mathbf{v}\times\mathbf{B}) - \partial \mathbf{B}'/\partial t \qquad (1)$$

relating the time variation of the magnetic induction **B** to the flow velocity **v** [e.g., *Merrill and McElhinny*, 1983]. Assuming that the anisotropy can be described as a laminar texture that is either parallel or perpendicular to the core–mantle boundary, the magnetic diffusivity tensor (or, equivalently, the reciprocal of the electrical conductivity tensor) can be represented by a 3x3 diagonal matrix, and the components of the vector **C** are given in tensor notation by $C_x = \eta_{xx}(B_{z,y} - B_{y,z})$, with corresponding permutations of the indices. To be specific, the index z will be identified with the vertical direction, perpendicular to the core–mantle boundary.

If the laminations are oriented vertically, and the (horizontal) y-axis is set perpendicular to the laminations, then $\sigma_{zz} \sim \sigma_{xx} >> \sigma_{yy}$. This implies that $\eta_{yy} >> \eta_{zz} \sim \eta_{xx} \sim 0$ and $C_y >> C_z \sim C_x \sim 0$ where, for simplicity, infinite conductivity (analogous to the frozen-flux approximation) has been assumed for the x- and z-directions in the last step [*Merrill and McElhinny*, 1983]. Therefore, the diffusive term in (1), given by the curl of **C**, has components of magnitude $|\partial_t B_x'| \sim |\partial_t B_z'| \sim |[\eta_{yy}(B_{x,z} - B_{z,x})]_{,x}| >> |\partial_t B_y'| \sim 0$.

Evidently, variations in both the horizontal (x) and vertical (z) components of the magnetic induction are inter-related, even for the highly simplified geometry assumed in the present example. The large vertical component of conductivity (σ_{zz}) assumed present in D" allows a strong toroidal field to be maintained and to couple with the poloidal field in the lowermost mantle. More complex geometries, including the presence of off-diagonal terms in the diffusivity matrix $\underline{\eta}$, lead even more readily to coupling between radial and tangential components of **B**.

Significantly, the effects of anisotropy parallel *Busse and Wicht's* [1992] findings for heterogeneous conductivity at the base of the mantle: the result is to strongly mix poloidal and toroidal components of the geomagnetic field. Toroidal–poloidal coupling raises the possibility of dynamo action due to anisotropy, but it seems unlikely that this phenomenon can be proven to occur at the core–mantle boundary with present observations.

Instead, it is worth considering horizontal laminations of the conductive and less conductive zones (Figure 1). Now, the horizontal and vertical components of the diffusive term separate: $\sigma_H >> \sigma_V$ implies $\eta_V >> \eta_H$ and $C_V >> C_H$, with indices V = z or zz, H = x or xx = y or yy. The result is that diffusion of the horizontal field is much more rapid than that of the radial field: $|\partial_t B_H'| >> |\partial_t B_V'| \sim 0$ (the last approximation is for the simplifying case $C_H \sim 0$). As long as η_V remains finite ($\sigma_V > 0$), electrical currents are possible in the vertical direction and a toroidal field can be maintained above the core–mantle boundary.

Therefore, the present model with horizontal conductivity laminations (Figure 1) would be supported by observations showing a correlation between relatively rapid diffusion of the horizontal geomagnetic field and seismological evidence of horizontal-transverse anisotropy in the D" layer. Hope that such a test of the model may be possible is offered by recent observations of possible seismic anisotropy in the lowermost mantle [*Vinnik et al.*, 1989; *Lay and Young*, 1991].

Acknowledgments. I have benefitted from discussions with J. Bloxham, B. A. Buffett, M. S. T. Bukowinski, F. H. Busse, R. Coe, A. M. Dziewonski, L. A. Kellogg, E. Knittle, T. Lay, R. Merrill, B. O'Neill and S. K. Runcorn. This work was supported by NASA.

REFERENCES

Anderson, D. L., *Theory of the Earth*, Blackwell Scientific Publications, Boston, 1989.

Bataille, K., and S. Flatté, Inhomogeneities near the core–mantle boundary inferred from short-period scattered PKP waves recorded at the Global Digital Seismographic Network, *J. Geophys. Res.*, *93*, 15057-15064, 1987.

Baumgardt, D. R., Evidence for a P-wave velocity anomaly in D", *Geophys. Res. Lett.*, *16*, 657-660, 1989.

Bloxham, J., Models of the magnetic field at the core–mantle boundary for 1715, 1777 and 1842, *J. Geophys. Res.*, *91*, 13954-13966, 1986.

Bloxham, J., D. Gubbins and A. Jackson, Geomagnetic secular variation, *Phil. Trans. R. Soc. Lond. A*, *329*, 415-502, 1989.

Bogue, S. W., and R. T. Merrill, The character of the field during geomagnetic reversals, *Ann. Rev. Earth Planet. Sci.*, *20*, 181-219, 1992.

Buffett, B. A., Estimates of magnetic energy and mantle conductivity based on observations of the Earth's forced nutations (abstract), *EOS, Trans. Am. Geophys. Union, 73 (supplement)*, *55*, 1992.

Buffett, B. A., T. A. Herring, P. M. Mathews and I. I. Shapiro, Anomalous dissipation in the Earth's forced nutations: Inferences on the electrical conductivity and magnetic energy at the core–mantle boundary (abstract), *EOS, Trans. Am. Geophys. Union*, *71*, 496, 1990.

Busse, F. H., Theory of the geodynamo and core–mantle coupling, in *Chaotic Processes in the Geological Sciences*, edited by D. A. Yuen, in press, Springer Verlag, New York, 1992.

Busse, F. H., and J. Wicht, A simple dynamo caused by conductivity variations, *Geophys. Astrophys. Fluid Dynamics*, *64*, 135-144, 1992.

Coe, R. S., and M. Prévot, Evidence suggesting extremely rapid field variation during a geomagnetic reversal, *Earth Planet. Sci. Lett.*, *92*, 292-298, 1989.

Cohen, R. E., First-principles predictions of elasticity and phase transitions in high pressure SiO_2 and geophysical implications, in *High-Pressure Research: Application to Earth and Planetary Sciences*, edited by Y. Syono and M. H. Manghnani, Am. Geophys. Union, Washington, DC, in press, 1992.

Constable, C., Link between geomagnetic reversal paths and secular variation of the field over the past 5 Myr., *Nature*, *358*, 230-233, 1992.

Courtillot, V., and J.-L. LeMouël, Geomagnetic secular variation impulses: A review of observational evidence and geophysical consequences, *Nature*, *311*, 709-716, 1984.

Creager, K. C., and T. H. Jordan, Aspherical structure of the core–mantle boundary from PKP travel times, *Geophys. Res. Lett.*, *13*, 1497-1500, 1986.

Davies, G. F., and M. Gurnis, Interaction of mantle dregs with convection: lateral heterogeneity at the core-mantle boundary, *Geophys. Res. Lett.*, *13*, 1517-1520, 1986.

Davis, J. P., and M. Weber, Lower mantle velocity inhomogeneity observed at GRF array, *Geophys. Res. Lett.*, *17*, 187-190, 1990.

Dziewonski, A. M., Mapping the lower mantle: Determination of lateral heterogeneity in P velocity up to degree and order 6, *J. Geophys. Res.*, *89*, 5929-5952, 1984.

Dziewonski, A. M., and D. L. Anderson, Preliminary reference Earth Model, *Phys. Earth Planet. Int.*, *25*, 297-356, 1981.

Dziewonski, A. M., and J. H. Woodhouse, Global images of the Earth's interior, *Science*, *236*, 37-48, 1987.

Gaherty, J. B., and T. Lay, Investigation of laterally heterogeneous shear velocity structure in D" beneath Eurasia, *J. Geophys. Res.*, *97*, 417-435, 1992.

Garnero, E., D. Helmberger and G. Engen, Lateral variations near the core–mantle boundary, *Geophys. Res. Lett.*, *15*, 609-612, 1988.

Goarant, F., F. Guyot, J. Peyronneau and J. P. Poirier, High-pressure and high-temperature reactions between silicates and liquid iron alloys, in the diamond cell, studied by analytical electron microscopy, *J. Geophys. Res.*, *97*, 4477-4487, 1992.

Gubbins, D., Implications of geomagnetism for mantle structure, *Phil. Trans. R. Soc. Lond. A*, *328*, 365-375, 1989.

Gwinn, C. R., T. A. Herring and I. I. Shapiro, Geodesy by radio interferometry: Studies of the forced nutations of the Earth 2. Interpretation, *J. Geophys. Res.*, *91*, 4755-4765, 1986.

Hansen, U., and D. A. Yuen, Dynamical influences from thermal-chemical instabilities at the core–mantle boundary, *Geophys. Res. Lett.*, *16*, 629-632, 1989.

Herrero-Bervera, E., and M. A. Khan, Olduvai termination: Detailed paleomagnetic analysis of a north central Pacific core, *Geophys. J. Int.*, *108*, 535-545, 1992.

Hide, R., and J. O. Dickey, Earth's variable rotation, *Science*, *253*, 629-637, 1991.

Hoffman, K. A., Long-lived transitional states of the geomagnetic field and the two dynamo families, *Nature*, *354*, 273-277, 1991.

Inoue, H., Y. Fukao, K. Tanabe and Y. Ogata, Whole mantle P-wave travel time tomography, *Phys. Earth Planet. Int.*, *59*, 294-328, 1990.

Jeanloz, R., Phase transitions in the mantle, *Nature*, *340*, 184, 1989a.

Jeanloz, R., Physical chemistry at ultrahigh pressures and temperatures, *Ann. Rev. Physical Chemistry*, *40*, 237-259, 1989b.

Jeanloz, R., The nature of the Earth's core, *Ann. Rev. Earth Planet. Sci.*, *18*, 357-386, 1990.

Kellogg, L. H., and S. D. King, Interaction of mantle plumes with the core–mantle boundary, *Geophys. Res. Lett.*, submitted for publication, 1992.

Kellogg, L. H., and D. L. Turcotte, Mixing and the distribution of heterogeneities in a chaotically convecting mantle, *J. Geophys. Res.*, *95*, 421-432, 1990.

Kellogg, L. H., and C. A. Stewart, Mixing by chaotic convection in an infinite Prandtl number fluid and implications for mantle convection, *Phys. Fluids A*, *3*, 1374-1378, 1991.

Knittle, E., and R. Jeanloz, High-pressure metallization of FeO and implications for the Earth's core, *Geophys. Res. Lett.*, *13*, 1541-1544, 1986.

Knittle, E., and R. Jeanloz, Simulating the core-mantle boundary: An experimental study of high-pressure reactions between silicates and liquid iron, *Geophys. Res. Lett.*, *16*, 609-612, 1989.

Knittle, E., and R. Jeanloz, Earth's core–mantle boundary: Results of experiments at high pressures and temperatures, *Science*, *251*, 1438-1443, 1991a.

Knittle, E., and R. Jeanloz, The high-pressure phase diagram of $Fe_{0.94}O$: A possible constituent of the Earth's core, *J. Geophys. Res.*, *96*, 16169-16180, 1991b.

Laj, C., A. Mazaud, R. Weeks, M. Fuller and E. Herrero-Bervera, Geomagnetic reversal paths, *Nature*, *351*, 447, 1991.

Lay, T., Structure of the core–mantle transition zone: A chemical and thermal boundary layer, *EOS, Trans. AGU, 70,* 54-55, 58-59, 1989.

Lay, T., and C. J. Young, The stably stratified core revisited, *Geophys. Res. Lett., 17,* 2001-2004, 1990.

Lay, T., and C. J. Young, Analysis of seismic SV waves in the core's penumbra, *Geophys. Res. Lett., 18,* 1373-1376, 1991.

Mathews, P. M., and I. I. Shapiro, Nutations of the Earth, *Ann. Rev. Earth Planet. Sci., 20,* 469-500, 1992.

Merrill, R. T., and M. W. McElhinny, *The Earth's Magnetic Field: Its History, Origin and Planetary Perspective,* Academic Press, New York, 1983.

Merrill, R. T., P. L. McFadden and M. W. McElhinny, Paleomagnetic tomography of the core–mantle boundary, *Phys. Earth Planet. Int., 64,* 87-101, 1990.

Morelli, A., and A. M. Dziewonski, Topography of the core–mantle boundary and lateral homogeneity of the liquid core, *Nature, 325,* 678-683, 1987.

Morrison, L. V., Re-determination of the decade fluctuations in the rotation in the period 1861-1978, *Geophys. J. Roy. Astron. Soc., 58,* 349-360, 1979.

Olson, P., and C. Kincaid, Experiments on the interaction of thermal convection and compositional layering at the base of the mantle, *J. Geophys. Res., 96,* 4347-4354, 1991.

Ottino, J. M. *The Kinematics of Mixing: Stretching, Chaos, and Transport,* Cambridge U. Press, Cambridge, 1989.

Poirier, J. P., and J. L. LeMouël, Does infiltration of core material into the lower mantle affect the observed geomagnetic field?, *Phys. Earth Planet. Int.,* in press, 1992.

Prévot, M., and M. Perrin, Intensity of the Earth's magnetic field since Precambrian from Thellier-type paleointensity data and inferences on the thermal history of the core, *Geophys. J. Int., 108,* 613-620, 1992.

Rekdal, T., and D. J. Dornboos, The times and amplitudes of core phases for a variable core–mantle boundary layer, *Geophys. J. Int., 108,* 546-556, 1992.

Runcorn, S. K., Polar path in geomagnetic reversals, *Nature, 356,* 654-656, 1992.

Schneider, D. A., and D. Kent, The paleomagnetic field from equatorial deep-sea sediments: axial symmetry and polar asymmetry, *Science, 242,* 252-256, 1988.

Shercliff, J. A., *A Textbook of Magnetohydrodynamics,* Pergamon Press, New York, 1965.

Sleep, N. H., Gradual entrainment of a chemical layer at the base of the mantle by overlying convection, *Geophys. J., 95,* 437-447, 1988.

Souriau, A., and G. Poupinet, A latitudinal pattern in the structure of the outermost liquid core revealed by the travel times of SKKS-SKS seismic phases, *Geophys. Res. Lett., 17,* 2005-2007, 1990.

Vinnik, L. P., V. Fara and B. Romanowicz, Observational evidence for diffracted SV in the shadow of the Earth's core, *Geophys. Res. Lett., 16,* 519-522, 1989.

Weber, M., and J. P. Davis, Evidence of a laterally inhomogeneous lower mantle from P- and S-waves, *Geophys. J. Int., 102,* 231-255, 1990.

Williams, Q., and R. Jeanloz, Melting relations in the iron-sulfur system at ultra-high pressures: Implications for the thermal state of the Earth, *J. Geophys. Res., 95,* 19299-19310, 1991.

Woodhouse, J. H., and A. M. Dziewonski, Seismic modelling of the Earth's large-scale three-dimensional structure, *Phil. Trans. R. Soc. Lond. A, 328,* 291-308, 1989.

Woodward, R. L., and A. M. Dziewonski, Planetary scale structures in the Earth's deep interior and radial variations in the pattern of lateral heterogeneity, *Nature,* in press, 1992.

Wysession, M. E., and E. A. Okal, Regional analysis of D" velocities from the ray parameters of diffracted P profiles, *Geophys. Res. Lett., 16,* 1417-1420, 1989.

Young, C. J., and T. Lay, The core–mantle boundary, *Ann. Rev. Earth Planet. Sci., 15,* 25-46, 1987

Young, C. J., and T. Lay, Multiple phases analysis of the shear velocity structure in the D" region beneath Alaska, *J. Geophys. Res., 95,* 17385-17402, 1990.

Zallen, R. *The Physics of Amorphous Solids,* Wiley, New York, 1983.

Raymond Jeanloz, Department of Geology and Geophysics, University of California, Berkeley, California 94720-4767.

Pressure-Temperature Regimes and Core Formation in the Accreting Earth

HORTON E. NEWSOM AND FREDERICK A. SLANE

Institute of Meteoritics, Department of Geology, University of New Mexico, Albuquerque, New Mexico

Known abundances of siderophile elements place constraints on models for the accretion of the Earth and core formation. Phase changes in the interior of the present Earth at 400 and 675 Km (15 and 25 GPa, respectively) are additional constraints to Earth accretion models, because the metal-silicate partitioning behavior may be affected. However, possible scenarios include core formation under high-, low-, and mixed high- and low-pressure conditions for metal-silicate equilibria. The segregation of metal in the mantle may require the existence of deep (high-pressure model) or shallow (low-pressure model) magma oceans, with implications for the accretion rate, the possibility of giant impacts, or the existence of a dense primitive atmosphere.

INTRODUCTION

The title of this monograph and the theme of the Jeffreys Symposium held at the International Union of Geodesy and Geophysics, Vienna, Austria, is the interrelation between geophysical structures and processes. The processes under discussion in this article are accretion and core formation in the Earth. The geophysical structures in question are the present major divisions of the Earth into the core, lower mantle and upper mantle, and also the geophysical structures that may have been present during the earliest history of the Earth, such as magma oceans, and a dense high-temperature atmosphere.

Clues to the processes of core formation and accretion may be provided by the abundances of siderophile elements and chalcophile elements, which have an affinity for Fe-Ni-metal, and sulfide, respectively. Recent articles on the constraints placed by the abundances of siderophile elements on core formation in the Earth have emphasized quantitative models based on metal/silicate partition coefficients determined at low pressures (for example, Newsom and Sims, [1991]). However, the partitioning behavior of the siderophile and chalcophile elements may be dramatically affected by the varying pressure, and temperature, regimes during the accretion of the Earth [Stevenson, 1990]. In addition, the physical process of core formation and the development of magma oceans, which may be needed to segregate metal from silicates, may also be affected by the high pressures that develop in the Earth at the later stages of accretion.

Relating Geophysical Structures and Processes: The Jeffreys Volume
Geophysical Monograph 76, IUGG Volume 16

SIDEROPHILE ELEMENTS IN THE EARTH

The segregation of metal from the silicate portion of the earth is assumed to have depleted the bulk silicate earth (primitive mantle) in the siderophile and chalcophile elements relative to the lithophile elements (Fig. 1). The original abundances of the siderophile and chalcophile elements in the Earth are assumed to be related to those measured in chondritic meteorites. The abundances of siderophile elements in the Earth are determined by analysis of upper mantle inclusions found in continental volcanics and by geochemical modelling of trace element abundances in mantle-derived volcanic rocks and in the continental crust [Newsom, 1990; Newsom and Sims, 1991]. Whether the composition of the lower mantle is different from the upper mantle, possibly more iron-rich [Jeanloz and Knittle, 1989] is controversial, but the moderately siderophile trace element compositions of basalts from hotspots, thought to be derived from the lower mantle, do not seem to be significantly different [Newsom et al., 1986]. The data for the highly siderophile platinum group elements indicate significant variations occur regionally in analyses of upper mantle inclusions [Spettel et al., 1991] and possibly in hotspot derived samples compared to mid-ocean ridge basalts [Greenough and Fryer, 1990]. Further work is needed to establish whether these variations reflect incomplete mixing of a meteorite component, or compositional differences between the upper and lower mantle.

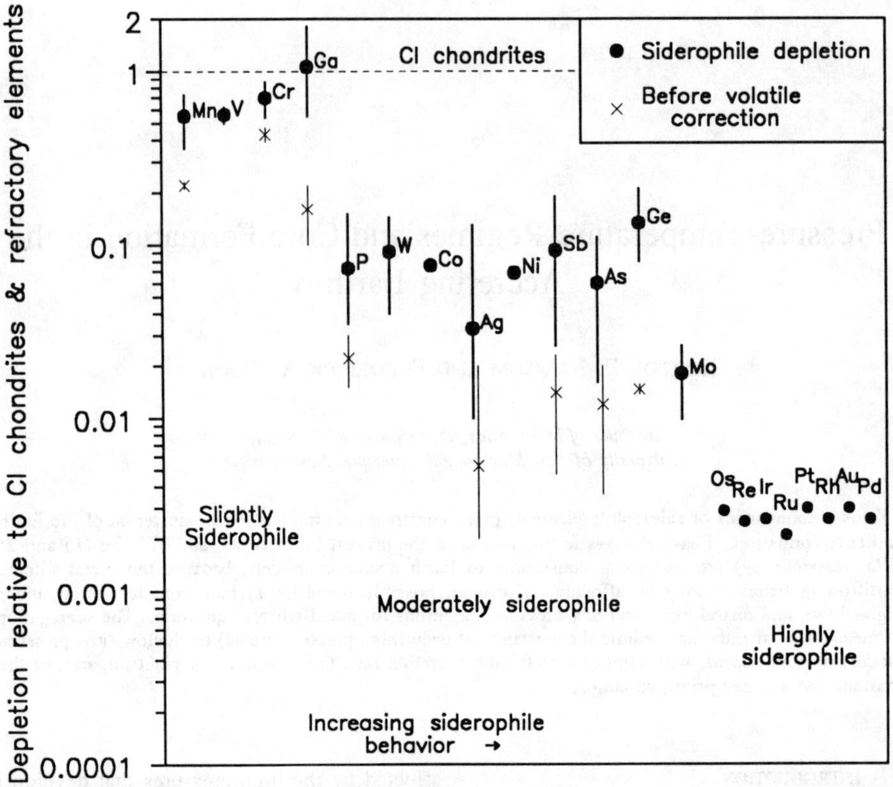

Fig. 1. Depletion of siderophile elements in the earth's primitive mantle (silicate part of the earth), normalized to mean CI chondrites, and refractory elements [Newsom, 1990]. The barely siderophile and moderately siderophile elements are arranged roughly in order of increasing siderophile behavior. The uncertainties in the depletions for the highly siderophile elements are not shown [Newsom, 1990]. The actual depletions of the volatile siderophile elements Mn, Cr, Ga, P, Ag, Sb, As, and Ge, in the primitive mantle relative to CI chondrites, are indicated by crosses. Since the depletions of these elements are due to both volatility and siderophility, their corrected siderophile depletions are indicated by filled circles.

The proposed models to explain the depletion of siderophile elements for the Earth can be easily divided into two classes: equilibrium models, in which the observed depletions of siderophile elements in the mantle are established by equilibrium processes, and additive (or disequilibrium) models, in which the abundances of siderophile elements are controlled by the addition of late veneers to a mantle that is highly depleted in siderophile elements. Both of these classes of models may be affected by metal-silicate partitioning at higher pressures and temperatures. At high pressures within the Earth, the stable mineral assemblage is different from the low pressure assemblage. Two pressures are of particular interest. In the present Earth, 15 GPa is roughly the pressure of the 400 km depth phase transition from α-olivine and pyroxene to the denser β -$(Mg,Fe)_2SiO_4$ phase plus garnet. 25 GPa is roughly the pressure of the 675 km depth phase transition in the Earth where β-phase spinel disproportionates to $MgSiO_3$ perovskite plus (Mg,Fe)O magnesiowüstite [Ringwood, 1991]. 25 GPa is also the pressure where oxygen becomes soluble in Fe-metal [Ringwood, 1990; Knittle and Jeanloz, 1989]. Evidence that such oxygen-rich alloys may be important in the Earth comes from the evidence that the core contains 10% of a light element [Stevenson, 1981]. Experimental evidence also suggests that a complete solid solution exists in the liquid Fe-FeO system above 70 GPa, and that this solid solution is metallic in character [Knittle and Jeanloz, 1991]. The partition coefficients for trace elements between an oxygen-rich alloy and silicates may be quite different from an oxygen-free alloy.

Recent work suggests that a large degree of melting is required to segregate metal from silicates. Metallic liquids do not wet silicate minerals, preventing the metal from aggregating into large masses that can sink. At high pressures, above 25 GPa, the interfacial angles of oxygen-rich metallic liquids may be reduced enough to allow percolation of metal, but this has not been confirmed [Stevenson, 1990]. Physical models of core formation and accretion may therefore involve the formation of magma oceans and the segregation of metal at both high and low pressures. We will discuss below models of core formation involving different pressure regimes, and then discuss some of the geophysical and chemical evidence bearing on the models.

Temperature Dependence

While different models for core formation depend on pressure, they also depend on temperature. Some work has been done on the early thermal history of the Earth [Abe and Matsui, 1985; Sasaki and Nakazawa, 1986; Turcotte and Pflugrath, 1985] but more study is needed. Considerations include the possibility of an existing or self-generated primitive atmosphere. Due to the strong dependence of all these models on the accretion rate of the Earth, thermal profiles vary significantly. However, a lower limit on the surface temperature of the proto-Earth seems to exist at about 400°K. Upper limits depend on the efficiency of convection at the surface, which in turn depends upon factors such as the existence of an atmosphere, fractionation, a magma ocean, a thin crust on the magma ocean.

The effects of temperature on metal-silicate partitioning of trace elements could be as extreme as those of pressure [Murthy, 1991]. However, Jones et al. [1991] have shown that thermodynamic extrapolation of lower temperature partition coefficients to high temperatures may not be straight forward. An additional complication is the new evidence that some siderophile elements, such as Ni, Co. and Ir may dissolve in silicate melts in the zero valence state at high temperatures and low oxygen fugacity [Morse et al., 1991; Colson and Steele, 1991; Borisov et al., 1991]. This phenomena may have profound implications for modelling of accretion and core formation.

High-pressure Models

Models involving high-pressure metal-silicate equilibrium can involve the formation of deep magma oceans, or percolation of metallic liquids at high-pressures. For example, high-pressure effects dominate a model in which core formation is delayed until late in the accretion of the Earth, when an event, such as a giant impact, melts the Earth, allowing core formation to proceed. Stevenson [1990] has discussed the possibility of metal-alloy segregating under high-pressure conditions from a viscous boundary layer at the base of such a magma ocean. This would correspond to a simple equilibrium siderophile element model if the high-pressure partition coefficients turn out to have the appropriate values. The evidence for core formation in the parent bodies of differentiated meteorites [Hewins and Newsom, 1988], however, suggests that metal segregation can occur easily, even on small asteroids, which argues against delayed core formation in the Earth.

For a continuous core formation model, a high pressure equilibrium siderophile signature could result if metal segregates at shallow depths, but reequilibrates with silicates at high pressure. Rapid convection during accretion would be necessary to transport upper mantle material to the lower mantle, where it can acquire a high pressure signature. Mantle material which accreted during the early low-pressure phase of accretion would also have to reequilibrate with metal-alloy at high pressures.

If percolation of metallic liquids occurs at high pressures, core formation could involve this process, combined with whole mantle convection to bring newly accreted upper mantle material into the lower mantle.

The heterogeneous accretion model (an additive model) assumes that most of the metal content of the Earth accreted by the time 80% to 90% of the Earth had accreted. During this first stage of accretion, the core formation process must result in significant depletion of the moderately siderophile elements below their present abundances. This could be consistent with high-pressure equilibration only if the high-pressure metal-silicate partition coefficients are large enough. Compared to the homogeneous accretion scenario, the pressure environment for metal segregation, in this model, will be on the average somewhat higher, because of the rapid accretion of the core. After accreting material to establish the abundance of the moderately siderophile elements, a second stage depletion is required, to lower the abundance of the highly siderophile elements (e.g., Pt, Ir, Au) below their observed abundance without significantly disturbing the abundances of the moderately siderophile elements (e.g., Co, Ni, W). This second stage depletion could involve the segregation of metal or sulfides at high pressures, for example at the base of a magma ocean. The heterogeneous accretion model usually invokes the third stage accretion of a late veneer of material to bring in the highly siderophile elements in their chondritic relative abundances. If the high-pressure partition coefficients for the highly siderophile elements are similar and of the right absolute magnitude, the highly siderophile element signature could be the result of high-pressure equilibrium during the second stage depletion phase.

Low-pressure Models

Models involving low-pressure metal-silicate equilibrium generally involve the formation of shallow magma oceans. During the initial stage of accretion, until the mass reaches 30% to 40% of the Earth's mass, the pressures within the proto-earth will be relatively low, below 25 GPa. Core formation probably occurs during this stage, based on the geophysical evidence for a core in Mars (11% of Earth's mass) and the evidence pointing to the differentiation of asteroids in the inner solar system [Hewins and Newsom, 1988]. In order to establish a low-pressure siderophile signature in the upper mantle, core formation must proceed in such a way that metal segregating at shallow depths cannot reequilibrate under high pressures. This could be accomplished, for example, by accumulation of metal from a shallow magma ocean into large masses, such that the metal will rapidly sink as large metallic diapirs through the lower mantle [Stevenson, 1981].

An important constraint on this type of model is the fraction of the mantle at high pressures during the accretion of the Earth. We have calculated the fraction of the mantle at pressures less than 25 GPa as a function of the accretion of the Earth in Figure 2. This calculation makes simple assumptions

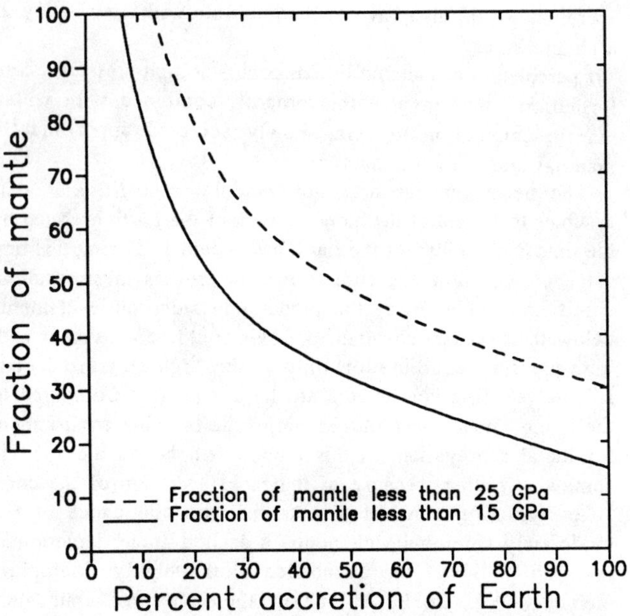

Fig. 2. The fraction of the mantle at pressures greater than 15 GPa and 25 GPa as a function of the accretion of the Earth. The calculation assumes: 1) Consistent (although not constant) core, lower mantle and upper mantle densities for each scale earth model and, 2) Core mass as a linear function of planetary mass. The resultant best-fit curves roughly approximate an inverse power relation.

about the density of the upper and lower mantle, but illustrates some important points. After the pressure at the base of the mantle exceeds 25 GPa, the relative mass of this region grows slowly and then increases rapidly during the later stages of accretion.

Another physical model that is consistent with a strictly low-pressure chemical signature in accessible reservoirs, is to infer a complete lack of communication between the lower mantle and the upper mantle since accretion. This would imply that we have no sample of silicates that reequilibrated with metal at high pressures. Metal segregation within the upper mantle in this model could also involve a shallow magma ocean. Arguments against this model include the possibility that hotspots originate in the lower mantle.

Quantitative siderophile element partitioning models that directly apply to these low-pressure scenarios have been discussed by Newsom [1990]. The most successful models are heterogeneous accretion, and metal retention, a model involving retention of metal and sulfide in the mantle.

Mixed High- and Low-pressure Models

Because the high-pressure environment of the Earth's mantle is restricted to the later stages of accretion, and to the lower mantle, the siderophile element signature of the mantle is likely to record effects due to both high and low pressure partitioning.

A simple homogeneous accretion model, with continuous core formation would result in early accreting mantle material that equilibrated with metal at low pressures. During the later stages of accretion, metal segregation in a deep magma ocean or percolation of metallic liquids at high pressures will begin to impart a high-pressure signature. Depending on the amount of convection, the depth of a magma ocean, etc., the effects of the two different regimes may vary.

To understand the development of high pressure regions during accretion we made some simple calculations. Using scaled Earth models the percent of the mantle mass above the 15 GPa and 25 GPa regions were calculated. First some assumptions were made about the densities of core, lower mantle and upper mantle. The initial data for these generalizations was obtained from the Basaltic Volcanism Study Project [1981]. Second, planetary masses, as a fraction of the earth mass, were picked at 11% (Mars), 20%, 30%, 50%, and 81% (Venus) for the calculations. Third, an initial assumption was made of the percent planetary mass for each region core, lower mantle and upper mantle). Finally, the calculations were made, iterations made on the masses of the regions and the results plotted. The resulting best-fit curves (Fig. 2) are roughly proportional to inverse power curves. The mass of the upper mantle above 15 and 25 GPa does not depend on the density of the upper mantle, but the depth to these pressures does depend on the density.

The heterogeneous accretion model assumes that most of the metal content of the Earth accreted by the time 80% to 90% of the Earth had accreted. During this first stage of accretion, the continuous core formation process must result in significant depletion of the moderately siderophile elements below their present abundances. This could be consistent with a significant role for high pressure equilibration if the high pressure metal-alloy partition coefficients are large, even if they are substantially different from the low pressure values. Compared to the homogeneous accretion scenario, the pressure environment for metal segregation, in this model, will be on the average somewhat higher, because of the rapid accretion of the core. After accreting material to establish the abundance of the moderately siderophile elements, a second stage depletion is required, to lower the abundance of the highly siderophile elements. This second stage depletion will probably involve high pressure effects.

Another accretion model that will involve high and low pressure partitioning effects involves the early accretion of a cold undifferentiated core. This is followed by formation of a magma ocean, due to a blanketing atmosphere, in which metal segregation can occur forming a metal layer on top of the cold core. In the quantitative treatment of Sasaki and Nakazawa, [1986], the lower portion of the magma ocean will reach pressures greater than 25 GPa after about 50% of accretion has occurred. The eventual melting and segregation of metal within the cold core will also probably involve high pressures.

Discussion

Unfortunately, current evidence does not allow us to discard any of the above models. Available geophysical data is ambiguous. The nature of the 670 km boundary (chemical difference or strictly phase change) between the upper and lower mantle is in doubt. There is some evidence that plumes are derived from the lower mantle, and seismic tomography strongly indicates penetration of subducting oceanic crust into the lower mantle, but the tomography data also indicates that the 670 km discontinuity is a major barrier to general mantle convection. The presence of the D" layer at the base of the lower mantle could be a reaction zone between the mantle and core indicating core-mantle disequilibrium, or the D" layer could be subducted material. The abundance of the siderophile elements in the mantle could provide clues to the importance of high pressure processes in the Earth, but partition coefficients at high pressures are only beginning to be measured.

Acknowledgements. Funding for this work was provided by National Science Foundation grants EAR 9005199, EAR 9209641 and the Institute of Meteoritics, University of New Mexico.

REFERENCES

Abe, Y., and T. Matsui, The Formation of an Impact-generated H_2O Atmosphere and Its Implications for the Early Thermal History of the Earth, *Proc. Lunar Planet. Sci. Conf. 15th, Part 2, J. Geophys. Res. 90*, suppl., C545-C559, 1985.

Basaltic Volcanism Study Project, Geophysical and Cosmochemical Constraints on Properties of Mantles of the Terrestrial Planets, *Basaltic Volcanism on the Terrestrial Planets,* 1286 pp., Pergamon Press, New York, 1981.

Borisov, A., D.B. Dingwell, H.St.C. O'Neill, and H. Palme, Experimental Determination of the Solubility of Iridium in Silicate Melts: Preliminary Results, in *LPI-LAPST Workshop on the Physics and Chemistry of Magma Oceans from 1 bar to 4 Mbar*, LPI, Houston, Texas, 7-8, 1991.

Colson, R.O., Nichel and Cobalt as Incompatible Elements in Olivine and Orthopyroxene, *EOS Transactions Amer. Geophys. Union 70,* 547, 1991.

Greenough, J.D., and B.J. Fryer, Distribution of Au, Pd, Pt, Rh, Ru, and Ir in ODP Leg 115 (Indian Ocean) Hot-spot Basalts: Implications for Magmatic Processes, in *Proc. Ocean Drilling Prog., Scientific Results, 115*, College Station, Texas, 71-84, 1990.

Hewins, R.H. and H. E. Newsom, Igneous activity in the early solar system, in *Meteorites and the Early Solar System*, edited by J.F. Kerridge and M. S. Matthews, pp. 73-101, University of Arizona Press, Tucson, 1988.

Jeanloz, R. and E. Knittle, Density and Composition of the Lower Mantle, *Phil. Trans. R. Soc. Lond., A328*, 377-389, 1989.

Jones, J.H., Calculation of Siderophile Element Partition Coefficients at Elevated Temperatures, *EOS Transactions Amer. Geophys. Union 70,* 548, 1991.

Knittle, E. and R. Jeanloz, Simulating the Core-Mantle Boundary: an Experimental Study of High-Pressure Reactions Between Silicates and Liquid Iron, *Geophys. Res. Lett. 16*, 609, 1989.

Knittle, E. and R. Jeanloz, The High Pressure Phase Diagram of $FeO_{0.94}O$: A Possible Constituent of the Earth's Core, *J. Geophys. Res. 96*, 16,169-16,180, 1991.

Morse, S.A., J.M. Rhodes, and K.M. Nolan, Redox Effect on the Partitioning of Nickel in Olivine, *Geochim. Cosmochim. Acta*, 55, 2373-2378, 1991.

Murthy, R.V., Early Differentiation of the Earth and the Problem of Mantle Siderophile Elements: A New Approach, *Science, 253*, 303-306, 1991.

Newsom, H.E., W.M. White, K.P. Jochum, and A.W. Hofmann, Siderophile and Chalcophile Element Abundances in Oceanic Basalts, Pb Isotope Evolution and Growth of the Earth's Core, *Earth Planet. Sci. Lett., 80*, 299-313, 1986.

Newsom, H.E., Accretion and Core Formation in the Earth: Evidence from Siderophile Elements, in *Origin of the Earth*, edited by H.E. Newsom and J.H. Jones, pp. 273-288, Oxford Press, New York, 1990.

Newsom, H.E., and K.W.W. Sims, Core Formation During Early Accretion of the Earth, *Science, 252*, 933, 1991.

Ringwood, A.E., Earliest History of the Earth-Moon System, in *Origin of the Earth*, edited by H.E. Newsom and J.H. Jones, pp. 101-134, Oxford Press, New York, 1990.

Ringwood, A.E., Phase Transformations and Their Bearing on the Constitution and Dynamics of the Mantle, *Geochim. Cosmochim. Acta, 55*, 2083-2110, 1991.

Sasaki, S., and K. Nakazawa, Metal-Silicate Fractionation in the Growing Earth: Energy Source for the Terrestrial Magma Ocean, *J. Geophys. Res. 91*, 9238, 1986.

Stevenson, D.J., Models of the Earth's core, *Science, 241*, 619, 1981.

Stevenson, D.J., Fluid Dynamics of Core Formation, in *Origin of the Earth*, edited by H.E. Newsom and J.H. Jones, pp. 231-250, Oxford Press, New York, 1990.

Turcotte, D.L., and Pflugrath, J.C., Thermal Structure of the Accreting Earth, *Proc. Lunar Planet. Sci. Conf. 15th, Part 2, J. Geophys. Res., 90*, suppl., C541-C544, 1985.

H. Newsom and F. Slane, Institute of Meteoritics, Department of Earth and Planetary Sciences, University of New Mexico, Albuquerque, NM 87131-1126

